VOLUME ONE HUNDRED AND FOURTY

PROGRESS IN MOLECULAR BIOLOGY AND TRANSLATIONAL SCIENCE

Genetics of Monogenic and Syndromic Obesity

VOLUME ONE HUNDRED AND FOURTY

PROGRESS IN MOLECULAR BIOLOGY AND TRANSLATIONAL SCIENCE

Genetics of Monogenic and Syndromic Obesity

Edited by

Ya-Xiong Tao

Department of Anatomy, Physiology and Pharmacology, College of Veterinary Medicine, Auburn University, Auburn, AL, USA

ELSEVIER

AMSTERDAM • BOSTON • HEIDELBERG • LONDON
NEW YORK • OXFORD • PARIS • SAN DIEGO
SAN FRANCISCO • SINGAPORE • SYDNEY • TOKYO
Academic Press is an imprint of Elsevier

Academic Press is an imprint of Elsevier
125 London Wall, London EC2Y 5AS, United Kingdom
525 B Street, Suite 1800, San Diego, CA 92101-4495, United States
50 Hampshire Street, 5th Floor, Cambridge, MA 02139, United States
The Boulevard, Langford Lane, Kidlington, Oxford OX5 1GB, UK

First edition 2016

Notices

Knowledge and best practice in this field are constantly changing. As new research and experience broaden our understanding, changes in research methods, professional practices, or medical treatment may become necessary.

Practitioners and researchers must always rely on their own experience and knowledge in evaluating and using any information, methods, compounds, or experiments described herein. In using such information or methods they should be mindful of their own safety and the safety of others, including parties for whom they have a professional responsibility.

To the fullest extent of the law, neither the Publisher nor the authors, contributors, or editors, assume any liability for any injury and/or damage to persons or property as a matter of products liability, negligence or otherwise, or from any use or operation of any methods, products, instructions, or ideas contained in the material herein.

ISBN: 978-0-12-804615-9
ISSN: 1877-1173

Publisher: Zoe Kruze
Acquisition Editor: Mary Ann Zimmerman
Editorial Project Manager: Helene Kabes
Production Project Manager: Magesh Kumar Mahalingam
Designer: Mark Rogers

Typeset by Thomson Digital

CONTENTS

CONTRIBUTORS

E. Burgio
European Cancer and Environment Research Institute (ECERI), Bruxelles, Belgium; ISDE International Society of Doctors for Environment (Scientific Office), Arezzo, Italy

M.G. Butler
Departments of Psychiatry & Behavioral Sciences and Pediatrics, University of Kansas Medical Center, Kansas City, KS, United States of America

D.D. Duan
Laboratory of Cardiovascular Phenomics, Center for Cardiovascular Research, Department of Pharmacology, and Center for Molecular Medicine, University of Nevada School of Medicine, Reno, NV, United States

J.C. Han
Department of Pediatrics, University of Tennessee Health Science Center and Children's Foundation Research Institute, Le Bonheur Children's Hospital, Memphis, TN, United States of America

I. Lindberg
Department of Anatomy and Neurobiology, University of Maryland, Baltimore, MD, United States of America

A. Lopomo
Department of Translational Research and New Technologies in Medicine and Surgery, Medical Genetics Laboratories, University of Pisa, Pisa, Italy; Doctoral School in Genetics, Oncology, and Clinical Medicine; University of Siena; Siena, Italy

M.G. Martin
Department of Pediatrics, Division of Gastroenterology and Nutrition, Mattel Children's Hospital and the David Geffen School of Medicine, Los Angeles, CA, United States of America

L. Migliore
Department of Translational Research and New Technologies in Medicine and Surgery, Medical Genetics Laboratories, University of Pisa, Pisa, Italy

B. Ramos-Molina
Department of Human Genetics, KU Leuven, Leuven, Belgium

M. Stachowiak
Department of Genetics, Animal Breeding, Poznań University of Life Sciences, Poznań, Poland

M. Switonski
Department of Genetics, Animal Breeding, Poznań University of Life Sciences, Poznań, Poland

I. Szczerbal
Department of Genetics, Animal Breeding, Poznań University of Life Sciences, Poznań, Poland

Y.-X. Tao
Department of Anatomy, Physiology and Pharmacology, College of Veterinary Medicine, Auburn University, Auburn, AL, United States of America

W. Wang
Department of Anatomy, Physiology, and Pharmacology, College of Veterinary Medicine, Auburn University, Auburn, AL, United States of America

Z. Yang
Department of Anatomy, Physiology and Pharmacology, College of Veterinary Medicine, Auburn University, Auburn, AL, United States of America

Y.-P. Zhang
Pediatric Heart Center, Beijing Anzhen Hospital, Capital Medical University, Beijing, China

Y.-Y. Zhang
Department of Cardiology, Changzhou Second People's Hospital, Changzhou, Jiangsu, China

PREFACE

Obesity is an epidemic in developed countries and becoming ever more prevalent in developing countries. Obesity is strongly associated with several metabolic diseases, including type 2 diabetes mellitus, non-alcoholic fatty liver disease, hypertension, arteriosclerosis, and cardiovascular disease. It is also associated with sleep apenea, and some types of cancers such as breast cancer in women and prostate cancer in men. It also has significant psychological effects on the obese. Therefore obesity has become a major public health problem worldwide, with tremendous health, economic and social costs. Childhood obesity presents an even bigger challenge, due to the increased chance of developing metabolic syndrome early in adulthood. Obesity significantly decreases the potential of life expectancy increase that could be achieved with reduced smoking and improved health care.

Obesity is caused by multiple factors, including genetic, environment, and social elements. Genetic contribution to obesity was clearly demonstrated with elegant studies in naturally occurring rodent obesity models (such as *ob/ob* and *db/db* mice and Zucker *Fatty* rat) as well as twin studies in humans. Indeed, mutations in several genes, such as leptin, leptin receptor, proopiomelanocortin, prohormone convertase 1/3, melanocortin-4 receptor, brain-derived neurotrophic factor, and Sim1, causing monogenic obesity have been identified. Some of these genes were discussed in previous chapters in this Series.

This volume reviews several aspects of the genetics of monogenic and syndromic obesity. Starting with Butler's introduction to several examples of monogenic and syndromic obesity, Ramos-Molina, Martin, and Lindberg discussed mutations in prohormone convertase 1/3 and human obesity. Han summarized the studies on brain-derived neurotrophic factor variants and human obesity. My students and I reviewed variants in melanocortin-3 receptor and ghrelin receptor and obesity. Epigenetic aspect of obesity was reviewed by Migliore and her colleagues. Duan and his colleagues provided a guided tour for us "From genome-wide association study to phenome-wide association study" as applied to obesity studies. Last but not least, Switonski and his colleagues demonstrated that genetics is also relevant in our best friend (dogs) and an important source of our meat (pigs). They also argued that these species could serve as large animal models for human obesity studies, complementing rodent studies.

I am immensely grateful to all the authors for their wonderful contributions to this volume. I also greatly appreciate the expert guidance and enormous patience of the colleagues at Elsevier, Mary Ann Zimmerman and Helene Kabes. It has been a great pleasure to work with you on this and other volumes. Finally, without the unwavering support of the Series Editor, Professor P. Michael Conn, this volume would not possible. I thank you all for the support.

Finally, as always, I thank my family for continuous support and unconditional love.

Ya-Xiong Tao
Auburn, Alabama

CHAPTER ONE

Single Gene and Syndromic Causes of Obesity: Illustrative Examples

M.G. Butler[1]

Departments of Psychiatry & Behavioral Sciences and Pediatrics, University of Kansas Medical Center, Kansas City, KS, United States of America

[1] Corresponding author: e-mail address: mbutler4@kumc.edu.

Contents

Abstract

Obesity is a significant health problem in westernized societies, particularly in the United States where it has reached epidemic proportions in both adults and children. The prevalence of childhood obesity has doubled in the past 30 years. The causation is complex with multiple sources, including an obesity promoting environment with plentiful highly dense food sources and overall decreased physical activity noted for much of the general population, but genetic factors clearly play a role. Advances in genetic technology using candidate gene approaches, genome-wide association studies, structural and expression microarrays, and next generation sequencing have

Progress in Molecular Biology and Translational Science, Volume 140
ISSN 1877-1173
http://dx.doi.org/10.1016/bs.pmbts.2015.12.003

led to the discovery of hundreds of genes recognized as contributing to obesity. Polygenic and monogenic causes of obesity are now recognized including dozens of examples of syndromic obesity with Prader–Willi syndrome, as a classical example and recognized as the most common known cause of life-threatening obesity. Genetic factors playing a role in the causation of obesity will be discussed along with the growing evidence of single genes and the continuum between monogenic and polygenic obesity. The clinical and genetic aspects of four classical but rare obesity-related syndromes (ie, Prader–Willi, Alström, fragile X, and Albright hereditary osteodystrophy) will be described and illustrated in this review of single gene and syndromic causes of obesity.

1. INTRODUCTION AND BACKGROUND

Obesity is an increasing serious health problem recognized worldwide and reaching epidemic status particularly in westernized societies. A role for genetic factors is now recognized as contributing to obesity and will be summarized in this review on single gene and syndromic causes of obesity. The United States currently leads as the most obese nation. The causation of obesity is clearly complex. The rising obesity prevalence is partially due to an obesity-promoting environment with highly dense, inexpensive, and plentiful food sources with a relatively sedentary lifestyle due to advances in modern technology leading to decreased energy expenditure for employees in the work place.[1–3] A survival advantage over time has led to a thrifty phenotype due to a more efficient use of calories with fat deposition thereby contributing to the obesity epidemic in the presence of increased caloric intake.

Major health concerns related to long term obesity status include insulin resistance, type 2 diabetes, fatty liver, sleep apnea, cancer, hypertension, cardiovascular disease, stroke, and physical limitations with disabilities. Obesity, as a global public health problem, is on the rise as evidenced in 1997 at 19.4%; 24.5% in 2004, 33.8% in 2008, and 35.7% of adults in 2010.[3] Childhood obesity is also present at 17% in the United States with 70% of obese adolescents becoming obese adults. The prevalence of childhood obesity has doubled in the past 30 years.[4] The medical costs for treating complications from an overweight and obesity status are inordinate with estimates for treating people for overweight and obesity estimated at $72 billion and $198 billion, respectively impacted by the current obesogenic environment.[3,5–10]

Body mass index (BMI) is a commonly used measure of obesity in adults and children and defined as weight measured in kilograms divided by the square of height in meters (kg/m^2), but this measure is not as accurate for determination of obesity in the pediatric age group. A BMI between 25 and 30 is considered overweight while BMIs greater than or equal to 30 are considered obese in adulthood. Since the 1980s, the mean BMI has increased by 0.4–0.5 kg/m^2 per decade in adults. In childhood a BMI from the 85th to 95th percentile may be considered overweight and equal to or greater than the 95th percentile for obesity.[3,11–13]

Twin and family studies strongly implicate genetic factors as playing an important role in the development of obesity. Genetic predisposition for obesity can also be influenced by ethnicity. Monozygotic twin studies show concordance for obesity at a range from 70 to 90%, while in dizygotic twins range from 35 to 45%. Heritability estimates derived from twin studies report a representative number of 77% for BMI.[14,15] Increased physical activity and exercise can lower the effects of genes impacting BMI and therefore obesity status. Hence, genes play an important central role in the determination of BMI and pathogenesis of obesity. In addition, strong genetic factors influence percentage body fat, waist circumference, energy expenditure, eating behavior, and level of physical activity. Specific gene mutations are found in 5–10% of obesity in childhood supported by a recent review of 370 recognized genes identified in the literature playing a role in obesity.[16] Several genes have been studied in relationship to obesity, particularly in mongenic causation; for example, MC4R, which is the single most commonly recognized gene causing childhood obesity and found in about 4% of cases.[6,7] Structural chromosomal anomalies involving genomic regions containing causative genes for obesity have been reported such as 3p25 duplications and 16p11.2 deletions.[17]

With continually improved and varied complex genetic techniques requiring smaller quantities of DNA (and RNA) such as next generation sequencing further supported by prometaphase chromosome analysis, fluorescence in situ hybridization, linkage and genome-wide association studies (GWAS), copy number variant (CNV), and single nucleotide polymorphism (SNP) probes utilized in high resolution microarrays have been instrumental in identifying structural, chromosomal, and DNA abnormalities in genomic regions and candidate genes causing both rare and common forms of obesity. Specifically, chromosome abnormalities (deletions, duplications) have been recorded in humans with dysmorphic and syndromic obesity including chromosome 1p36 deletion; chromosome 2q37.3 deletion; chromosome 3p23 duplication; 3p25.3 duplication (contains the GHRL gene); chromosome 4q32.1 duplication; 4q35.1 duplication; chromosome 5p13.1

duplication; chromosome 6q16.2 deletion (SIM1 gene); 6q22.2 deletion; 6q24.3 duplication; 6q15-q21 deletion (SIM1 gene); 6q16-q21 deletion (SIM1 gene); 6q16.1-q16.3 deletion (SIM1 gene); chromosome 7q36 deletion; chromosome 9p23 deletion; 9q34 deletion; 9q34.3 deletion; 9q33.3q34 duplication; chromosome 10q22.3q23.2 duplication; chromosome 11p12-p14 deletion; 11p13-p14.2 deletion (BDNF gene); 11p11.2 deletion; chromosome 12p13.1 duplication; 12qter deletion; chromosome 14q32.2 hypomethylation status (maternal disomy 14); chromosome 16q13 duplication; 16p11.2 deletion (SH2B1 gene); 16q11.2-q13 duplication (FTO gene); chromosome 18q12.2-q21.1 deletion; chromosome 19q12-q13.2 duplication; 19q13.2 deletion; chromosome 20q13.13-q13.32 deletion; chromosome 22q11.2 deletion; chromosome Xq26.3-q27.3 deletion (FMR1 gene); Xq23q25 duplication; Xp11.3p21.1 duplication; Xp11.4q11.2 inversion; and Xq27.1-q28 deletion (FMR1 gene)[17] (Table 1). Small deletions involving the chromosome 16p11.2 band have been reported in 0.5–0.7% of individuals with severe obesity and not found in nonobese healthy individuals. It includes the SH2B1 gene, an obesity-related gene that impacts leptin sensitivity and hence caloric intake.[18] Chromosome microarray studies have shown rare CNVs which are greater than 2 Mb in size and found in 1.3% of individuals with obesity and evidence of disrupting multiple candidate genes for obesity (eg, POMC, UCP1, GHRL). Hence, the research of CNVs in the study of obesity has merits. Recently, Butler et al.[16] summarized the literature and authoritative computer websites and found 370 clinically relevant and known genes reported for obesity and plotted the genes on chromosome ideograms to represent a visual display of gene distribution (Fig. 1A and B).

2. GENETIC VARIANT AND GENOME-WIDE ASSOCIATION STUDIES IN OBESITY

Identification of gene loci or markers for disease states became available in late 2005 and is used to discover a number of loci for obesity. The first study of its type to use DNA markers in childhood obesity was published in 2010 and two loci were found (ie, SDCCAG8 and TNKS/MSRA).[19] Now larger, more advanced genetic loci studies have utilized 14 existing GWAS data sets to identify additional loci for obesity.[3] The GWAS method is considered high-throughput with the advantage to assay millions of DNA markers as SNPs that span the entire human genome. When combined with

Table 1 List of Chromosome Locations and Obesity Genes in Relationship to Obesity Phenotypes.

Genetic Factors and Relationship to Obesity Phenotypes	Chromosome Location and Candidate Gene for Obesity
Obesity, early onset	1p36.11 (NROB2); 2p23.3 (POMC); 18q21.32 (MC4R)
Obesity, severe	3p25.2 (PPARG); 6q16.3 (SIM1); 11q13.4 (ULP3); 16p11.2 (SH2B1)
Obesity, generalized	1p35.2 (SDC3); 3p25.3 (GHRL); 4q31.1 (UCP1); 5q13.2 (CART); 5q32 (ADRB2; PPARGC1B); 6q23.2 (ENPP1); 8p11.23 (ADRB3); 11p14.1 (BDNF); 13q14 (OLFM4); 14q32.2 (MEG-DMR); 16p11.2 (SH2B1); 16q12.2 (FTO); 16q22.1 (AGRP); 17q21 (HOXB5); 17q21.31 (PYY); 18q21.32 (MC4R)
Description	**Gene of Interest**
Genetic loci associated with child and/or adult BMI levels including extreme obesity identified by genome-wide studies (GWAS), linkage or with CNV/SNP microarrays	FTO; TMEM18; GNPDA2; INSIG2; MC4R; NEGR1; BDNF; KCTD15; PCSK1; CTNNBL1; MTCH2; NPC1; MAF; PTER; PRL; FAIM2; TFAP2B; SEC16B; ETV5; AIF1; GPRC5BB; MAP2K5; GIPR; FANCL; SDCCAG8; TNKS-MSRA; TNN13K; LRRN6C; NRXN3; FLJ35779; SLC39A8; TMEM160; CADM2; LRP1B; PRKD1; MTIF3; ZNF608; PTBP2; TUB; HMGA1; PPYR1
Associated with childhood obesity—novel loci	SH2B1; EDIL3; S1PR5; FOXP2; TBCA; ABCB5; ZPLD1; KIF2B; ARL15; EPHA6-UNQ6114
Associated with childhood and/or adult waist to hip ratio	LYPLAL1; C12orf51; LY86; HOXC13; RSPO3; VEGFA; TBX15; NFE2L3; GRB14; DNM3; PIGC; ITPR2; ADAMTS9; ZNRF3; NISCH; CPEB4

powerful statistical approaches, the use of SNPs will identify additional loci for specific conditions including obesity.

Common variants in a number of genes encoding proteins involved with obesity include leptin and leptin receptors which regulate caloric intake and have been found to be associated with BMI and obesity status identified in

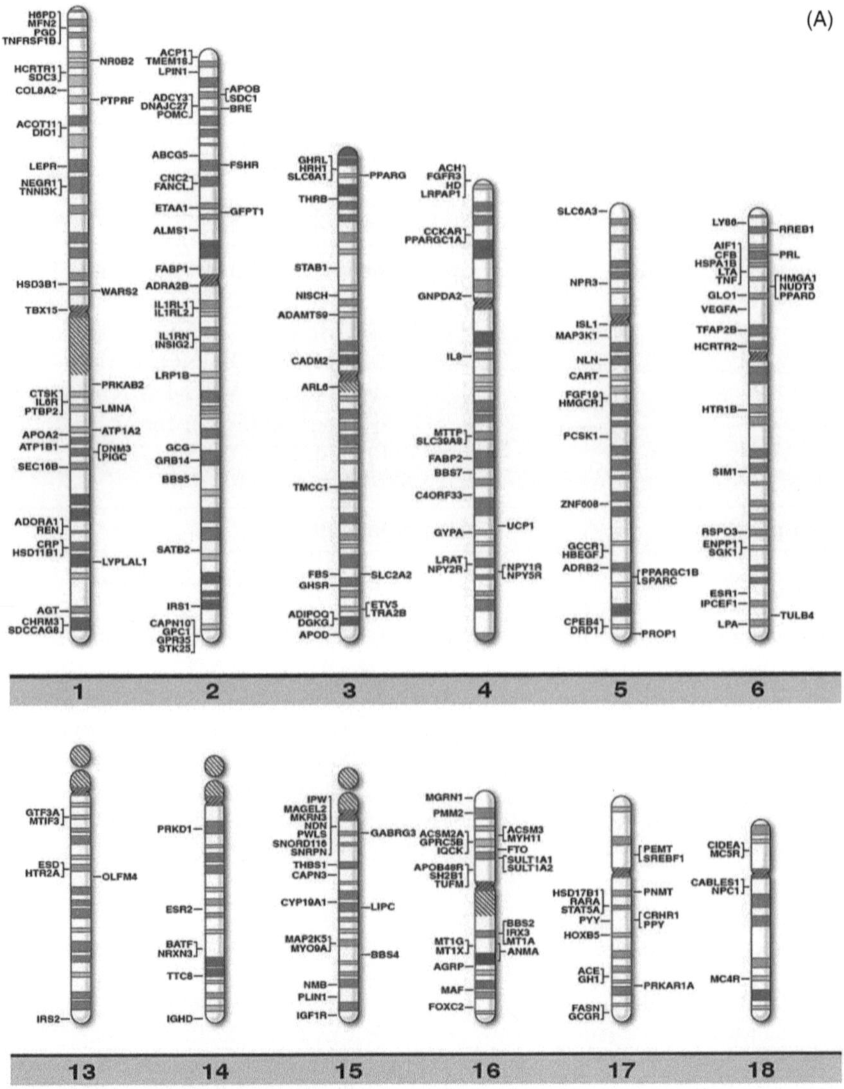

Figure 1 (A) Obesity gene ideogram part A. High resolution human chromosome ideograms (850 band level) with symbols representing recognized genes for obesity positioned at the chromosome band location. The upper "p" and lower "q" arms for each chromosome are separated by the centromere area.[16] (B) Obesity gene ideogram part B. High resolution human chromosome ideograms (850 band level) with symbols representing recognized genes for obesity positioned at the chromosome band location. The upper "p" and lower "q" arms for each chromosome are separated by the centromere area.[16]

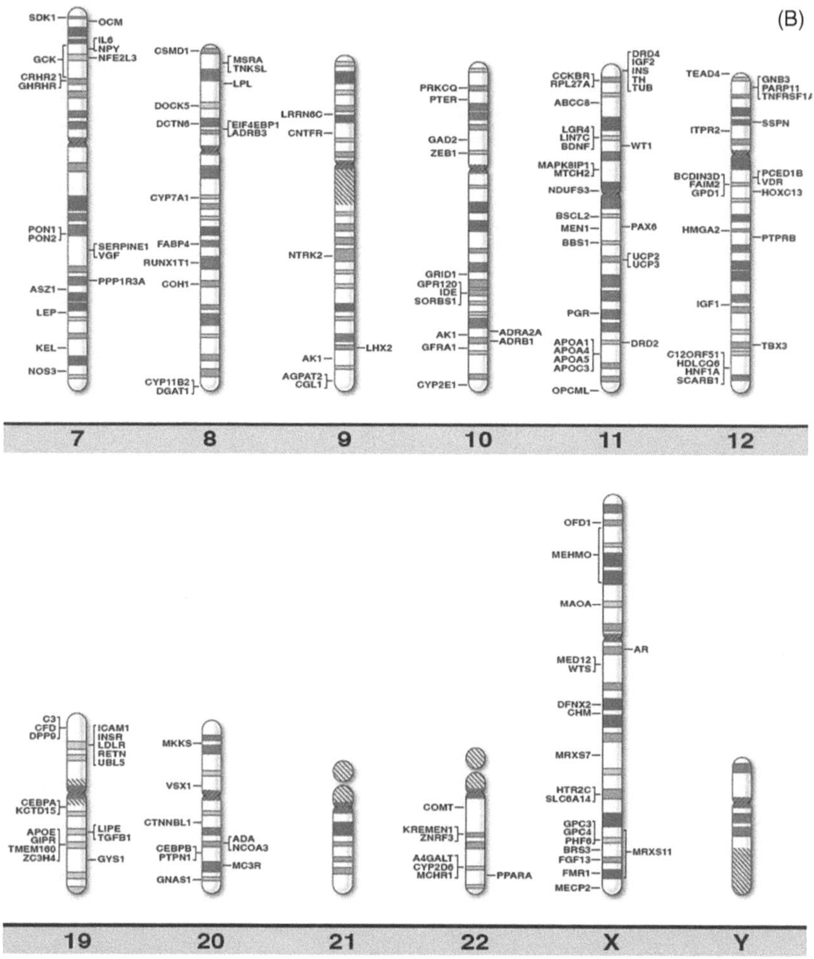

Figure 1 (Continued)

several human populations. Another hormone that plays a role in glucose and fatty acid regulation and production is adiponectin which is coded by a specific gene and known to be lower in obese individuals and in those with type 2 diabetes. Other variants include the cannabinoid receptor 1 (CNR1), dopamine receptor 2 (DRD2), Fat mass and obesity associated (FTO), insulin induced gene 2 (INSIG2), serotonin receptor 2C (HTR2C), and SLC6A4. Another important obesity gene is peroxisome proliferator-activated receptor gamma (PPARγ) found to be associated with both obesity and type 2 diabetes along with the FTO gene which is considered the most robust common obesity–susceptibility locus found to date. The frequency for a minor allele (ie, rs9939609) of the FTO gene ranges from 38 to 44% in the Caucasian population and clearly correlates with weight gain by producing higher BMI levels and obesity.[3,6,7,20–25]

Using the GWAS approach, larger metaanalysis studies have been carried out by the Genetic Investigation of Anthropometric Traits (GIANT) consortium in Caucasians and continues to uncover obesity-related genes including transmembrane protein 18 (TMEM18), potassium channel tetramerization domain containing 15 (KCTD15), glucosamine-6-phosphate deaminase 2 (GNPDA2), SH2B adaptor protein 1 (SH2B1), mitochondrial carrier 2 (MTCH2), and neuronal growth regulator 1 (NEGR1). Furthermore, novel loci reported involve chromosome regions 1q25, 3q27, and 12q13. A more recent, expanded GIANT metaanalysis study involving 249,796 individuals has uncovered 32 BMI-associated genetic loci including 10 loci previously reported and associated with BMI and 4 associated with weight gain and/or increased waist–hip ratio (SEC16B, TFAP2B, FAIM2, NRXN3). Eighteen loci were reported to be novel (RBJ-ADCY3-POMC, GPRC5B-IQCK, MAP2K5- LBXCOR1, QPCTL-GIPR, TNNI3K, SLC39A8, FLJ35779-HMGCR, LRRN6C, TMEM160-ZC3H4, FANCL, CADM2, PRKD1, LRP1B, PTBP2, MTIF3-GTF3A, ZNF608, RPL27A-TUB, and NUDT3- HMGA1). Common CNVs were also found including a 21 kb deletion which was 50 kb upstream to the GPRC5B gene. Continuation of analytical studies by this consortium group have confirmed previously identified obesity loci such as FTO, SEC16B, MC4R, GIPR-QPCTL, ADCY3-DNAJC27, BDNF, and MAP2K5 and new additional gene loci (CDKAL1, PCSK1, GP2, GNPDA2, TFAP2B, PAX6, CDKAL1, and KLF9) associated with BMI.[3] Hence, candidate gene, DNA linkage, and GWAS approaches have led to the identification of large sets of genes and involved genomic regions found to be associated with obesity and obesity-related endophenotypes. Evidence now exists for the presence of 370 genes playing a role and 153 genes associated with known reproduction and infertility with 21 of these genes in common in both obesity and infertility.[16]

3. RECOGNIZED SINGLE GENE CAUSES OF OBESITY

Genetic forms of obesity can be grouped into Mendelian or single gene and syndromic or multifactorial. These include recessive gene inheritance, partial gene deficiencies or duplications, genomic structural variations or CNVs, and polygenic causes. Monogenic forms or single gene conditions causing obesity have been reported for at least eight genes including leptin (LEP), leptin receptor (LEPR), proopiomelanocortin (POMC), prohormone convertase 1 (PCSK1), melanocortin 4 receptor (MC4R), single-minded homolog 1 (SIM1), brain-derived neurotrophic factor (BDNF), and the neurotrophic tyrosine kinase receptor type 2 gene (NTRK2). The hypothalamic leptin–melanocortin system is critical for regulating energy balance with disturbances leading to severe obesity disorders.[3,6,7] These obesity genes are known to primarily affect common pathways involving lipid metabolism, deposition or transport, food seeking behavior and calorie selection types (ie, fat, protein, carbohydrates), level of physical activity, and forms of energy expenditure related to employment or recreation.

Several obesity-related syndromic genetic disorders are identified in humans, both common and rare and will be discussed later. Monogenic causes of morbid obesity are uncommon but marked obesity and extreme eating behavior (hyperphagia) are key features of several rare genetic syndromes including Prader–Willi, Alström, Bardet–Biedl, Albright hereditary osteodystrophy (AHO), Cohen, and fragile X syndromes. Recognized genes playing a role in these disorders are known [eg, SNRPN for Prader–Willi syndrome (PWS), GNAS for AHO, FMR1 for fragile X syndrome (FXS)],[26] which will be discussed later. Understanding the molecular basis of these rare disorders and their genetic mechanisms involving the control of food intake and energy balance in the general population will be important to address the obesity epidemic. Coding and noncoding RNA expression patterns, specifically microRNAs and small nucleolar RNAs (snoRNAs) having important regulatory roles in biological processes, need to be better characterized including their impact on appetite regulation, gene–environment interaction, adipocyte differentiation, and biochemical pathways.[16] Genetic variants have been reported near the MC4R and FTO genes which increase body weight in those carrying these variants, with mutations of the MC4R gene present in about 2% of all obese individuals. Male and female heterozygous carriers weigh 15–30 kg more, respectively compared with their relatives without the MC4R gene changes or mutations.[3,6,7] More research is needed to examine genetic differences among obese and nonobese individuals, particularly rare CNVs to give novel insight into the genetic causation and architecture of obesity and associated infertility in the general population.

The consequences of increased weight and obesity can shorten life expectancy as well as affecting reproduction with dysfunction in ovulation, spontaneous abortions, and overall infertility. Adverse pregnancy outcomes are also noted including preeclampsia, fetal growth failure with premature delivery, and gestational diabetes. About 15% of all women in the United States are also infertile with advanced age playing a role. The prevalence of maternal obesity in the US population is increased with more women having obesity-related reproductive problems. Gradual and sustained maternal weight loss is needed to improve menstrual cycles and ovulation and thus reproductive rates and outcomes.[27,28] Weight loss is considered the first line of treatment in those women with reproductive failure and obesity-related infertility. One of the most common causes of subfertility in women with obesity is polycystic ovarian syndrome which is associated with the androgen receptor (AR) gene.[29,30] The AR gene produces a steroid hormone-activated transcription factor important in regulating androgen activity and sensitivity to sex hormones in both males and females involved in weight and body composition. This X-linked gene contains a polymorphic CAG trinucleotide repeat length that is inversely correlated with gene expression.[31]

Butler and Manzardo[29] examined the polymorphic AR gene with CAG repeats and measures of weight and BMI in a cohort of nonsyndromic obese and lean controls, compared with PWS subjects. PWS is a rare obesity-related genetic disorder with growth hormone (GH) deficiency, hypogonadism, and natural sex hormone deficits. The effects of the AR gene CAG repeat length was examined in relationship to androgen-mediated response and obesity-related factors relevant to human infertility and reproduction. The average CAG repeat length in base pair size did not significantly differ by subject group but was significantly positively correlated with height among lean and obese males, but not present in PWS males. A negative correlation was also observed for weight among females when grouped together. The summary results support the role of sex hormones and the AR gene interaction in obesity and infertility, both cardinal features seen in PWS. The AR gene CAG repeat length is a marker for increased androgen sensitivity and shorter CAG repeat length in this study predicted smaller stature in non-PWS adult males. This process may accelerate fusion of bone growth plates and reduce the length of the growth phase. Hence, increased androgen effects from shorter CAG repeat lengths in non-PWS females could impact pregnancy-related weight gain and possibly pregnancy outcomes.

4. OBESITY GENES AND THEIR ENCODED PROTEINS

One of the key obesity-related proteins is leptin which is a 16-kDa secreted protein and encoded by the leptin (LEP) gene. This gene is expressed and secreted by adipocytes, but its receptor is primarily expressed in the hypothalamus. Leptin plays a major role in food intake regulation, energy balance, and body weight. Mutations and polymorphisms of the leptin and leptin receptor genes are associated with obesity in Caucasians as well as POMC with the gene located on chromosome 2. Its protein is produced by the hypothalamus and thus POMC plays a role in feeding behavior. The POMC gene expression level is positively regulated by leptin. Alterations in the POMC gene including a frame shift mutation have been reported to cause loss of function. This results in early-onset obesity, adrenal insufficiency, and red hair pigmentation. POMC also produces β-MSH and β-endorphin peptides via a proteolytic process involving the enzyme, prohormone convertase (PC), and when altered produces an aberrant fusion protein with lower binding affinity to the MC4R. This can have devastating effects because MC4R plays an important role in both food intake and body weight by controlling leptin effects and shown to cause severe obesity when disrupted in mice.[6,7]

Multiple nonsense and missense mutations in genes have been identified in humans and strongly associated with many obesity-related traits including the BDNF gene located on chromosome 11. Its receptor is a tyrosine receptor kinase B and encoded by the NTRK2 gene which regulates eating behavior and energy balance with growing evidence in humans as playing a role in obesity. In addition, the conditional knockout ofBDNF gene in mice will develop obesity and hyperactivity. The SIM1 gene is located on chromosome 6 and also important in the differentiation and function of the central nervous system. When this gene is disturbed, early-onset obesity occurs along with increased linear growth, hyperinsulinemia, and hyperleptinemia in humans. The FTO gene in humans consists of nine exons and is located on chromosome 16. FTO gene expression is most abundant in the brain based on mouse studies, particularly in hypothalamic nuclei, an area involved in energy balance. The mRNA levels for the FTO gene are found specifically in the arcuate nucleus which regulates feeding patterns.[3] Current research and supporting evidence directly identifies human disorders of energy balance and obesity such as Prader–Willi, Alström, and Albright hereditary osteodystrophy syndromes as targets for study in understanding the genetics of obesity. PWS and AHO are both due to errors in genomic imprinting or epigenetics. Epigenetics refers to changes in gene expression without altering the DNA code and influenced by the parent of origin. Epigenetics is an emerging field of study which impacts the development of multiple diseases and disorders including obesity and will require additional testing to gain a better understanding of pathogenesis.[26]

Genetic factors which predispose weight gain and development of obesity in humans can also impact the response to intervention in terms of weight loss. For example, those individuals with MC4R or POMC gene mutations appear to respond to a reduced caloric diet and exercise program but those with MC4R mutations fail to maintain weight lost after intervention. Hence, the gene–environment interaction needs to be considered when treating those affected with obesity. Access to a multidisciplinary team may be helpful in achieving weight loss including genetic evaluations and screening for obesity-related disorders or monogenic causes, dietary consultation, and behavior/pharmaceutical therapy. There is a growing list of pharmaceutical agents and drugs under clinical trial development to target known genetic causes of obesity, for example, PWS. Information learned through this process may have direct application on obesity in the general population. Bariatric surgery may also be considered. However, this is typically used as a final option for treating and managing exogenous obesity but with only limited experience in individuals with syndromic obesity.

Progress in clinical evaluations, awareness, and genetic testing will help to identify DNA factors contributing to obesity specific for each individual based on their family history, lifestyle, and obesogenic environment common in modern society. Other contributing factors may include environmental chemicals, heavy metals, and additive supplements. Next generation DNA sequencing including exome and whole-genome approaches will generate extensive genetic data and comprehensive genetic maps of potentially predisposing causative factors for obesity. This detailed picture of biological and molecular mechanisms and pathways will be key in understanding the development of childhood obesity and eating behavior.

Pediatric obesity in the United States is on the rise with about one-fifth of children being affected.[4] Most or greater than 50% of cases of childhood obesity are exogenous in nature and result from excessive energy intake in relationship to energy expenditure for extended periods of time. Some cases of obesity are associated with hormonal or genetic factors, or syndromes. Hormone imbalances such as hypothyroidism, Cushing disease, GH deficiency, defective leptin signaling, and insulin resistance are influenced by genetic factors and account for a subset of cases including those with MC4R or POMC mutations. Examples of syndromic obesity include Prader–Willi, Alström, and Bardet–Biedl syndromes.

Critical features can be distinguished between the more rare endogenous obesity genetic disorders when compared with the more common exogenous forms of obesity. Those with endogenous obesity may have an earlier onset of obesity, lack of satiety, inadequate linear growth patterns, dysmorphic features, and cognitive/behavior problems. To identify differences between the two obesity forms a clinical genetics evaluation may be required to rule out dysmorphic/syndromic causes, obtain family history data, and request dietitian consultations for reviewing food records and calorie intake. In addition, appropriate hormonal and genetic laboratory testing would be ordered and results interpreted applicable for individual medical care.

5. SYNDROMIC CAUSES OF OBESITY

Syndromic obesity may result from a single gene condition (eg, Cohen, Alström, Bardet–Biedl syndromes) or errors in imprinting (ie, epigenetics) as seen in PWS involving chromosome 15 or AHO involving chromosome 20. Other syndromes may have more than one cause, leading to the collection of findings such as obesity found in Down syndrome or the Prader–Willi phenotype (PWP) of FXS males. We will review the genetics of syndromic obesity and clinical presentation of classical and underreported rare causes of obesity. There are several syndromic causes of obesity in humans that are currently recognized; some are common such as Down syndrome while others are rare [eg, Alström syndrome (ALMS)] (Table 2).

PWS is the first example of syndromic obesity to be discussed. It was first described in 1956 with nine cases and since that time well over 1000 cases have been reported.[32] PWS is a neurodevelopmental disorder resulting from errors in genomic imprinting (epigenetics) with loss of only paternally expressed genes located on chromosome 15. It results usually from a de novo deletion of the 15q11-q13 region.[33–36] A deletion of the 15q11-q13 region from the mother causes a different clinical disorder known as Angelman syndrome. About 70% of those with PWS will have the 15q11-q13 deletion while the remaining (25–30% of cases) will have both 15s from the mother or a defect of the imprinting center controlling the activity of the imprinted

Table 2 Syndromic Causes of Obesity in Humans.

Prader–Willi
Alstrom
Albright hereditary osteodystrophy
Bardet–Biedl
Cohen
Fragile X
Down
Carpenter
Kabuki
Turner
Borjeson-Forssman-Lehmann
Klinefelter
1p36 deletion
Killian/Teschler-Nicola

genes in the region found in only a few subjects. Those with an imprinting defect may have a high recurrence risk (up to 50%) for subsequently affected children with PWS in the family.[33,35]

PWS shows a range of mild learning deficits and characteristic behavior problems including self-injury (skin picking), outbursts, obsessive compulsions, temper tantrums, and food seeking with hyperphagia beginning in childhood. Additional clinical findings include growth and other hormone deficiencies leading to short stature, small hands and feet, hypogonadism, hypogenitalism, hypotonia, a poor suck, feeding difficulties during infancy, and a particular facies (Fig. 2). Infertility is present in both PWS males and females. Obesity can result if not controlled in early childhood and be life threatening.[37] PWS is found in about 1 in 10,000–30,000 babies with a lower prevalence of 1 in 30,000–50,000 individuals; PWS affects between 350,000 and 400,000 people worldwide.[38] Most cases are sporadic but PWS is considered one of the most common known genetic causes of life threatening obesity affecting all races and ethnic groups[34] (Table 3).

FXS is the most common cause of intellectual disability that runs in families. It is due to a triplet repeat mutation of the FMR1 gene located at chromosome Xq27.3 while a carrier status or premutation form of this gene can be seen in females at risk of developing a full mutation in offspring in the next generation. FXS is associated with obesity but a subset has features in common with PWS referred to as the Prader–Willi phenotype (PWP). FXS generally affects males and was first reported in 1969.[39] It is found in about 1 in 4000 males in the general population.[40] Prominent clinical features

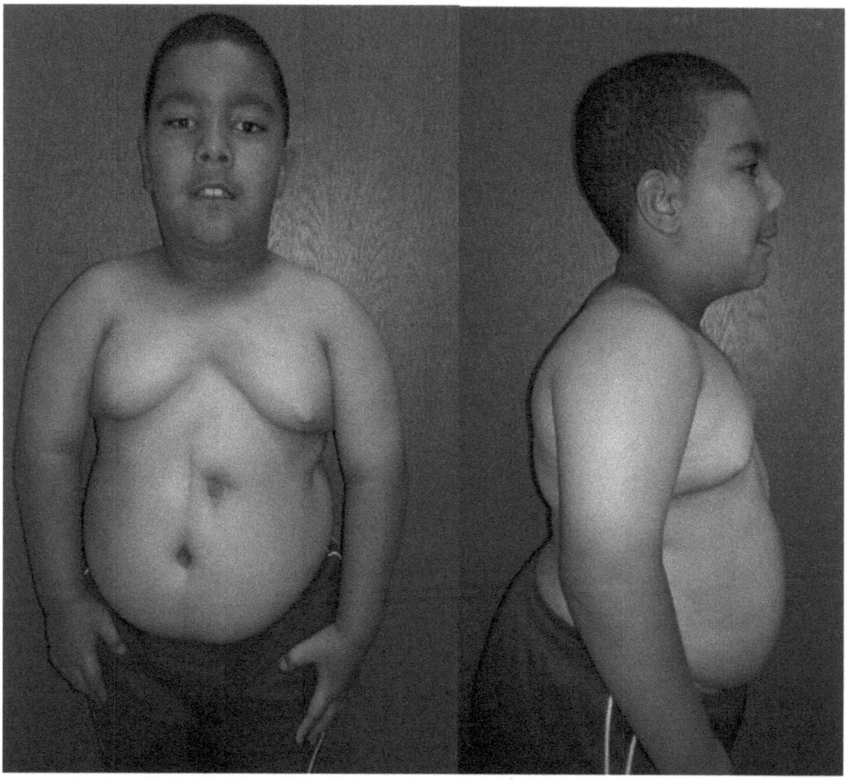

Figure 2 Frontal and profile view of an 8-year-old male with PWS showing the typical facial features, central obesity, and gastrostomy-tube site located on the abdomen.

include intellectual disability, prominent large ears, a narrow head, an elongated face with a flattened midface, and prognathism. Joint laxity, mitral valve prolapse, and macroorchidism are also common. Autism spectrum disorder can also be a common finding in those with classical FXS. As noted, a subset of males with this gene mutation have features similarly seen in PWS including hypotonia, developmental delay, short stature, a small penis, behavioral problems, excessive eating, and marked obesity[41–43] (Table 3).

A third rare separate genetic obesity-related disorder is ALMS. This syndrome is due to mutations of a single gene (ALMS1) located on chromosome 2p13. ALMS occurs in about 1 in 1,000,000 individuals[44] and is characterized by multiorgan involvement with fibrosis, progressive vision, and hearing loss, accompanied by obesity in childhood. Insulin resistance, type 2 diabetes mellitus, and high lipid levels are noted, along with

Table 3 Clinical, Cognitive, Behavior, and Genetic Findings Seen in PWS, PWP in FXS, ALMS, and AHO.

Syndromes	Facial Features	Physical Features	Cognitive/Behavior Features	Genetics
Prader–Willi Syndrome (PWS)	Narrow forehead, almond-shaped eyes, strabismus, short nose with thin upper lip, downturned corners of mouth, dry sticky saliva, enamel hypoplasia	Severe hypotonia, short stature, obesity, osteoporosis, small hands and feet, scoliosis, hypopigmentation, head tilt forward, hypogenitalism	Mild learning impairment, hyperphagia, skin and rectal picking, difficulty with transitions, stubbornness, temper tantrums, perseverative speech, autism, obsessive compulsions, unusual skill with jigsaw puzzles, high pain tolerance	Paternally derived 15q11-q13 deletion (70% of cases), maternal disomy 15 (about 25% of cases), imprinting defects (1–3% of cases)
Prader–Willi Phenotype (PWP) in Fragile X Syndrome (FXS)	Round face, almond-shaped eyes, prominent ears	Obesity, delayed puberty, small penis, hypotonia	Developmental delay, food seeding behavior and hyperphagia, difficulty with transitions, perseverative speech, hand flapping, poor eye contact, autism, obsessive compulsions	*FMR1* gene triplet repeat mutations (at chromosome Xq27.3)
Alström Syndrome (ALMS)	Round face, deep-set eyes, thickened skull, thick ears, frontal hair loss	Wide, flat feet with brachydactyly, scoliosis, dental anomalies, truncal obesity, short suture, hypogonadism, cardiomyopathy, vision (cone-rod dystrophy) and hearing loss, type 2 diabetes, progressive pulmonary, renal and hepatic problems with fibrosis	Developmental delay, balance disturbances and neurosensory deficits, depression, autism, obsessive compulsions	*ALMS1* gene mutations (at chromosome 2p13)

Albright Hereditary Osteodystrophy (AHO)	Rounded face with short nose and short neck, delayed dental eruption or enamel hypoplasia	Moderate obesity, short fourth and fifth metacarpals and metatarsals, short distal phalanx of thumb, small stature, osteoporosis, subcutaneous mineral deposits and basal ganglia, thickened calvarium, variable hypocalcaemia and hyperphosphatemia, occasionally hypothyroidism, hypogonadism and infertility, lens opacity or cataracts, optic atrophy, scoliosis and vertebral anomalies	Mental deficiency (average IQ of 60)	Defects of the *GNAS* gene or imprinting center domain associated with different forms of pseudohypoparathyroidism (PHP) and pseudopseudohypoparathyroidism (PPHP) depending on parent of origin. Maternal inheritance leads to PHP-Ia (AHO plus hormone resistance) while paternal inheritance leads to PHPP or AHO without evidence of resistance to parathyroid hormone (PTH)

hypogonadism; short stature; and cardiac, liver, lung, and kidney problems. The organ systems are often complicated by fibrotic changes noted throughout life. Obesity is observed in most children with ALMS and they are noted to have significant and rapid weight gain beginning with the first year of life.[45,46] The obesity is predominantly distributed in both subcutaneous and visceral compartments. Several endocrine disturbances, including hypogonadism and hypothyroidism, are known to occur in this rare syndrome, as well as decreased growth velocity with advancing age. Most adolescents and adults are reported with short stature and low levels of insulin-like growth factors and GH. Chronic respiratory tract infections are common beginning in early childhood with some becoming severe and leading to chronic bronchitis, asthma, and obstructive pulmonary disease. Pulmonary hypertension is also common with severe interstitial obliterating fibrosis reported, consistent with the fibrotic changes occurring in other major organs.[44,45]

Most individuals with ALMS are reported to have normal intelligence, although mild to moderate delay in reaching major milestones may be noted, such as speech problems. Major depression, obsessive compulsions, and psychotic disturbances may occur particularly during adulthood in ALMS. The protein encoded by the ALMS1 gene is known to cause ALMS when disturbed and is related to ciliary function as similarly seen in an another genetic obesity disorder known as Bardet–Biedl syndrome which consists of more than one type.[47] The ALMS1 gene contains 23 exons with several alternatively spliced transcripts (isoforms).[48] ALMS occurs as a result of mutations of this protein coding gene and inherited as an autosomal recessive disorder (Table 3).[45,46,49]

A second obesity-related genetic disorder due to epigenetics is AHO. AHO was first reported in 1942[50] and is caused by errors in genomic imprinting found on chromosome 20.[26] AHO results from an end-organ resistance to PTH and other associated hormones. Obesity is also a major manifestation along with small stature, mild mental deficiency, a particular facial appearance, short fourth and fifth metacarpals, skeletal anomalies, and delayed dental eruption. Osteoporosis, subcutaneous calcium deposits, hypothyroidism, hypogonadism, ocular problems, and cataracts are occasionally seen.[42,51] Two major clinical variants are known for this complex disorder and referred to as PHP (PHP-Ia or PHP-Ib) and PPHP depending on the presence or absence of additional hormone resistance, along with the presence of the AHO phenotype and the specific pattern of inheritance (Table 3).

6. PRADER–WILLI SYNDROME

6.1 Genomic Imprinting Defects in Prader–Willi Syndrome

Most genes and transcripts on the chromosome 15q11-q13 region are subject to genomic imprinting with disturbances leading to PWS, a rare obesity-related disorder.[35] Loss or deletion of gene alleles that are only active on the chromosome 15 received from the father and controlled by an imprinting center within the 15q11-q13 region leads to PWS. These same alleles when present on the maternal chromosome 15 are silenced by epigenetic factors usually through methylation. Loss of activity of the maternally expressed gene (ie, UBE3A) in the 15q11-q13 region, usually by the same deletion but from the mother leads to Angelman syndrome, an entirely different clinical disorder. PWS and Angelman syndrome were the first examples of errors in genomic imprinting causing disease in humans. It is now recognized that multiple syndromes affect growth, for example, Beckwith–Wiedemann syndrome with overgrowth as a manifestation and Silver–Russell syndrome with growth retardation as a feature are due to errors in genomic imprinting.[26] About 1% of human genes are thought to be imprinted (estimated at about 150 in number); many code for growth factors, related proteins, and receptors and are expressed on only one chromosome depending on the parent of origin. Paternally expressed imprinted genes are thought to enhance the growth of the fetus while maternally expressed imprinted genes are more likely to inhibit growth. Genomic imprinting is now thought to play a role in other diseases such as malignancies, diabetes, and the aging process.[26,52,53]

Genomic imprinting is related to the methylation of cytosine bases in the CpG dinucleotides of the DNA molecule which are key regulatory elements of genes. Almost all imprinted genes have a CpG-rich differentially methylated region (DMR) which usually relates to allele repression. Many imprinted genes are arranged in clusters (imprinted domains) on different chromosomes under control of an imprinting center affecting animal growth, development, and viability. Imprinted genes may also contribute to behavior and language development, alcohol dependency, schizophrenia, and possibly bipolar affective disorders. In addition, the phenomena of genomic imprinting with abnormal imprinting and loss of heterozygosity (LOH) contributes to a wide range of malignancies.[52–54]

The expression of imprinted genes may be tissue- and stage specific with one of the parental alleles being differentially expressed only at a certain developmental stage or in certain cells. However, the monoallelic expression of an imprinted gene is not absolute. Thus, a potential role of genomic imprinting in the differentiation of tissue types may be to determine the transcription rate of genes that influence growth through a fine balance between the expression of the two parental alleles.[55]

Experimental evidence suggests that genomic imprinting evolved about 150 million years ago in a common live-born mammalian ancestor after divergence from egg-laying animals.[56] Imprinting genes provide the paternal and maternal genomes the ability to exert counteracting growth effects during embryonic development.[57] Approximately 1% of all mammalian genes are thought to be imprinted with the first gene (H19) reported to be imprinted in humans in 1992.[58] Since then, many imprinted genes are now candidates for human disease including cancer, obesity, and diabetes.[56]

Imprinted genes are targets for environmental factors to influence expression through epigenetics whereby the expression level is altered without changing the DNA nucleotide coding structure. Imprinting disturbances have been reported in classical genetic disorders such as Beckwith–Wiedemann, Angelman, and Prader–Willi syndromes while the incidence of these disorders are increased in those individuals conceived with the use of assisted reproductive technology (ART). Hence, ART may increase imprinting defects by changing the regulation of imprinted genes.[59]

Epigenetics involve various processes altering gene activity without changing the primary nucleotide sequence of the DNA molecule. A common process for controlling gene activity is methylation. A gene that is methylated (inactivated) can be reactivated in male or female gametogenesis for the next generation. For example, a maternally imprinted gene (inactivated by methylation) may be unmethylated by male gametogenesis and transmitted as an active gene in the sperm.

A genome-wide search for imprinted genes in the human genome has identified over 150 candidate imprinted genes involving 115 chromosome bands.[60] The number of human diseases or disorders due to genomic imprinting maybe greater than 100 conditions as a consequence of an inappropriate genetic alteration, such as a deletion or uniparental disomy involving a gene or chromosome region. Humans are predicted to have fewer imprinted genes than mice, but the types of human genes involved are markedly different from mice. Therefore, questions have been raised about the use of mice as models for human diseases, particularly those involved with imprinted genes, and assessing environmental factors that may impact on genes and their activity. Examples of classical human disorders related to alterations of genomic imprinting, besides Prader–Willi and Angelman syndromes, include Silver–Russell syndrome, Beckwith–Wiedemann syndrome, AHO, and more recently, uniparental disomy 14 (both paternal and maternal forms).[26,61,62]

Genes clustered together under the regulation of a single imprinting-controlling element suggest possible involvement of higher order regulatory elements showing allelic specific DNA replication. Genes contributed by the mother generally replicate or express at different rates than genes contributed by the father. However, inappropriate methylation may contribute to tumor formation by silencing tumor-suppressing genes or by activating growth-stimulating genes. In mammals, DNA methylation patterns are established and maintained during development by three distinct DNA cytosine methyltransferases (Dnmt1, Dnmt3a, and Dnmt3b). In mammalian somatic cells, cytosine methylation occurs in 60–80% of all CpG dinucleotides that are not randomly distributed in the genome. Heavily methylated heterochromatin and repetitive sequences contribute to gene silencing. Most CpG islands located at the promoter regions of many active genes are methylation free. Understanding the functions of DNA methylation and its regulation in mammalian development will help to elucidate how epigenetic mechanisms play a role in human diseases such as neurobehavioral problems and cancer.[26,63]

6.2 Genetic and Clinical Aspects of Prader–Willi Syndrome

The chromosome 15q11-q13 region contains several million DNA base pairs with dozens of imprinted genes and/or transcripts clustered and under the control of two imprinting-controlling centers (one for PWS and one for Angelman syndrome) in this cytogenetic region. There is also a nonimprinted domain or cluster of genes which are equally expressed from either of the parental chromosome 15s.[33,35] Low copy DNA repeats called duplicons are located at the end of this chromosome region and are novel with designation at chromosome 15q11-q13 breakpoint sites.[64,65] These repetitive DNA areas contain a functional HERC2 gene which is located distally at the chromosome 15q11-q13 region found at breakpoint BP3. Two HERC2 pseudogenes which are located at two proximal breakpoint sites (BP1 and BP2) are separated by about 500 kb size DNA segment[66] (Fig. 3). Because these chromosome 15 breakpoint sites contain similar DNA sequences, they can contribute to nonhomologous or mispairing of the chromosome 15s during meiosis I leading to aberrant recombination or crossing-over events.

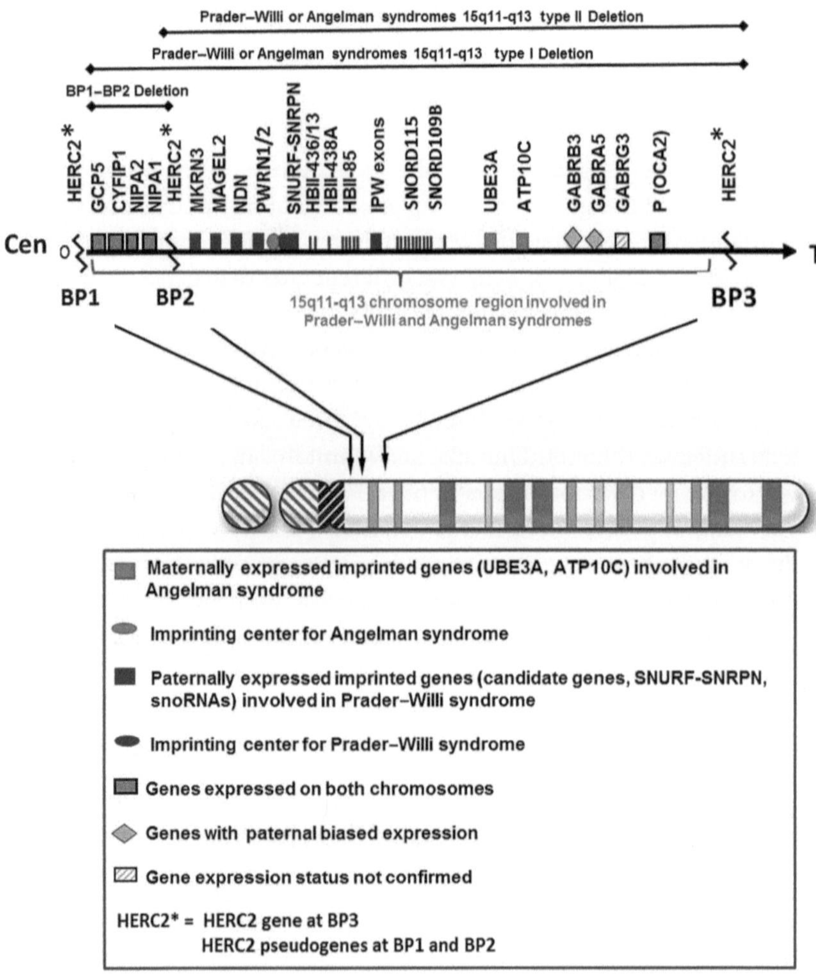

Figure 3 Chromosome 15 ideogram showing the location of genes in the 15q11-q13 region. BP1, BP2, and BP3 are the three common chromosome 15 breakpoints in the region at the site of breakage leading to the larger typical Type I deletion between BP1 and BP3 and the smaller Type II deletion between BP2 and BP3. The *dark gray* (blue in the web version) *colored rectangle*-shaped symbols represent paternally expressed genes (eg, MAGEL2) which when disturbed leads to PWS. The *gray* (red in the web version) *colored square*-shaped symbols represent maternally expressed genes and the UBE3A gene when disturbed causes Angelman syndrome. The *light gray* (green in the web version) *colored rectangle*-shaped symbols represent genes (eg, CYFIP1) that are expressed on both the maternal and paternal chromosome 15s.

This homologous chromosome 15 mispairing can therefore lead to deletions in the offspring and hence PWS when the deletion is from the father's chromosome 15 or Angelman syndrome when the deletion is present on the chromosome 15 donated by the mother.

About 70% of individuals with PWS will show the typical paternal deletion of the 15q11-q13 region and will be de novo in origin and not inherited. There are two types of the typical 15q11-q13 deletion, Type I and Type II with the Type I deletion being larger (about 6.5 Mb in size) and involving chromosome breakpoints BP1 and BP3.[33,35] This is detectable by high resolution chromosomal microarray analysis (Fig. 4A). The smaller Type II deletion (about 5.3 Mb in size) involves breakpoints BP2 and BP3.[67] Four genes (TUBGCP5, CYFIP1, NIPA1, and NIPA2) are located between BP1 and BP2 (Fig. 3) and are not imprinted. They show biallelic or normal expression from either the maternal or paternal chromosome 15. Individuals recently reported with speech delay and autistic characteristics are shown to have deletions and/or duplications of only the four genes between BP1 and BP2. In about 5% of individuals with PWS, an unusual or atypical deletion can be present that is either larger or smaller in size than the typical Type I or Type II deletion.[68–70]

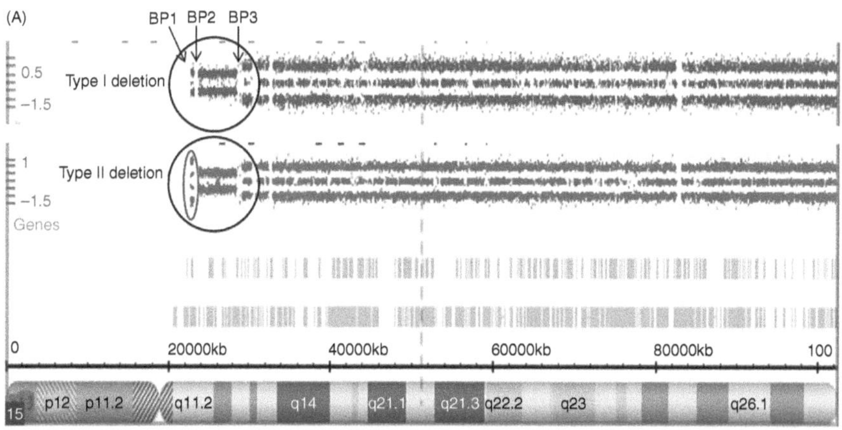

Figure 4 (A) Chromosome microarray results for PWS deletion subtypes. High resolution chromosome microarray using CNV and SNP probes to identify typical 15q11-q13 deletions in PWS classified as Type I involving breakpoints BP1 and BP3 and Type II involving breakpoints BP2 and BP3. (B) Chromosome Microarray Results for Uniparental Maternal Disomy 15. High resolution chromosome microarray using CNV and SNP probes to identify the maternal disomy 15 subtype (segmental isodisomy 15, isodisomy 15, and heterodisomy 15).

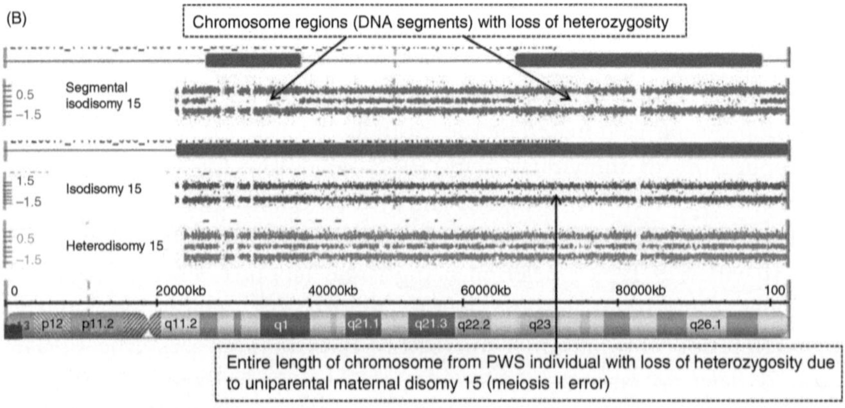

Figure 4 (Continued)

Several studies have shown that individuals with the larger typical 15q11–q13 Type I deletion have more severe neurodevelopmental symptoms, as compared to those with PWS or Angelman syndrome with the smaller typical Type II deletions.[71,72] Bittel et al.[73] later reported findings from mRNA isolated from lymphoblastoid cell lines established from males with PWS for four genes (ie, NIPA1, NIPA2, CYFIP1, and TUBGCP5) in the genomic area between BP1 and BP2 in the 15q11.2 chromosome band. They reported that 24–99% of the phenotypic variability in behavioral and academic measures obtained in their subjects could be explained by the individual gene expression levels. Dykens and Roof[74] later examined and reported behavior findings in a mixed cohort of young and old subjects with PWS and showed a relationship between their genetic subtypes and age. They found negative associations between age and behavior in the 15q11–q13 Type I deletion subtype only which implicated nonimprinted genes between breakpoints BP1 and BP2, specifically the CYFIP1 gene. Disturbed expression of CYFIP1 is seen in other developmental disabilities including those with 15q disorders without PWS.

Chai et al.[75] showed that these four genes are highly conserved and biallelically expressed. NIPA1 or nonimprinted in Prader–Willi/Angelman syndrome 1 gene is the best studied gene and associated with autosomal dominant hereditary spastic paraplegia.[76] NIPA1 also mediates Mg^{2+} transport and is highly expressed in neuronal tissue. Jiang et al.[77] later reported childhood absence epilepsy when mutations were found in the NIPA2 gene. The TUBGCP5 gene or tubulin gamma complex associated protein 5 gene is involved in neurobehavioral disorders including ADHD and OCD [24].

CYFIP1 or cytoplasmic fragile X mental retardation 1 (FMR1) interacting protein 1 gene interacts with FMRP in a ribonucleoprotein complex. FMRP is the product of the FMR1 gene which is associated with FXS, the most common cause of familial intellectual disability that primarily affects males.[40] Hence, the importance of these four genes in causation of neurological development and functions but not all individuals with defects within the 15q11.2 band (ie, microdeletions or microduplications) share a clinical phenotype or are clinically affected. Therefore, this region contains genetic material that shows incomplete or low penetrance along with variable expressivity. Population surveys show that about 0.25% of controls will have the 15q11.2 BP1–BP2 microdeletion with penetrance estimated at 10.4% representing a twofold increase over the general population.[78,79]

In summary, studies of the 15q11.2 BP1–BP2 microdeletion or the Burnside Butler syndrome found that affected individuals will show developmental and language delay, neurobehavioral disturbances, and psychiatric problems that can vary from person to person. However, this emerging syndrome is now recognized with a prevalence ranging from 0.57–1.27% of patients presenting for high resolution microarray analysis accounting for a two- to fourfold increase compared with controls.[79] Autism is reported in this disorder and this chromosome anomaly is considered one of the more common findings in those present for microarray analysis. Seizures, schizophrenia, and mild dysmorphic features are less commonly seen in this condition but are at risk.

The second most common genetic cause of PWS is maternal disomy 15 where both chromosome 15s come from the mother and is found in about 25% of affected individuals.[35] There are three recognized forms of maternal disomy 15. These include maternal heterodisomy 15 with two different chromosome 15s from the mother due to errors in the first stage of meiosis (meiosis I) from nondisjunction and without cross-over events or shuffling of genes from the two maternal chromosome 15s. A second form is maternal isodisomy 15 with two identical chromosome 15s from the mother due to errors in the second division of meiosis (meiosis II) or the equational phase due to nondisjunction. The third form is segmental maternal isodisomy 15 with two partially different chromosome 15s received in the offspring from the mother due to errors in meiosis I from nondisjunction with cross-over events leading to segments of isodisomy or DNA sequence regions with identical gene alleles. High resolution CNV/SNP microarray studies have shown that if DNA segments of 10 Mb size area or larger are seen with LOH on a specific chromosome with a normal copy number, then uniparental disomy is present. For example, if this LOH occurs on chromosome 15 in the presence of DNA methylation testing showing a PWS genetic pattern, then this represents maternal disomy 15 (Fig. 4B). An individual with PWS and maternal disomy 15 can be at risk for a second genetic condition, if the mother is a carrier of an autosomal recessive gene mutation on chromosome 15 for a disorder if the gene is located in the LOH or isodisomic region with identical DNA sequences representing two copies of the same gene allele.[33,35]

Maternal disomy 15 is thought to arise from an error in gametogenesis in the female with the egg containing two chromosome 15s from the mother and if fertilized by a normal sperm with a single chromosome 15 then a trisomy 15 zygote results.[26,80] Trisomy 15 is lethal and a relatively common cause of spontaneous abortions. In an event of trisomy 15 rescue, the extra chromosome 15 is not passed in the next cell division in the developing embryo. Thus, a normal 46 chromosome number is now established in the embryo from an abnormal 47 chromosome count and leads to viability of the fetus. If the father's chromosome 15 is lost then the two remaining chromosome 15s from the mother will lead to maternal disomy 15 and the fetus is born with PWS due to his genetic subtype.

Another genetic phenomenon can occur in females with PWS due to maternal disomy 15 which involves the X chromosome. Females have two X chromosomes (one from the father and one from the mother) while males have only one X chromosome; however, the number of active X-linked genes remains constant in both sexes due to gene dosage compensation in females with or without PWS. Females generally inactivate one of their X chromosomes at random which then equals the number of X-linked genes found in the male. This process of X chromosome inactivation occurs very early in pregnancy. Occasionally, this process is not random and skewness occurs which can allow for expression of X-linked conditions in females. Hence, the trisomy 15 rescue event in the early pregnancy of a developing female with PWS and maternal disomy 15 may allow for a small number of cells to survive and to populate embryo development. These small number of cells rescued by the trisomy event may have the same X chromosome active leading to X chromosome inactivation skewness. This allows for the presence of an X-linked condition if the mother is a carrier of an X-linked gene and PWS due to maternal disomy 15.[35]

The third major category of genetic subtypes in PWS is an imprinting defect. These defects may be due to microdeletions of the imprinting center or due to epimutations through DNA methylation errors in gametogenesis.

If the father carries an imprinting defect (microdeletion) that he inherits from his mother's chromosome 15, he is unaffected due to the presence of his father's normal chromosome 15. However, when he passes the imprinting defect in his chromosome 15 onto his offspring, that offspring will then have PWS. The risk for him to pass the defect to his offspring would be 50%.[37]

6.3 Molecular Genetics and Prader–Willi Syndrome

There are dozens of genes and transcripts located in the 15q11-q13 region with most under an imprinting center control. About 10 genes are imprinted and all but 2 (UBE3A and ATP10C) are paternally expressed and regulated by DNA methylation (methylation equates to gene inactivity and unmethylation equates to gene activity). The maternally expressed UBE3A gene causes Angelman syndrome.[35,36]

The SNRPN (small nuclear ribonucleoprotein N) and a second protein coding sequence (SNURF, or SNRPN upstream reading frame) are located in the 15q11-q13 region. Exons 4–10 of the complex bicistronic SNURF–SNRPN gene encode a core spliceosomal protein (SmN) involved in mRNA splicing in the brain, whereas exons 1–3 encode a 71-amino-acid protein enriched in arginine residues. A disruption of this complex locus will cause loss of function of paternally expressed genes in this region, leading to PWS.[35,66] Multiple copies of noncoding C/D box snoRNAs or SNORDs involved in RNA processing are embedded within the long SNURF-SNRPN transcript. These include SNORD64, SNORD107, SNORD108, SNORD109A, and SNORD109B (previously referred to as HBII-13), HBII-436, HBII-437, HBII-438A, and HBII-438B, respectively. Deletions of other snoRNAs also located in the same region have been implicated in causing a PWS phenotype,[81] specifically SNORD115 (HBII-52) and SNORD116 (HBII-85). Other imprinted genes that are not components of the SNURF-SNRPN gene complex locus and located proximally are MKRN3, MAGEL2, NDN, and C15orf2. They are involved in brain development and function.[35,36]

Necdin (NDN) is a paternally expressed gene from the melanoma-associated protein (MAGE) family and required for cell cycle proliferation and differentiation. It is expressed in the hypothalamus, thalamus, and pons suggesting a role in brain development and axon growth. Mice lacking this gene show delayed migration of the sympathetic neurons, neonatal lethality, and respiratory problems. A second gene in this region is the MAGEL2 gene which is paternally expressed in various brain regions including the hypothalamus. It appears to play a role in circadian rhythm, brain structure, behavior, and maintenance of fertility recently reported to be associated with autism.[82] The MKRN3 gene is a member of the makorin (MKRN) RING finger protein gene family that encodes a specific group of proteins (makorins) and present in a wide variety of eukaryotes. The MKRN3 is abundantly expressed (paternal only) in the developing brain and nervous system. Mutations of this gene have recently been found in individuals with precocious puberty.[83]

Another class of genes located in the distal area of the 15q11-q13 region are not imprinted and include three gamma aminobutyric acid (GABA) receptor subunits (GABRB3, GABRA5, and GABRG3) with evidence of unequal expression (paternal bias).[66] The disturbances of receptor subunit genes for GABA, a major inhibitory neurotransmitter, have been implicated in a number of symptoms associated with PWS including hunger, obsessive–compulsive disorder, and altered visual perception and memory. The P gene encodes a protein required for pigment production and is located in the distal end of the chromosome 15q11-q13 region and noted to have equal expression from each parental allele. Mutations of this gene are known to cause oculocutaneous albinism II.[33,35,36]

The UBE3A and ATP10C genes are located distal to the paternally expressed SNURF–SNRPN complex gene locus and imprinted with maternal expression only. A maternal 15q11-q13 deletion including the UBE3A gene or mutations of this gene causes Angelman syndrome. Several transcripts in the 15q11-q13 region are thought to read in an antisense direction, which is complementary to DNA sequences of other genes in the region, but in a reverse direction thereby impacting on the regulation or control of gene activity including the UBE3A antisense transcript.

Butler et al.[68] reported a submicroscopic deletion of the 15q11-q13 region approximately 100–200 kb in size in a 5-year-old female with PWS in 1996 as one of the first PWS subjects with an atypical small deletion detectable only with molecular genetic or cytogenetic techniques and included the imprinting center controlling element with adjoining genetic regions now recognized as the snoRNAs (SNORDs). Later, Sahoo et al.[81] described a male child with features of PWS and a paternal deletion involving the SNORDs, particularly SNORD109A, the SNORD116 gene cluster, and part of the snoRNA SNORD115 cluster. De Smith et al.[84] also reported a 19-year-old male with a PWS phenotype and a 187 kb microdeletion of chromosome 15q11-q13 encompassing SNORD116. Therefore, growing evidence including genetic data in mice support that paternal loss of

SNORD116 may contribute to energy homeostasis, growth pattern, and reproduction; all disturbed in PWS. Larger deletions have also been reported which include the 15q14 band and in individuals with features of PWS but with additional findings. These include hearing deficits, cutaneous ear tags, and congenital heart defects. The expanded 15q11-q14 deletion contains about 60 protein-coding genes and making it larger (about twice the size) than the typical 15q11-q13 deletion seen in PWS.[69]

6.4 Clinical Features in Prader–Willi Syndrome

PWS is a complex genomic imprinting disorder characterized by mental, behavior, and physical findings with obesity as the most significant health problem. Key neuroendocrine peptides produced by the gastrointestinal system are known to play a role in feeding and eating behaviors that are disturbed in PWS.[85,86] For example, ghrelin stimulates eating and peptide YY inhibits eating, but when disturbed can lead to abnormal eating patterns and obesity. Plasma ghrelin levels are elevated in PWS as noted during infancy and early childhood.[71] This elevation potentially contributes significantly to hyperphagia seen as a cardinal feature in PWS and usually begins in early childhood, if not controlled. In addition, the noncoding snoRNA SNORD115 is thought to regulate alternative splicing of the human serotonin 5-HT2C receptor gene, an important receptor that contributes to normal eating behavior in humans.[87] When this snoRNA expression pattern is disturbed through loss of function (eg, deleted in PWS) then an altered receptor results leading to excessive eating behavior and obesity.

Bittel et al.[88] also reported on gene expression patterns in males with PWS and found several disturbed genes when compared with control males. Several of these genes were involved in eating behavior and obesity (ADIPOR2, MC2R, SAG, HCRT, STAR, OXTR) and serotonin receptors (eg, HTR2B). Hence, evidence to date supports both disturbed neuropeptide and gene/transcript products playing a role in the PWS phenotype requiring further studies.

PWS is generally divided into two major stages of clinical course development. The first stage is characterized by infantile hypotonia, temperature instability, a weak cry and poor suck, feeding difficulties, developmental delay, and hypogonadism/hypogenitalism. The second stage occurs in early childhood (2–5 years of age) and is characterized by developmental delay, behavioral problems (skin picking, tantrums, obsessive compulsions), speech delay, an insatiable appetite, food seeking with rapid weight gain, and subsequent obesity, if diet is not controlled. There are now several recognized nutritional phases and subphases referred to as 0, 1a, 1b, 2a, 3, and 4. These occur from the time of pregnancy with decreased fetal movements (ie, phase 0) then followed by severe hypotonia and feeding difficulties in infancy. Transitions then occur to an increased weight phase with or without increased caloric intake and followed by food seeking with hyperphagia in early childhood and progressed into adulthood (ie, phases 3 and 4).[89]

Short stature and small hands/feet also occur in PWS due to GH deficiency, along with rumination, sleep problems, physical inactivity, and decreased pain sensitivity. Hypopigmentation, hypogenitalism/hypogonadism, scoliosis, sleep apnea, enamel hypoplasia, and decreased saliva are more apparent later in life and accompanied by almond-shaped palpebral fissures, down-turned corners of the mouth, a narrow bifrontal diameter, a short nose, and a small chin.[34] Several of these features may be present in infancy but become more evident in childhood into adolescence and adulthood. Hypothalamic dysfunction is implicated in many manifestations of this genetic obesity-related syndrome with decreased growth velocity, endocrine disturbances accompanied by behavioral and intellectual disabilities (an average IQ of 65).

Due to developmental delay, PWS infants sit independently by 1 year of age, crawl at 16 months, walk at about 2 years, and talk (10 words) at 39 months but generally have a weak or absent cry with little spontaneous activity.[34,90] This is due to decreased muscle tone and strength. Excessive sleepiness with diminished swallowing and sucking reflexes are common and often necessitate gavage feedings with use of special nipples or gastrostomy tube placement to address the feeding problems. Growth parameters should be assessed regularly using recently established standardized growth charts developed for PWS infants.[91] Calories are adjusted accordingly, but fats required for brain growth should not be restricted as PWS infants usually require fewer calories than recommended to avoid rapid weight gain. Vitamin and mineral intake such as calcium should be added, as well as caloric intake monitored closely by a dietitian. Developmental assessments, early stimulation programs, and occupational and physical therapy services are recommended.[37] Endocrine disturbances including thyroid, growth, and sex hormones and cortisol levels should be monitored from infancy. Myopia and impaired vision may be recognized in early childhood.

Many children with PWS will be mainstreamed in the school setting, but special education and support services are often required. PWS children will generally have strengths related to reading, and with visual and long-term memory skills, but may have weaker math, sequential processing, and short-term memory skills. Verbal skills may be relative strengths particularly in those with PWS having maternal disomy 15. Unusual skill of working with jigsaw puzzles are also noted and more common in those with the 15q11-q13 deletion.[37,92]

During adolescence, hypogonadism and hypogenitalism becomes more pronounced in the vast majority of individuals with PWS due to hypothalamic hypogonadism leading to low testosterone and estrogen levels. About 90% of males will have cryptorchidism along with a micropenis and an underdeveloped scrotum.[34] A hypoplastic labia majora and minora with a small clitoris are often seen in PWS females and puberty is generally absent or delayed in both males and females. Menarche in females may occur but delayed until about 30 years of age but in rare cases, pregnancies in PWS females have been reported.[37] PWS males are infertile.

About 90% of individuals with PWS without GH treatment will develop short stature by adulthood. The average adult male will have a height of 155 cm and the adult female will be 147 cm tall.[34,37] Small hands and feet are common during adolescence and adulthood and often accompanied by scoliosis and kyphosis. Bracing or surgery may be required. PWS adolescents may weigh more than 300 pounds if caloric intake is not controlled and if GH was not administered at a younger age. Eating related deaths do occur including choking on gorged food and gastric necrosis with rupture.[93] Therefore, to avoid these complications, locking the refrigerator and food cabinets to prevent access to food and excessive eating are often required and accompanied by close supervision by all care givers inside and outside of the home setting.[94] Although psychotropic drugs are often prescribed to help control behavior problems, no specific medication has been consistently effective. Rigorous control of the food environment which should be secured is recommended to avoid hyperphagia and include regular exercise programs (about 30 min/day) to manage weight and the excessive eating behavior and associated health complications.[37,95] Mild mental deficiency is common for the family background and obsessive compulsions, outbursts, and temper tantrums may be triggered by withholding food. Self-injury (eg, skin picking) is also common beginning in childhood which correlates with the PWS genetic subtype, particularly those with the 15q11-q13 deletion.[71] Behavioral problems may begin by 3 years of age and manifest as poor peer relationships, immaturity, and inappropriate social behavior, which are reminiscent of autism.

As short stature is a main feature of PWS individuals due to GH deficiency, recombinant human GH which is FDA approved is now prescribed for PWS beginning in infancy and throughout childhood to achieve maximum adult height ultimately within the normal range.[33,37] Beneficial effects are noted on stature, weight, and body composition (increased muscle and decreased fat mass) with GH treatment in PWS.[96–98] Positive effects on motor development, strength, and cognitive effects have also been reported in both children and adults during GH treatment.[99] The quality of life is also improved in adults.[100] The prevalence of scoliosis in PWS is high (30–80%),[33,37] and thus a concern in those treated with GH and needs to be closely monitored. Recommended starting GH dose in children is 0.18–0.3 mg/kg per week given as daily subcutaneous injections with careful evaluation of clinical status, bone age measures, surveillance for scoliosis, and serum IGF-1 levels at regular intervals should be performed as well as monitoring regular caloric intake records regularly.[101] Plotting the individual growth parameters (height, weight, head circumference) on syndrome-specific standardized PWS growth charts is recommended for standard of care.[91,102]

Weight control requires a strict dietary plan and coordination with a dietitian. Increased physical activity is also encouraged and recorded. Caloric intake is restricted to 6–8 cal/cm of height for weight loss beginning in early childhood and to 10–12 cal/cm of height to maintain weight for nongrowth hormone treated individuals with PWS particularly during childhood. A general recommendation for dietary intake for PWS adolescents or adults may include 800 cal/day to lose weight or 1000–1200 cal/day to maintain weight if food seeking is controlled and exercise program in place.[95] A diet plan should be adjusted to include calories from protein at 30–35%; 45% for carbohydrates; and about 30% for fat.[37] Calcium, vitamin supplements, and fish oil are recommended throughout life.

6.5 Clinical Findings Associated with Typical 15q11-q13 Deletions (Type I and Type II) in Prader–Willi Syndrome

Differences in cognitive, psychological, and behavioral findings in young adults with PWS was first reported by Butler et al.[71] in 2004 with the longer 15q11-q13 Type I deletion involving breakpoints BP1 and BP3 versus those with the shorter Type II deletion involving breakpoints BP2 and BP3. PWS individuals with the longer Type I deletion score significantly worse in self-injurious (skin picking) assessments and in maladaptive behavior findings when compared to those with the smaller Type II deletion. Obsessive–compulsive behavior was also more frequently seen in PWS subjects with the Type I deletion. Academic achievement scores did differ between PWS subjects with the shorter Type II versus longer Type I typical 15q11-q13 deletion. This study supported that the loss of genes between BP1 and BP2 when deleted would increase the severity of the behavioral and psychological problems seen in PWS. For example, the adaptive behavior scores were generally worse in PWS individuals having the Type I deletion, such as obsessive–compulsive behaviors. Those with Type I deletion generally had poorer math and reading skills. Additional problems included poorer visual-motor integration, worse adaptive behavior, and more compulsions in those subjects with Type II deletion, particularly relating to grooming and bathing skills and compulsions that disrupted daily living. Intellectual ability and academic achievement were also poorer in those PWS subjects with Type I deletions as well as visual processing, compared with those with Type II deletions.[71]

7. FRAGILE X SYNDROME AND THE PRADER–WILLI PHENOTYPE

As obesity is a cardinal feature of PWS, it is also common in a subset of individuals with the FXS.[103] A recent survey of BMI data collected from 718 children with FXS showed a prevalence rate of obesity (31%) which was higher than that found in age matched control children (18%).[43,104] FXS is considered the most common cause of familial intellectual disability and due to a CGG triplet repeat expansion greater than 200 in size in the 5' untranslated region of the fragile X mental retardation 1 (FMR1) gene.[40,43,104] Those with a smaller repeat size (50–200) would be considered as carriers or having a premutation form while the normal number of CGG repeats is less than 50 and usually less than 30. The expanded full mutation usually leads to methylation and little to no mRNA is transcribed; thus, resulting in a lack of production of the fragile X mental retardation protein (FMRP) encoded by the FMR1 gene.[40,104] Lack of this protein is correlated with an increased childhood growth rate in FXS patients and supported by evidence from the FXS knockout mouse model showing enhanced growth rate for the mice and obesity. Interestingly, about 10% of individuals with FXS will have severe obesity, hyperphagia, hypogonadism, or delayed puberty as seen in PWS.[43]

This subset of FXS patients, termed the Prader-Willi phenotype (PWP), do not have chromosome 15q11-q13 gene deletions or abnormalities.[103]

Mild to profound intellectual disability is seen in affected males with an IQ in the 30–55 range but can extend into the borderline normal range. Hand flapping or biting (self-injury) is frequently seen along with poor eye contact. The speech is often repetitive (echolalia) and clustered. Complete lack of speech is seen in those with more severe and profound range of intellectual disability. Attention problems, hyperactivity, and overstimulation are common leading to abnormal behavioral outcomes, hyperkinetic activity, and emotional instability. Autism is seen in most FXS patients and may be recorded as high as 60%.[42] Females may also have the full mutation resulting in the same learning and behavioral problems seen in males. Females who carry a premutation may have social or generalized anxiety, and math skill deficits. They are at risk of having premature ovarian failure and twin pregnancies. Other occasional clinical findings seen in individuals with FXS include eye problems such as strabismus, nystagmus, and myopia. Seizures may also be present along with torticollis, pectus excavatum, kyphoscoliosis, and a submucous cleft palate. Some individuals also present cerebral gigantism. Expansion of the carrier gene status (premutation) to a full mutation occurs only in females and correlates with the size of the premutation. Males with the premutation are prone to anxiety and deficits in executive function and tremors (Parkinson-like) with advancing age.[42,104,105]

The FMR1 gene is located at the chromosome Xq27.3 band consisting of 17 coding exons that span 38 kb in size.[40,104] FMRP is transported back and forth between the nucleus and the cytoplasm with binding and transportation of mRNAs in the neurons at the synapse level in the brain. FMRP stabilizes mRNA or can enhance degradation of mRNA. It appears to shape the pattern of mRNA regulation throughout human fetal development. Interestingly, a lower expression of a gene located in the 15q11-q13 region involved with PWS that encodes the cytoplasmic FMR1 interacting protein 1 (CYFIP1) also works in concert with FMRP and is associated with PWP.[106]

Hence, the classic features seen in FXS patients include an elongated face with a prominent forehead and ears, flat feet, mitral valve prolapse, soft fleshy hands, and hyperextensible joints. These indicate the presence of a connective tissue disorder. For example, individuals with FXS can have macroorchidism (two to four times larger testicular volume than seen in normal adult males). Hand callouses from self-injury and a plantar crease between the first and second toes are helpful indicators for the FXS. Characteristic behavioral findings include anxiety, attention deficit hyperactivity disorder, autistic tendencies such as hyperarousal to sensory stimuli for sound and food texture, self-injury, and often obsessive–compulsive disorder.[42,104,105] The family history often supports an X-linked inheritance pattern with intellectual disability and/or autism in affected males (uncles and nephews on the maternal side). The diagnosis of FXS and molecular genetic testing to identify the number of CGG repeats on the FMR1 gene would be indicated. The obsessions seen in males with FXS when focused on food can lead to overeating, food seeking at night, and obesity. This can become severe and, thus, lead to the PWP status. PWP individuals are described as having a lack of satiation leading to hyperphagia by 5 years of age and truncal obesity involving the torso and abdomen as seen in PWS by 10 years of age. Other PWP features in common with PWS include a round face, hypotonia, short stature, delayed puberty, and hypogenitalism with a ravenous appetite including the consumption of inedible food such as raw meat and lack of satiety requiring locking refrigerators.[43]

8. ALSTRÖM SYNDROME: A RARE OBESITY-RELATED SINGLE GENE DISORDER

ALMS is another obesity-related disorder due to a single gene defect with an autosomal recessive inheritance pattern. It has a prevalence of about 1 per 1,000,000 individuals and considered very rare.[44] Due to multiorgan involvement and fibrosis, individuals with ALMS have a reduced life expectancy of less than 50 years. This obesity syndrome is caused by mutations in the ALMS1 gene located on chromosome 2p13. The symptoms typically begin during infancy. Affected individuals have visual impairment; cone-rod dystrophy; hearing loss; childhood truncal obesity; developmental delay; insulin resistance and type 2 diabetes mellitus; hypertriglyceridemia; short stature; scoliosis and kyphosis; dilated cardiomyopathy; and progressive pulmonary, hepatic, and renal dysfunction. Fibrosis can also develop in multiple organs.[45,46] Photoreceptor dystrophy which is present in 100% of subjects, begins at birth to 15 months of age and is accompanied by nystagmus leading to early blindness. Progressive sensorineural hearing loss is severe and occurs in the first 10 years of life in about three-fourths of affected individuals.

During the first 5 years of life, truncal obesity develops along with insulin resistance, acanthosis nigricans, hyperinsulinemia, and hyperlipidemia.[44] Other endocrine problems include hypothyroidism, GH alterations, polycystic ovarian syndrome, acute pancreatitis, and type 2 diabetes mellitus. About 45% of individuals with ALMS will have developmental delay and impaired learning skills along with sleep disturbances.

The protein coded by the ALMS1 gene when disturbed involves organs throughout the body. Its function is unclear but a role in ciliary function, intracellular trafficking, and adipocyte differentiation is reported.[46,47] This is similar to what is found in another obesity-related genetic disorder, Bardet–Biedl syndrome. The diagnosis of ALMS is usually made based on clinical features and confirmed by molecular genetic testing of disease causing mutations in the ALMS1 gene.

Cardiomyopathy and congestive heart failure (CHF) are seen in about 70% of individuals with this syndrome and account for major causes of morbidity and mortality. The onset of CHF can occur as early as infancy or develop later leading to fibrosis and myocardial hypertrophy with dilation and restrictive impairment of both ventricles.[46] Hepatic changes are seen including cirrhosis, steatosis, and hepatosplenomegaly followed by gastrointestinal bleeding, inflammation, and fibrosis. End-stage liver disease is the cause of death in about 10% of individuals with ALMS. Renal disease is a major cause of morbidity in this syndrome. Progressive renal impairment and varying degrees of glomerular disease, albuminuria, and interstitial fibrosis may be found with end-stage renal disease occurring as early as the teenage years. Hypogonadism is seen in both sexes with ALMS. Low–normal levels of testosterone and elevated gonadotropins are seen in males along with a small penis and testicular atrophy. Secondary sex characteristics may be normal in males. Hypogonadism in females becomes evident by puberty with delays in secondary sex characteristics and in menarche. As seen in other obesity-related disorders, infertility is a consistent finding in ALMS.[45,47,49]

Heterozygous carriers of mutations in the ALMS1 gene are generally asymptomatic. Exon 8 contains a large tandem repeat domain which encodes 34 imperfect repeats of about 50 amino acids constituting 40% of the protein.[48,49] Over 100 different mutations have been reported in the ALMS1 gene with the majority being of the nonsense or frameshift type. The coding regions of exons 8, 10, and 16 are most commonly affected. Population studies suggest founder effects in certain ethnic backgrounds with lack of disease causing mutations in the 5' half of the coding region of ALMS1 which include exons 1 through 7, seen as lethal to the developing embryo.[45,47,49]

Hyperphagia is seen in several rare but classical genetic obesity-related syndromes such as PWS, ALMS, and FXS, specifically the PWP which is seen in a subset of males (about 10%) with FXS. Using advanced genetic microarray methods to compare coding and noncoding gene expression patterns in adult males with PWS, ALMS, and nonsyndromic obesity relative to nonobese males using readily available lymphoblastoid cells, we attempted to identify disease-specific molecular signature patterns and disturbed mechanisms in obesity.[107] Through this study, we found 231 genes that were upregulated in ALMS males compared with nonobese males using stringent classification criteria, but no genes were upregulated in either obese or PWS adult males. We found 124 genes downregulated in ALMS using this high resolution expression microarray approach. Interestingly, the metallothionein gene (MT1X) which is involved in metal toxicity was significantly downregulated in ALMS and also in obese males but not in PWS males. Only the complex SNRPN gene locus was disturbed or downregulated in PWS along with several downregulated snoRNAs in the 15q11-q13 region including SNORD116, SNORD109B, SNORD109A, and SNORD107. Several snoRNAs that were disturbed in ALMS males and found outside chromosome 15 targeted multiple genes that impacted rRNA processing, developmental pathways, and associated diseases. Fifty-two miRNAs were upregulated in ALMS, and four were shared with obese males but not PWS males. Our data supported that the cell cycle (eg, PPP3CA), transcription (eg, POLE2), and developmental processing may be impacted by the upregulated genes found in ALMS. The downregulated genes found in ALMS were involved with metabolic processes (eg, FABP3), immune responses (eg, IL32), and cell signaling (eg, IL1B) when examined using computer generated pathway analyses. The high number of gene and noncoding RNA disturbances found in ALMS did contrast the observations seen in PWS and in males with nonsyndromic obesity. This contrast may reflect the progressing multiorgan pathology seen in ALMS and the ongoing disease process related to a shortened life expectancy.

9. ALBRIGHT HEREDITARY OSTEODYSTROPHY

AHO was first reported in 1942 and is due to an end-organ resistance to PTH and other hormones.[50] Obesity is a major manifestation. In addition to obesity, other clinical features of AHO consist of small stature (adult height of 54–60 in.), mild mental deficiency with an average IQ of 60 (range of 20–99), a round face with a low nasal bridge, a short nose and neck, delayed dental eruption with enamel hypoplasia, short metacarpals and metatarsals especially of the fourth and fifth digits, and short distal phalanges of the thumb.[42] These skeletal changes represent early epiphyseal fusion and may not be evident until several years of age. In addition, arm span is decreased for height. Reproductive dysfunction is common due to partial resistance to gonadotropins. Other features seen in this disorder include osteoporosis, areas of subcutaneous mineralization with calcium deposits including the basal ganglia, a thickened calvarium, variable low plasma calcium and/or high phosphate levels, and seizures secondary to low calcium levels. Hypogonadism and hypothyroidism are seen along with abnormal eye findings including lens opacity or cataracts, optic atrophy, diplopia, microphthalmia, ocular, and macular degeneration. Pancreatic dysfunction, vertebral anomalies, and advanced bone age are reported in affected individuals.[42,51,108,109]

There are two major clinical variants of this disorder referred to as PHP (PHP-Ia or PHP-Ib) and PPHP. The clinical presentation depends on the presence or absence of hormonal resistance and the AHO phenotype with the pattern of inheritance. Individuals with PHP-Ia have features of AHO, along with hypocalcemia, hyperphosphatemia, mild hypothyroidism, and hypogonadism with an abnormal response to the GH releasing hormone at the brain level. Affected individuals also have high serum PTH levels. Individuals with PPHP have the characteristic features of AHO, but without identified resistance to parathyroid and thyroid stimulating hormones and gonadotropins. PHP-Ia and PPHP have been reported in the same families, but depend on the parent of origin thereby indicating the role of the genomic imprinting status in causation of this disorder. Both disease variant types result from decreased activity of the alpha subunit of the membrane bound trimeric G subunit-regulatory signaling protein (GNAS).[42,51]

GNAS is involved in several complex cellular pathways and mechanisms by stimulating cellular adenyl cyclase activity. PHP individuals present PTH-resistance but lack features of AHO including obesity and are defined as having the PHP-Ib subtype. Most PHP-Ib cases are sporadic in origin and typically lack GNAS gene mutations. PHP-la and PPHP are caused by heterozygous mutations in exons of the GNAS gene which leads to gene inactivity. GNAS encodes the alpha subunit of the stimulatory guanine nucleotide-binding protein (G-protein) subunit alpha (Gs-alpha) and the autosomal dominant form of PHP-lb. This condition is caused by heterozygous mutations that disrupt an imprinting controlling element of the GNAS gene and not within the exons of the gene.[110,111]

Gs-alpha is a trimeric regulatory protein which functions by coupling to membrane receptors for adenyl cyclase and thus stimulates cyclic adenosine monophosphate. GNAS is expressed biallelically on both maternal and paternal alleles of chromosome 20 in most tissues but expressed monoallelically from only the maternal GNAS allele in a selected number of tissues (eg, the pituitary and thyroid glands, proximal tubules of the kidneys and gonads). Other gene transcripts of the GNAS are known and produced or expressed only from the maternal or paternal GNAS imprinted gene allele. The promoter of the gene lacks methylation. Individuals with PHP-Ia and features of AHO also have mutations of the GNAS gene or cytogenetic deletions of chromosome 20q13 including the complex imprinted GNAS gene locus. Those with PPHP (or those AHO patients without evidence of hormone resistance) also carry heterozygous inactivating GNAS gene mutations. Maternal inheritance of a mutation can lead to PHP-Ia (AHO with hormone resistance) while paternal inheritance of the same mutation leads to PPHP or AHO alone. When the altered gene is inherited from the affected father with either PHP-Ia or PPHP, then PPHP occurs in the offspring. If the same GNAS mutation is present in the mother affected with either PHP-Ia or PPHP, then the offspring will be affected with PHP-Ia.[42,110] Hence, the parent of origin of GNAS gene disturbances lead to different clinical disorders and inheritance patterns.

GNAS is a complex imprinted locus on chromosome 20q13.11. Several transcripts are produced using alternative promoters and splice sites. As illustrated when altered, these transcripts can lead to several clinical disorders or presentations. The best characterized GNAS transcript encodes the G protein Gs-alpha subunit. This transcript is biallelically expressed in most tissues, but expressed on one allele (maternal) in the gonads, pituitary, and thyroid glands and in the proximal renal tubules. Other GNAS transcripts are expressed exclusively from either the paternal or the maternal GNAS allele. Hence, the GNAS locus encodes four main transcripts: Gs-alpha, XLAS, NESP55, and the A/B transcript, and an antisense transcript (GNASAS). Gs-alpha is found ubiquitously and stimulates adenyl cyclase when coupled with a membrane receptor that generates the second messenger cyclic AMP (cAMP). Many hormones and neurotransmitters exert their actions through receptors coupled to Gs-alpha. Gs-alpha is involved in AHO, XLAS is paternally expressed, and NESP55 is exclusively maternally expressed and encodes a chromogranin-like neuroendocrine secretory protein. NESP55 does not share any amino acid sequences with Gs-alpha because of a stop codon in its unique first exon sequence. The A/B transcript uses an alternative first exon A/B or exon 1A and the antisense GNAS transcript is expressed from the paternal GNAS allele; all involved in cellular processes and when disturbed can lead to disease.[111,112] The XLAS transcript contains a previously undescribed large variant of the Gs-alpha subunit and expressed exclusively from the paternal GNAS allele that is located upstream of exon 1. It occurs most often in neuroendocrine and neurological tissue.

Reproductive- and health-related problems are common in individuals with AHO having GNAS gene disturbances. Clinical and genetic evaluations with testing are required to identify the specific genetic subtype and to supply appropriate and accurate genetic counseling leading to better medical care for affected or at-risk individuals.

In summary, as common or exogenous obesity represents an interaction of complex multifactorial agents including susceptibility genes with an obesogenic environment characterized by increased consumption of high caloric foods and a sedentary lifestyle, genetic mutations and variants are becoming more recognized as playing a role in response to our emerging environment. Research with known dietary components (high-fat food or high saturated-fat food) and population-based studies are needed to identify and develop targeted treatment protocols for personalized nutritional or pharmaceutical therapy for individuals at an increased risk for genetic factors and developing childhood obesity. In addition, the growing evidence learned about the role of genes from single gene obesity-related disorders such as Alström, Prader–Willi, and fragile X syndromes (with or without the PWP) and mutations or gene variants present in the general population need to be better characterized and information applied in the clinical setting beginning in early childhood to avoid obesity and its manifestations.

REFERENCES

1. Friedman JM. Modern science versus the stigma of obesity. *Nat Med.* 2004;10(6): 563–569.
2. Lyon HN, Hirschhorn JH. Genetics of common forms of obesity: a brief overview. *Am J Clin Nutr.* 2005;82(1):215S–217S.
3. Xia Q, Grant S. The genetics of human obesity. *Ann NY Acad Sci.* 2013;1281(1): 178–190.
4. Mason K, Page L, Balikcioglu PG. Screening for hormonal, monogenic, and syndromic disorders in obese infants and children. *Pediatr Ann.* 2014;43(9):e218.
5. Behan DF, Cox SH, Lin Y, et al. Obesity and its Relation to Mortality and Morbidity Costs. Schaumburg, Illinois: Society of Actuaries: Committee on Life Insurance Research; 2010: 1–78
6. Choquet H, Meyre D. Genetics of obesity: what have we learned? *Curr Genomics.* 2011;12(3):169.

7. Choquet H, Meyre D. Molecular basis of obesity: current status and future prospects. *Curr Genomics.* 2011;12(3):154–168.
8. Must A. Does overweight in childhood have an impact on adult health? *Nutr Rev.* 2003;61:139–142.
9. Nicklas TA, Baranowski T, Cullen KW, et al. Eating patterns, dietary quality and obesity. *J Am Coll Nutr.* 2001;20(6):599–608.
10. Parsons TJ, Power C, Logan S, et al. Childhood predictors of adult obesity: a systematic review. *Int J Obes Relat Metab Disord.* 1999;23(suppl 8):S1–S107.
11. Daniels SR, Arnett DK, Eckel RH, et al. Overweight in children and adolescents: pathophysiology, consequences, prevention, and treatment. *Circulation.* 2005;111:1999–2012.
12. Flegal KM, Wei R, Ogden C. Weight-for-stature compared with body mass index-for-age growth charts for the United States from the centers for disease control and prevention. *Am J Clin Nutr.* 2002;75:761–766.
13. Himes JH, Dietz WH. Guidelines for overweight in adolescent preventive services: recommendations from an expert committee. The Expert Committee on Clinical Guidelines for Overweight in Adolescent Preventive Services. *Am J Clin Nutr.* 1994;59:307–316.
14. Farooqi IS, O'Rahilly S. New advances in the genetics of early onset obesity. *Int J Obes.* 2005;29:1149–1152.
15. Stunkard AJ, Foch TT, Hrubec Z. A twin study of human obesity. *JAMA.* 1986;256:51–54.
16. Butler MG, McGuire AM, Manzardo AM. Clinically relevant known and candidate genes for obesity and their overlap with human infertility and reproduction. *J Assist Reprod Genet.* 2015;32(4):495–508.
17. Dasouki MJ, Youngs EL, Hovanes K. Structural chromosome abnormalities associated with obesity: report of four new subjects and review of literature. *Curr Genomics.* 2011;12(3):190.
18. Walters RG, Jacquemont S, Valsesia A, et al. A new highly penetrant form of obesity due to deletions on chromosome 16p11. *Nature.* 2010;463:671–675.
19. Scherag A, Dina C, Hinney A, et al. Two new Loci for body-weight regulation identified in a joint analysis of genome-wide association studies for early-onset extreme obesity in French and German study groups. *PLoS Genet.* 2010;6:e1000916.
20. Boissel S, Reish O, Proulx K, et al. Loss-of-function mutation in the dioxygenase-encoding FTO gene causes severe growth retardation and multiple malformations. *Am J Hum Genet.* 2009;85:106–111.
21. Deliard S, Panossian S, Mentch FD, et al. The missense variation landscape of FTO, MC4R, and TMEM18 in obese children of African Ancestry. *Obesity.* 2013;21(1):159–163.
22. Loos RJ, Lindgren CM, Li S, et al. Common variants near MC4R are associated with fat mass, weight and risk of obesity. *Nat Genet.* 2008;40:768–775.
23. Meyre D, Proulx K, Kawagoe-Takaki H, et al. Prevalence of loss-of-function FTO mutations in lean and obese individuals. *Diabetes.* 2009;59:311–318.
24. Thorleifsson G, Walters GB, Gudbjartsson DF, et al. Genome-wide association yields new sequence variants at seven loci that associate with measures of obesity. *Nat Genet.* 2009;41:18–24.
25. Willer CJ, Speliotes EK, Loos RJF, et al. Six new loci associated with body mass index highlight a neuronal influence on body weight regulation. *Nat Genet.* 2009;41:25–34.
26. Butler MG. Genomic imprinting disorders in humans: a mini-review. *J Assist Reprod Genet.* 2009;26(9–10):477–486.
27. Jungheim ES, Travieso JL, Carson KR, et al. Obesity and reproductive function. *Obstet Gynecol Clin North Am.* 2012;39:479–493.

28. Venkatesh T, Suresh PS, Tsutsumi R. New insights into the genetic basis of infertility. *Appl Clin Genet.* 2014;7:235–243.
29. Butler MG, Manzardo AM. Androgen receptor (AR) gene CAG trinucleotide repeat length associated with body composition measures in non-syndromic obese, non-obese and Prader–Willi syndrome individuals. *J Assist Reprod Genet.* 2015;32:1–7.
30. Saute JA, Silva AC, Souza GN, et al. Body mass index is inversely correlated with the expanded CAG repeat length in SCA3/MJD patients. *Cerebellum.* 2012;11:771–774.
31. Barber TM, Franks S. Genetics of polycystic ovary syndrome. *Front Horm Res.* 2013;40:28–39.
32. Prader A, Labhart A, Willi H. Ein sydnrom von adipositas, kleinwuchs, kryptorchismus und oligophrenie nach myatonieartigem zustand im neugeborenenalter. *Schweizerische Medizinishce Wochenschrift.* 1956;6(3):1260–1261.
33. Angulo MA, Butler MG, Cataletto ME. Prader–Willi syndrome: a review of clinical, genetic, and endocrine findings. *J Endocrinol Invest.* 2015;38:1–15.
34. Butler MG. Prader–Willi syndrome: current understanding of cause and diagnosis. *Am J Med Genet.* 1990;35(3):319–332.
35. Butler MG. Prader–Willi syndrome: obesity due to genomic imprinting. *Curr Genomics.* 2011;12:204–215.
36. Cassidy SB, Schwartz S, Miller JL, et al. Prader–Willi syndrome. *Genet Med.* 2012;14(1):10–26.
37. Butler MG, Lee PDK, Whitman BY, eds. In: *Management of Prader–Willi Syndrome 3rd ed. New York, NY: Springer Science+Business Media; 2006:.*
38. Butler MG, Thompson T. Prader–Willi syndrome: clinical and genetic finding. *Endocrinology.* 2000;10:3S–16S.
39. Lubs HA. A marker X chromosome. *Am J Hum Genet.* 1969;21(3):231.
40. Hagerman RJ, Hagerman PJ. Fragile X syndrome: model of gene-brain-behavior relationships. *Mol Genet Metab.* 2001;74(1):89–97.
41. Chudley AE, Hagerman RJ. Fragile X syndrome. *J Pediatr.* 1987;110(6):821–831.
42. Jones KL, ed. In: *Smiths Recognizable Patterns of Human Malformation 6th ed. Philadelphia, PA: WB Saunders Company; 2006:.*
43. Lozano R, Rosero CA, Hagerman RJ. Fragile X spectrum disorders. *Intractable Rare Dis Res.* 2014;3(4):134.
44. Álvarez-Satta M, Castro-Sánchez S, Valverde D. Alström syndrome: current perspectives. *Appl Clin Genet.* 2015;8:171–179.
45. Marshall JD, Bronson RT, Collin GB, et al. New Alström syndrome phenotypes based on the evaluation of 182 cases. *Arch Intern Med.* 2005;165(6):675–683.
46. Marshall JD, Maffei P, Collin GB, Naggert JK. Alström syndrome: genetics and clinical overview. *Curr Genomics.* 2011;12(3):225–235.
47. Waters AM, Beales PL. Ciliopathies: an expanding disease spectrum. *Pediatr Nephrol.* 2011;26(7):1039–1056.
48. Collin GB, Marshall JD, Ikeda A, et al. Mutations in ALMS1 cause obesity, type 2 diabetes and neurosensory degeneration in Alstrom syndrome. *Nat Genet.* 2002;31(1):74–78.
49. Marshall JD, Beck S, Maffei P, et al. Alström syndrome. *Eur J Hum Genet.* 2007;15(12):1193–1202.
50. Albright F, Burnett CH, Smith PH, et al. Pseudohypoparathyroidism—an example of "Seabright-Bantam syndrome": report of three cases. *Endocrinology.* 1942;30:922–932.
51. Levine MA. Clinical spectrum and pathogenesis of pseudohypoparathyroidism. *Rev Endocr Metab Disord.* 2000;1(4):265–274.
52. Bartolomei MS, Tilghman SM. Genomic imprinting in mammals. *Annu Rev Genet.* 1997;31:493–525.
53. Walter J, Paulsen M. Imprinting and disease. *Semin Cell Dev Biol.* 2003;14:101–110.

54. Delaval K, Wagschal A, Feil R. Epigenetic deregulation of imprinting in congenital diseases of aberrant growth. *Bioessays.* 2006;28(5):453–459.
55. Platonov ES, Isaev DA. Genomic imprinting in the epigenetics of mammals. *Genetika.* 2006;42(9):1235–1249.
56. Murphy SK, Jirtle RL. Imprinting evolution and the price of silence. *Bioessays.* 2003;25 (6):577–588.
57. Haig D, Graham C. Genomic imprinting and the strange case of the insulin-like growth factor II receptor. *Cell.* 1991;64(6):1045–1046.
58. Zhang Y, Tycko B. Monoallelic expression of the human H19 gene. *Nat Genet.* 1992;1 (1):40–44.
59. Niemitz EL, Feinberg AP. Epigenetics and assisted reproductive technology: a call for investigation. *Am J Hum Genet.* 2004;74(4):599–609.
60. Luedi PP, Dietrich FS, Weidman JR, et al. Computational and experimental identification of novel human imprinted genes. *Genome Res.* 2007;17(12):1723–1730.
61. Falk MJ, Curtis CA, Bass NE, et al. Maternal uniparental disomy chromosome 14: case report and literature review. *Pediatr Neurol.* 2005;32(2):116–120.
62. Temple K, Shrubb V, Lever M, et al. Isolated imprinting mutation of the DLK1/GTL2 locus associated with a clinical presentation of maternal uniparental disomy of chromosome 14. *J Med Genet.* 2007;44:637–640.
63. Zakharova IS, Shevchenko AI, Zakian SM. Monoallelic gene expression in mammals. *Chromosoma.* 2009;118(3):279–290.
64. Nicholls RD, Knepper JL. Genome organization, function, and imprinting in Prader–Willi and Angelman syndromes. *Annu Rev Genomics Hum Genet.* 2001;2:153–175.
65. Christian SL, Robinson WP, Huang B, et al. Molecular characterization of two proximal deletion breakpoint regions in both Prader–Willi and Angelman syndrome patients. *Am J Hum Genet.* 1995;57(1):40–48.
66. Bittel DC, Butler MG. Prader–Willi syndrome: clinical genetics, cytogenetics and molecular biology. *Expert Rev Mol Med.* 2005;7(14):1–20.
67. Butler MG, Fischer W, Kibiryeva N, et al. Array comparative genomic hybridization (aCGH) analysis in Prader–Willi syndrome. *Am J Med Genet A.* 2008;146(7):854–860.
68. Butler MG, Christian SL, Kubota T, et al. A 5-year-old white girl with Prader–Willi syndrome and a submicroscopic deletion of chromosome 15q11q13. *Am J Med Genet.* 1996;65(2):137–141.
69. Butler MG, Bittel DC, Kibiryeva N, et al. An interstitial 15q11-q14 deletion: expanded Prader–Willi syndrome phenotype. *Am J Med Genet A.* 2010;152A(2):404–408.
70. Kim SJ, Miller JL, Kuipers PJ, et al. Unique and atypical deletions in Prader–Willi syndrome reveal distinct phenotypes. *Hum Mol Genet.* 2012;20(3):283–290.
71. Butler MG, Bittel DC, Talebizadeh Z. Plasma peptide YY and ghrelin levels in infants and children with Prader–Willi syndrome. *J Pediatr Endocrinol Metab.* 2004;17: 1177–1184.
72. Zarcone J, Napolitano D, Peterson C, et al. The relationship between compulsive behaviour and academic achievement across the three genetic subtypes of Prader–Willi syndrome. *J Intellect Disabil Res.* 2007;51(pt 6):478–487.
73. Bittel DC, Kibiryeva N, Butler MG. Expression of 4 genes between chromosome 15 breakpoints 1 and 2 and behavioral outcomes in Prader–Willi syndrome. *Pediatrics.* 2006;118(4):e1276–e1283.
74. Dykens EM, Roof E. Behavior in Prader–Willi syndrome: relationship to genetic subtypes and age. *J Child Psychol Psychiatry.* 2008;49(9):1001–1008.
75. Chai JH, Locke DP, Greally JM, et al. Identification of four highly conserved genes between breakpoint hotspots BP1and BP2 of the Prader–Willi/Angelman syndromes deletion region that have undergone evolutionary transposition mediated by flanking duplicons. *Am J Hum Genet.* 2003;73(4):898–925.

76. Svenstrup K, Moller RS, Christensen J, et al. NIPA1 mutation in complex hereditary spastic paraplegia with epilepsy. *Eur J Neurol.* 2011;18(9):1197–1199.
77. Jiang Y, Zhang Y, Zhang P, et al. NIPA2 located in 15q11. 2 is mutated in patients with childhood absence epilepsy. *Hum Genet.* 2012;131(7):1217–1224.
78. Burnside RD, Pasion R, Mikhail FM, et al. Microdeletion/microduplication of proximal 15q11.2 between BP1 and BP2: a susceptibility region for neurological dysfunction including developmental and language delay. *Hum Genet.* 2011;130(4): 517–528.
79. Cox DM, Butler MG. The 15q11.2 BP1-BP2 microdeletion syndrome: a review. *Int J Mol Sci.* 2015;16(2):4068–4082.
80. Cassidy SB, Lai LW, Erickson RP, et al. Trisomy 15 with loss of the paternal 15 as a cause of Prader–Willi syndrome due to maternal disomy. *Am J Hum Genet.* 1992;51:701–708.
81. Sahoo T, del Gaudio D, German JR, et al. Prader–Willi phenotype caused by paternal deficiency for the HBII-85 C/D box small nucleolar RNA cluster. *Nat Genet.* 2008;40 (6):719–721.
82. Schaaf CP, Gonzalez-Garay ML, Xia F, et al. Truncating mutations of MAGEL2 cause Prader–Willi phenotypes and autism. *Nat Genet.* 2013;45(11):1405–1408.
83. Macedo DB, Abreu AP, Reis AC, et al. Central precocious puberty that appears to be sporadic caused by paternally inherited mutations in the imprinted gene makorin ring finger 3. *J Clin Endocrinol Metab.* 2014;99(6):E1097–E1103.
84. de Smith AJ, Purmann C, Walters RG, et al. A deletion of the HBII-85 class of small nucleolar RNAs (snoRNAs) is associated with hyperphagia, obesity and hypogonadism. *Hum Mol Genet.* 2009;18(17):3257–3265.
85. Cummings DE, Clement K, Purnell JQ, et al. Elevated plasma ghrelin levels in Prader–Willi syndrome. *Nat Med.* 2002;8(7):643–644.
86. Irizarry K, Bain J, Butler MG, et al. Metabolic profiling in Prader–Willi syndrome and non-syndromic obesity: sex differences and the role of growth hormone treatment. *Clin Endocrinol (Oxf).* 2015;83(6):797–805.
87. Kishore S, Stamm S. The snoRNA HBII-52 regulates alternative splicing of the serotonin receptor 2C. *Science.* 2006;311(5758):230–232.
88. Bittel DC, Kibiryeva N, Sell SM, et al. Whole genome microarray analysis of gene expression in Prader–Willi syndrome. *Am J Med Genet.* 2007;143A(5):430–432.
89. Miller JL, Lynn CH, Driscoll DC, et al. Nutritional phases in Prader–Willi syndrome. *Am J Med Genet A.* 2011;155A(5):1040–1049.
90. Butler MG, Meaney FJ, Palmer CG. Clinical and cytogenetic survey of 39 individuals with Prader–Labhart–Willi syndrome. *Am J Med Genet.* 1986;23(3):793–809.
91. Butler MG, Sturich J, Lee J, et al. Growth standards of infants with Prader–Willi syndrome. *Pediatrics.* 2011;127(4):687–695.
92. Dykens EM. Are jigsaw puzzle skills "spared" in persons with Prader–Willi syndrome? *J Child Psychol Psychiatry.* 2002;43(3):343–352.
93. Stevenson DA, Heinemann J, Angulo M, et al. Gastric rupture and necrosis in Prader–Willi syndrome. *J Pediatr Gastroenterol Nutr.* 2007;45(2):272–274.
94. McCandless SE. Clinical report-health supervision for children with Prader–Willi syndrome. *Pediatrics.* 2011;127(1):195–204.
95. Butler MG. Management of obesity in Prader–Willi syndrome. *Nat Clin Pract Endocrinol Metab.* 2006;2(11):592–593.
96. Angulo M, Castro-Magana M, Mazur B, et al. Growth hormone secretion and effects of growth hormone therapy on growth velocity and weight gain in children with Prader–Willi syndrome. *J Pediatr Endocrinol Metab.* 1996;9(3):393–400.
97. Carrel AL, Myers SE, Whitman BY, et al. Long-term growth hormone therapy changes the natural history of body composition and motor function in children with Prader–Willi syndrome. *J Clin Endocrinol Metab.* 2010;95:1131–1136.

98. Eiholzer U, l'Allemand D. Growth hormone normalizes height, prediction of final height and hand length in children with Prader–Willi syndrome after 4 years of therapy. *Horm Res.* 2000;53(4):185–192.
99. Höybye C, Thorén M, Böhm B. Cognitive, emotional, physical and social effects of growth hormone treatment in adults with Prader–Willi syndrome. *J Intellect Disabil Res.* 2005;49:245–252.
100. Butler MG, Smith BK, Lee J, et al. Effects of growth hormone treatment in adults with Prader–Willi syndrome. *Growth Horm IGF Res.* 2013;23(3):81–87.
101. Deal CL, Tony M, Höybye C, et al. Growth hormone research society workshop summary: consensus guidelines for recombinant human growth hormone therapy in Prader–Willi syndrome. *J Clin Endocrinol Metab.* 2013;98:E1072–E1087.
102. Butler MG, Lee J, Manzardo AM, et al. Growth charts for non-growth hormone treated Prader–Willi syndrome. *Pediatrics.* 2015;135(1):e126–e135.
103. De Vries BB, Fryns JP, Butler MG, et al. Clinical and molecular studies in fragile X patients with a Prader–Willi-like phenotype. *J Med Genet.* 1993;30(9):761–766.
104. Hagerman PJ, Hagerman RJ. Fragile X–associated tremor/ataxia syndrome. *Ann NY Acad Sci.* 2015;1338(1):58–70.
105. Niu YQ, Yang JC, Hall DA, et al. Parkinsonism in fragile X-associated tremor/ataxia syndrome (FXTAS): revisited. *Parkinsonism Relat Disord.* 2014;20(4):456–459.
106. Nowicki ST, Tassone F, Ono MY, et al. The Prader–Willi phenotype of fragile X syndrome. *J Dev Behav Pediatr.* 2007;28(2):133–138.
107. Butler MG, Wang K, Marshall JD, et al. Coding and noncoding expression patterns associated with rare obesity-related disorders: Prader–Willi and Alström syndromes. *Adv Genomics Genet.* 2015;5:53.
108. Bastepe M. The GNAS locus and pseudohypoparathyroidism. *Adv Exp Med Biol.* 2008;626:27–40.
109. Fitch N. Albright's hereditary osteodystrophy: a review. *Am J Med Genet.* 1982;11 (1):11–29.
110. Jüppner H. Genetic and epigenetic defects at the GNAS locus cause different forms of pseudohypoparathyroidism. *Annales d'endocrinologie.* 2015;76(2):92–97.
111. Mantovani G, Elli FM, Spada A. GNAS epigenetic defects and pseudohypoparathyroidism: time for a new classification? *Horm Metab Res.* 2012;44(10):716–723.
112. Bastepe M, Jüppner H. GNAS locus and pseudohypoparathyroidism. *Horm Res Paediatr.* 2005;63(2):65–74.

CHAPTER TWO

PCSK1 Variants and Human Obesity

B. Ramos-Molina[1], M.G. Martin[2], I. Lindberg[3,*]

[1] Department of Human Genetics, KU Leuven, Leuven, Belgium
[2] Department of Pediatrics, Division of Gastroenterology and Nutrition, Mattel Children's Hospital and the David Geffen School of Medicine, Los Angeles, CA, United States of America
[3] Department of Anatomy and Neurobiology, University of Maryland, Baltimore, MD, United States of America
* Corresponding author. E-mail address: iLindberg@som.umaryland.edu

Contents

Abstract

PCSK1, encoding prohormone convertase 1/3 (PC1/3), was one of the first genes linked to monogenic early-onset obesity. PC1/3 is a protease involved in the biosynthetic processing of a variety of neuropeptides and prohormones in endocrine tissues. PC1/3 activity is essential for the activating cleavage of many peptide hormone precursors implicated in the regulation of food ingestion, glucose homeostasis, and energy homeostasis, for example, proopiomelanocortin, proinsulin, proglucagon, and

Progress in Molecular Biology and Translational Science, Volume 140
ISSN 1877-1173
http://dx.doi.org/10.1016/bs.pmbts.2015.12.001

proghrelin. A large number of genome-wide association studies in a variety of different populations have now firmly established a link between three *PCSK1* polymorphisms frequent in the population and increased risk of obesity. Human subjects with PC1/3 deficiency, a rare autosomal-recessive disorder caused by the presence of loss-of-function mutations in both alleles, are obese and display a complex set of endocrinopathies. Increasing numbers of genetic diagnoses of infants with persistent diarrhea has recently led to the finding of many novel *PCSK1* mutations. *PCSK1*-deficient infants experience severe intestinal malabsorption during the first years of life, requiring controlled nutrition; these children then become hyperphagic, with associated obesity. The biochemical characterization of novel loss-of-function *PCSK1* mutations has resulted in the discovery of new pathological mechanisms affecting the cell biology of the endocrine cell beyond simple loss of enzyme activity, for example, dominant-negative effects of certain mutants on wild-type PC1/3 protein, and activation of the cellular unfolded protein response by endoplasmic reticulum–retained mutants. A better understanding of these molecular and cellular pathologies may illuminate possible treatments for the complex endocrinopathy of *PCSK1* deficiency, including obesity.

1. INTRODUCTION

Neuropeptides and prohormones require proteolytic activation prior to release for subsequent action in various tissues. This activation occurs within the regulated secretory pathway of both neurons and endocrine cells. The enzymes responsible for the initial proteolytic cleavages are known as *prohormone* convertases (PCs). PCs are calcium-dependent serine endoproteases that undergo final maturation in the acidic environment of the secretory granules, where they acquire full catalytic activity.[1] PC1/3 (also known as PC1, PC3, and SPC3) and PC2 are abundantly expressed in neuroendocrine cells, although the two enzymes do not always colocalize within the same cells.[2] The processing of protein precursors in nonendocrine cells is mediated by other widely expressed enzymes (in this case *proprotein* convertases), which act within the constitutive secretory pathway. The best characterized proprotein convertase is furin, which has been reported to process a variety of substrates such as growth factors and receptors, plasma proteins, proteases, bacterial toxins, and viral coat glycoproteins.[3] The other mammalian members included within this family are PC4, PC5/6, PACE4, and PC7.

Although PC1/3 usually initiates precursor cleavage, either PC1/3 or PC2 can process prohormones and neuropeptides by cleaving at conserved dibasic sites, most frequently Lys-Arg or Arg-Arg.[2] Genetically modified mice lacking convertase expression[4,5] have provided considerable information on convertase-mediated processing of substrates. Human studies have also provided considerable information on PC1/3 function, beginning with early studies of an individual with compound heterozygous mutations in the gene that encodes PC1/3, *PCSK1*.[6] This patient exhibited multiple endocrinopathies as well as morbid obesity.[6] *PCSK1* has subsequently become recognized as one of the first genes causing rare forms of monogenic obesity.[7] More recently, many index cases have been described with severe biallelic variants of the *PCSK1* gene that result, in most cases, in the total lack of PC1/3 activity.[8,9] In addition, numerous genome-wide association studies (GWAS) show a convincing association between certain polymorphisms in the *PCSK1* gene that are exceedingly common in the population and increased risk of obesity.[10,11] In this chapter, we review the biochemistry and cell biology of PC1/3 and its prohormone substrates related to obesity and glucose metabolism. A specific goal of the chapter is to address the possible mechanism(s) by which the total or partial lack of PC1/3 activity in humans may contribute to obesity.

2. PROHORMONE CONVERTASE 1/3: GENERAL PROPERTIES

PC1/3 was the third member of the proprotein convertase family to be cloned from mammalian organisms, after furin and PC2.[12,13] The PC1/3 protein is encoded by the *PCSK1* gene, which is located on chromosome 5q15–21 in humans, and chromosome 13c in the mouse.[12,14] PC1/3 mRNA is translated by ribosomes located on the rough endoplasmic reticulum (ER) into a 753-residue protein, the signal peptide-bearing inactive precursor preproPC1/3. Like other members of the subtilisin superfamily (which includes bacterial subtilisin and the yeast subtilase kex2), the catalytic triad is formed by the amino acids Asp, His, and Ser, arranged in this order (as opposed to the Ser, His, Asp order present in trypsin-like proteases). The domain structure of proPC1/3 consists of four well-defined domains: a prodomain, a catalytic domain, a P domain specific to this superfamily (including yeast kexin), and a carboxy-terminal domain. The catalytic domain is highly conserved among various species and paralogs. The propeptide domain is essential to the proper folding and ER exit of the protein.[15] The P domain plays an important role in the regulation of the calcium and pH dependence of PCs,[16] while the carboxy-terminal domain participates in the sorting of

PC1/3 into secretory granules.[17] The carboxy-terminal domain may also assist the oligomerization of PC1/3, since the truncated form of PC1/3 is less prone to oligomerization.[18]

Fig. 1 depicts the cellular maturation of pre-proPC1. After signal peptide removal in the ER, the 94 kDa proPC1/3 zymogen is multiply N-glycosylated. While the PC1/3 prodomain is known to undergo early and rapid cleavage at the primary cleavage site by autocatalysis,[19,20] it is still unclear whether the cleaved prodomain remains bound to the catalytic domain, to act as an intramolecular inhibitor, until the enzyme reaches the trans-Golgi network (TGN), as occurs for the closely related enzyme furin.[21] The furin prodomain remains associated with the catalytic domain until a Golgi compartment of lower pH is reached, at which time a "histidine switch" is activated which results in secondary site cleavage, dissociation of propeptide fragments, and enzyme activation.[22,23] This type of study has not yet been performed for PC1/3, though it is clear that PC1/3 can also operate within TGN[24]; indeed, the initial cleavages of POMC are known to occur in the Golgi.[25] Within the TGN, proPC1/3 undergoes terminal glycosylation, and also becomes sulfated.[19,20,26]

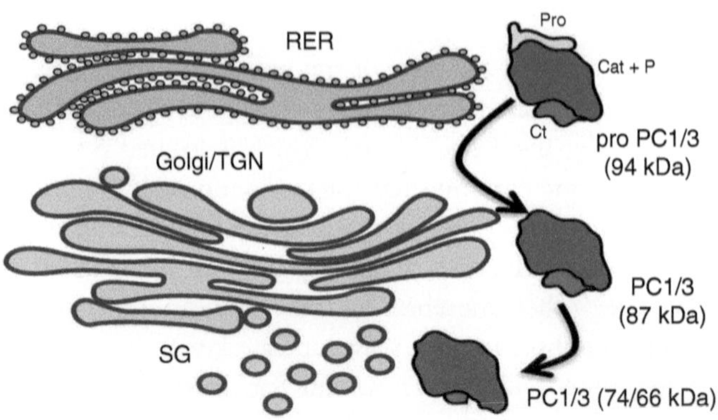

Figure 1 PC1/3 trafficking in the regulated secretory pathway. During translocation the signal peptide is removed, resulting in the inactive zymogen proPC1/3. In the ER, proPC1/3 is folded, N-glycosylated and undergoes rapid initial prodomain cleavage. In the Golgi compartments, the protein undergoes additional posttranslational modifications, including further addition glycosylation, sugar trimming; and sulfation. Once in the TGN, PC1/3 is sorted into dense-core granules, and the Ct domain is autocatalytically cleaved, resulting in complete maturation. *RER*, rough endoplasmic reticulum; *TGN*, trans-Golgi network; *SG*, secretory granules; *Pro*, prodomain; *Cat*, catalytic domain; *Ct*, carboxy-terminal domain.

The 87 kDa active PC1/3 form undergoes further intermolecular autocatalytic cleavage of the carboxyl terminal domain within dense-core secretory granules[27] (Fig. 1). These truncated forms (74 and 66 kDa PC1/3) are much more catalytically active than the parent 87 kDa species, and also require higher calcium concentrations and lower pH (5–5.5) to exhibit maximal activity.[27,28] Although both of these smaller forms are much more unstable than the 87 kDa form,[27] impaired processing to the smaller forms is associated with weaker tissue activity against various substrates.[29]

PC1/3 expression is restricted to the central and peripheral nervous systems and endocrine and neuroendocrine organs. In brain, PC1/3 levels are particularly high in certain regions of the hypothalamus, such as the arcuate nucleus,[30] an area known to be involved in the regulation of food intake and body weight. In the periphery, PC1/3 is also highly expressed in the pancreas, pituitary, stomach, and intestine, as well as in the adrenal and thyroid glands.[31–34] PC1/3 expression is limited to specific ghrelin-expressing endocrine cells in the stomach; the α- and β-cells of the islets of Langerhans in the pancreas; and various enteroendocrine cells present in the intestine.[31,32,35] As discussed later, these specialized cells play an important role in appetite, glucose homeostasis, and nutrient assimilation by secreting several PC1/3 products including ghrelin, insulin, and proglucagon-derived peptides such as GLP-1.

3. PC1/3 SUBSTRATES INVOLVED IN THE REGULATION OF ENERGY HOMEOSTASIS AND FOOD BEHAVIOR

Many if not all circulating and hypothalamic peptide hormones involved in appetite regulation require PC processing, for example, the production of ACTH, β-endorphin, and α-MSH from POMC.[36–38] Both cell studies and peptidomic analyses of tissues from mice lacking PC1/3 activity support the idea that there are a large number of PC1/3 substrates (see Table 2[2] and Table 1[39]). In this section we will describe only the PC1/3 substrates most likely to be involved in the control of appetite, glucose homeostasis, and nutrient assimilation. Table 1 shows a set of such anorexigenic and orexigenic peptide hormones whose biosynthesis is thought to involve PC1/3 action; this should not be taken as a comprehensive list of PC1/3-synthesized peptides involved in these processes, as the biosynthesis of many peptides (for example, neuromedin U) is not yet understood.

Table 1 Main PC1/3 Substrates Involved in Feeding Behavior and Energy Homeostasis.

Hormone/ Neuropeptide	Peptide Precursor	Feeding Regulation	Tissue Distribution
α-MSH	POMC	Anorexigenic	Hypothalamus
CART	proCART	Anorexigenic	Hypothalamus
NPY	proNPY	Orexigenic	Hypothalamus
AgRP	proAgRP	Orexigenic	Hypothalamus
Orexin/hypocretin	proOrexin/hypocretin	Orexigenic	Hypothalamus
Ghrelin	proghrelin	Orexigenic	Stomach
Insulin	proinsulin	Anorexigenic	Pancreas
CCK	proCCK	Anorexigenic	Small intestine
GLP-1	proglucagon	Anorexigenic	Small intestine

The processing of POMC and proinsulin to α-MSH and insulin requires the sequential action of PC1/3 and PC2. The bioactive peptides CART, NPY, AgRP, ghrelin, CCK, and GLP-1 are generated by PC1/3-mediated processing of their respective peptide precursors; proorexin/hypocretin processing has not yet been investigated, although PC1/3 is expressed in orexin-positive neurons.

3.1 PC1/3 Activation of Neuropeptides in the Hypothalamus: A Key Role in the Melanocortin Pathway?

The hypothalamus is an important brain region that contains distinct neuronal populations which control energy homeostasis and feeding behavior. As mentioned earlier, hypothalamic PC1/3 expression is especially high in the arcuate nucleus, where it resides in two leptin-sensitive neuronal populations: proopiomelanocortin (POMC)-expressing neurons, and neuropeptide Y (NPY) and agouti-related peptide (AgRP)-expressing neurons.[40,41]

Alpha-melanocyte stimulating hormone (α-MSH) is a potent POMC-derived anorexigenic hormone that reduces food intake, increases energy expenditure, and regulates glucose metabolism through the activation of melanocortin 4 receptor (MC4R) in the hypothalamic paraventricular nucleus[42,43] (Fig. 2), among other target areas in the central nervous system. α-MSH is produced within POMC-expressing neurons by a proteolytic process consisting of PC1/3-mediated generation of adrenocorticotropic hormone (ACTH) followed by PC2-specific cleavage to ACTH 1-18, and then terminal modification (dibasic trimming, acetylation, and amidation) to result in α-MSH.[44] While pituitary α-MSH levels were not altered in a peptidomics study of PC1/3 null mice,[45] reduced hypothalamic α-MSH levels were detected in a PC1/3-deficient mouse model[29] (discussed later). The mechanism underlying this decrease is not yet clear and represents an intriguing area for further study.

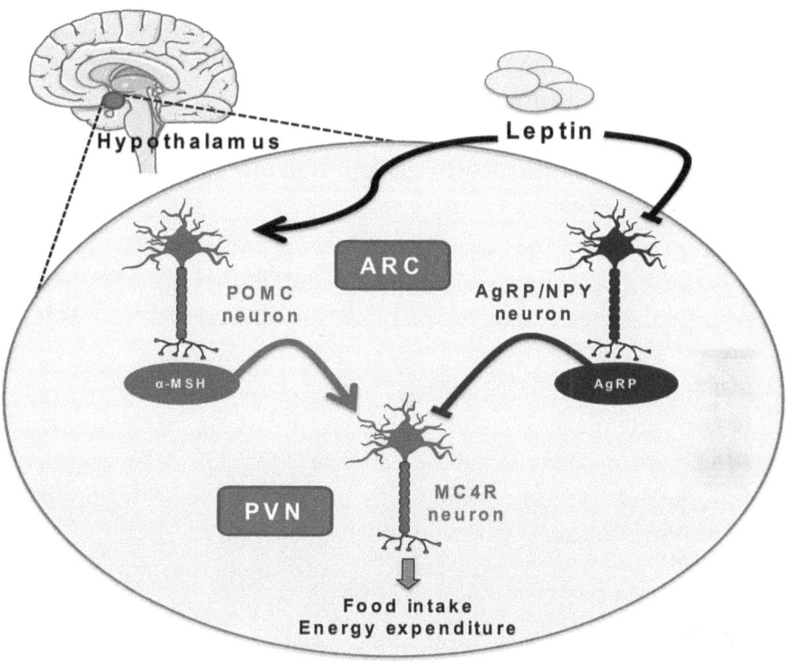

Figure 2 PC1/3 activity is essential to the leptin–melanocortin pathway. The hypothalamus is a key regulator of food intake and energy expenditure, responding to the action of peripheral hormones. Leptin is an anorexigenic hormone, produced and secreted by white adipose tissue, whose circulating levels are proportional to the mass of body fat. Leptin binds to leptin receptors on AgRP/NPY- and POMC-expressing neurons in the ARC of the hypothalamus. Leptin binding inhibits the synthesis of AgRP and stimulates the synthesis of POMC. Acting together, PC1/3 and PC2 accomplish the proteolytic processing of POMC to α-MSH within POMC-expressing neurons. PC1/3 is specifically required for the processing of proAgRP in NPY/AgRP-producing neurons. α-MSH and AgRP compete for the MC4R in the paraventricular nucleus *(PVN)* of the hypothalamus. Whereas increased receptor activity by α-MSH binding generates an anorexigenic signal, AgRP reduces receptor activity, generating an orexigenic signal.

POMC processing is clearly impaired in cases of biallelic human *PCSK1* deficiency, with the generation of large ACTH-containing fragments rather than mature ACTH.[6,8,9,46] While in some cases this lack of authentic ACTH results in adrenocortical insufficiency,[8] other subjects exhibit normal cortisol levels.[46] Jackson et al. provided evidence for the presence of authentic ACTH in the plasma of a human *PCSK1* null patient, suggesting redundancy in the enzymatic processing of POMC.[46] It seems likely that further processing of large ACTH-containing intermediates can occur in plasma. Whether *PCSK1*-deficient patients exhibit reduced circulating α-MSH levels has not yet been examined; this might be expected based on results obtained in mice.

3.2 Other PC1/3-Generated Peptide Hormones Involved in Feeding Behaviors

In addition to α-MSH, the cocaine- and amphetamine-regulated transcript (CART) is an anorexigenic peptide expressed in POMC-positive hypothalamic neurons that clearly requires PC1/3 action for its synthesis.[47] AgRP is a PC1/3-synthesized[48] orexigenic neuropeptide that stimulates appetite and decreases energy expenditure, thereby antagonizing the effects of POMC within the arcuate nucleus. AgRP suppresses MC4R and MC3R activity by competing with α-MSH for these receptors (Fig. 2). NPY is a second orexigenic peptide hormone known to be involved in the hypothalamic control of food intake[49] whose synthesis depends on PC1/3.[50] Lastly, PC1/3 colocalizes with orexin-positive neurons in the hypothalamus,[51] though no biochemical studies have been performed to demonstrate whether its synthesis is PC1/3-dependent. Alterations in the processing of proCART, proorexin, proAgRP, and proNPY have not yet been examined in cases of human *PCSK1* deficiency, for example, by measuring CSF levels.

3.3 PC1/3 Processing of Proinsulin to Insulin in Pancreatic β-Cells

Given the key role of peripheral insulin in energy balance and glucose homeostasis it is important to understand how this anabolic hormone is produced in the pancreas. Like the majority of secreted proteins, insulin is initially synthesized as an inactive precursor (proinsulin) within pancreatic β-cells. When proinsulin reaches the TGN and is packaged into the secretory granules, it undergoes maturation to active insulin via the joint action of PC1/3 and PC2; here, PC1/3 plays the larger role.[52,53] The absence of either PC1/3 or PC2 causes incomplete processing from proinsulin to insulin, as demonstrated by the severe deficiency of active insulin found in both *Pcsk1* and *Pcsk2* null mice.[4,54] Human studies show enormous increases in circulating proinsulin in *PCSK1* deficiency.[6,8,9,46,55–58] Interestingly, while neither hyperglycemia nor diabetes mellitus has been reported in either humans or mice bearing severe loss-of-function *PCSK1* variants, common polymorphisms are associated with increased risk of these conditions (see discussion later).

3.4 PC1/3 Activity in the Gastrointestinal System

Proglucagon, encoded by the *Gcg* gene, is expressed in the α-cells of the pancreatic islets, in certain brain areas, and in enteroendocrine cells within the small intestine; it is cleaved in a tissue-specific manner by PC1/3 and PC2, resulting in different profiles of bioactive peptides. For example, in pancreatic α-cells, which express PC2, proglucagon is specifically processed to glucagon.[59–61] In contrast, PC1/3-mediated processing of proglucagon in intestinal L cells, where PC2 is lacking, results in the production of glucagon-like peptides such as GLP-1 and GLP-2, but not of glucagon itself.[62–64] Whereas GLP-1 regulates food intake behavior and glucose and energy homeostasis, GLP-2 is mostly involved in the regulation of intestinal absorption of nutrients and gut permeability.[65,66] While studies of *Pcsk1* null mice clearly show a profound deficit in proglucagon processing,[4] human studies have been ambiguous in this regard,[46,56] and further human studies may be warranted.

Glucose-dependent insulinotropic polypeptide (GIP) is a 42-amino acid gut hormone produced by PC1/3[67] in K-cells, a population of enteroendocrine cells mainly located in the upper intestine. GIP is secreted by K cells in response to food ingestion and plays an important role as a modulator of energy homeostasis. GIP receptors are widely distributed in peripheral tissues, including fat, gut, pancreas, bone, and brain.[68] While it is well accepted that GIP stimulates glucose-dependent insulin secretion, other proposed GIP functions, for example, the induction of GLP-1 secretion in L cells, or the regulation of food intake through specific hypothalamic circuits, are still controversial.[69]

Ghrelin is a potent orexigenic hormone secreted under fasting conditions from specialized endocrine cells within the stomach. Proghrelin is processed to ghrelin only by PC1/3.[70,71] In the hypothalamus, ghrelin activates NPY/AgRP neurons, inducing food intake.[72] Ghrelin is also associated with increased GLP-1 secretion by enteroendocrine L cells[73]; thus, loss of PC1/3 activity could conceivably affect GLP-1 secretion both directly and indirectly.

PC1/3 also mediates the processing of several other gut peptide hormone precursors, for example, procholecystokinin (proCCK)[74] and progastrin[75] and likely others.[76,77]

In summary, PC1/3 activity is required for the biosynthesis of a variety of hypothalamic, pancreatic, and gut peptide hormones implicated in the control of food intake, both orexigenic and anorexigenic, as summarized in

Table1. While direct links between a specific PC1/3-synthesized peptide and alterations in food intake have not yet been demonstrated, the known involvement of PC1/3 in the synthesis of so many different appetitive peptides supports the idea that PC1/3 deficiency alters the balance of anorexigenic/orexigenic peptidergic signals. Of note, appetitive changes could arise either as the direct loss of a PC1/3-made anorexigenic peptide, or an indirect mechanism, that is, the loss of negative peptidergic input (presumably a PC1/3 product) to a tonic orexigenic system. Indeed, the influence of other hypothalamic peptides, for example, β-endorphin, corticotropin-releasing hormone (CRH) and/or neurotensin, on the PC1/3-mediated obesity phenotype remains unexplored.

4. PC1/3 DEFICIENCY AND DISEASE

4.1 PC1/3 Knockout Mouse and the PC1/3-Deficient N222D Mouse

Mouse models of PC1/3 deficiency illustrate the critical importance of PC1/3-mediated processing of peptide hormones. A *Pcsk1* null mouse was constructed in which the *Pcsk1* gene lacks the promoter and exon 1.[4] While these targeted *Pcsk1* null mice exhibit highly increased proinsulin levels as well as greatly impaired processing of proglucagon to glucagon-like peptide-1 and -2, they do not show differences in glucose tolerance or develop diabetes; nor are these mice obese.[4] The *Pcsk1* null mouse has a high postnatal mortality rate, with only one-third surviving beyond 7 days of life, and most succumbing by the second day (Ref. [4]; M. Martin, unpublished observations). In the few animals that survive into adulthood, despite grossly impaired processing of POMC to ACTH, corticosterone levels are normal.[4] The small size of these mice was ascribed to the lack of processed growth hormone-releasing hormone (GHRH); in agreement, growth hormone and insulin-like growth factor 1 (IGF1) levels are very low. While this was postulated to represent a species-specific effect,[4] impaired growth is also seen in a subset of *PCSK1*-deficient humans (see further).

A second *Pcsk1* null was constructed which involved a 32.7-kb chromosomal deletion with insertion of a neomycin cassette and produced aberrant *Pcsk1* gene products. While no homozygous mice were obtained, female heterozygote mice exhibited stunted growth under a low fat diet, and catch-up growth under a high-fat diet.[78]

A *Pcsk1*-deficient mouse was created by chemical mutagenesis of C57BL/6 inbred mice, which resulted in a PC1/3 N222D missense mutation.[29] Homozygotes show glucose intolerance and become obese; obesity is associated with increased food intake. Heterozygotes show an intermediate phenotype. Despite the fact that proinsulin processing is impaired, neither diabetes nor insulin resistance were observed. Processing of 87 kDa PC1/3 to the 66 kDa form is clearly reduced in these mice.[29] In contrast to the proteomics results showing no change in pituitary α-MSH in the *Pcsk1* null mouse,[45] hypothalamic α-MSH levels are clearly reduced in the *Pcsk1* N222D mouse.[29] Heterozygous mutant N222D mice also show a robust obesity phenotype that is completely lacking in heterozygous PC1/3 null mice. These differences support the idea that the N222D mutation may produce alterations in hypothalamic neuronal cell biology that are not seen in haploinsufficient PC1/3 null mice. In support of this idea, we have observed a dominant-negative deleterious effect of the N222D PC1/3 mutant protein on wild-type PC1/3 trafficking and on proinsulin processing in cell lines and intact islets.[79] More recent data from our laboratory show that the expression of human PC1/3 mutant proteins which are ER-retained (eg, G593R and G209R) reduces wild-type PC1/3 trafficking to secretory granules, and also results in the induction of ER stress markers.[80] While N222D PC1/3 was not examined in this latter work, it is likely that this mutant mouse protein induces similar ER stress, potentially explaining the decrease in hypothalamic α-MSH in the N222D mutant mouse. In agreement, another group has shown that both N222D PC1/3 and certain human ER-retained PC1/3 forms coimmunoprecipitate with the ER stress marker BiP (J. W. Creemers, personal communication).

The modulation of PC1/3 message levels by glucose, and in response to feeding has been previously reviewed.[2]

4.2 Common and Rare Human Haploinsufficient PCSK1 Polymorphisms are Also Linked to Obesity

The strong involvement of *PCSK1* in obesity was established by a variety of GWAS that have shown that several common nonsynonymous polymorphisms in this gene constitute a large risk factor for obesity. Indeed, *PCSK1* constitutes the third most prevalent monogenic contributor to the risk of obesity in a variety of different human populations.[81–84] The minor allele frequency of N221D (rs6232) ranges from 3–5% in different ethnic populations, while that of the linked Q665E/S690T variants (which nearly always appear together; rs6234/6235) is 24%.[85] The presence of an N221D allele has been correlated with decreased insulin sensitivity, decreased oral glucose tolerance, and increased proinsulin levels, while the Q665E/S690T variation has been found to be associated with decreased fasting glucose and an increased insulin level.[86,87] However, obesity risk data regarding the Q665E/S690T allele are not always consistent for different populations.[83,88,89] Aside from ethnicity, this may also have to do with group sizes analyzed; a larger effect in children and younger adults[83]; sex differences, with larger effects in men versus women[89,90]; and/or other factors as yet unknown, for example, varying diets, activity, or hormonal status. Additional recent data in the form of two very large metaanalyses show that the N221D polymorphism is more strongly associated with childhood obesity than with adult obesity[10]; and confirm that both N221D and Q665E/S690T are significantly associated with the risks of both common and extreme obesity-risks that are strongly modulated by both age and ethnicity.[11]

The biochemical basis for effects of common polymorphisms on PC1/3 activity is unclear. The linked Q665E/S690T polymorphisms are found in the carboxyl terminal tail of PC1/3, a domain linked to protein targeting as well as enzyme inhibition.[2] Mbikay et al. have proposed that the secondary structure of this domain is significantly affected by these two mutations.[91] Although this group found that the joint presence of these three common polymorphisms (N221D and Q665E/S690T) conferred a significant increase in PC1/3 carboxy-terminal processing, no differences in POMC processing could be identified in these cell culture studies.[91] This leads to the question of whether this paired polymorphism biologically affects PC1/3-mediated peptide processing; produces cellular effects by another route; or requires intact endocrine tissue (rather than a tumor cell line) in order to manifest a biological effect. In contrast, the N221D mutation was proposed to affect catalysis due to its probable location adjacent to a calcium binding site[81]; however, early studies showed very little effect on enzyme activity.[85] We have recently reexamined the activity of human PC1/3 proteins containing the common polymorphisms N221D or the linked Q665E/S690T pair in cell culture studies. While the linked polymorphisms were not able to influence PC1/3 activity, the N221D polymorphism decreased the specific activity of the secreted protein by about 30%.[80] This large difference was not detected in the prior study[85] that employed a background vector containing an additional S357G mutation, which represents an activating mutation.[92]

In addition to the common mutations, nearly 1% of obese populations exhibit rare *PCSK1* mutations, and some of these are associated with profound effects on metabolism.[81] Eight rare heterozygous mutations resulting in partial loss of PC1/3 activity have been identified in obese patients.[81] Despite not displaying the severe endocrine phenotype of PC1/3-lacking subjects (described further), carriers of these mutations exhibit an 8.7-fold increased risk of obesity over subjects lacking these mutations.[81] When analyzed in cell culture, seven of the eight mutant PC1/3 proteins were found to exhibit impaired maturation and secretion, while one propeptide mutation seemed benign; again, this may be due to the inherent limitations of transfection analyses in constitutively-secreting cells rather than endocrine tissues.

The rare and common human *PCSK1* polymorphisms that have been genetically linked to obesity are shown on the top of Fig. 3. While most of the rare heterozygous mutations found in obese populations clearly impact the cell biology and enzymatic activity of PC1/3,[81] the physiological mechanisms underlying the obesity phenotype of the Q665E/S690T polymorphisms are as yet unclear. Obesity may result from changes in appetite; differences in energy metabolism; and alterations in glucose handling subsequent to peptide precursor processing effects; or, most likely, a combination of these factors. Somewhat paradoxically, the S690T polymorphism was found to be associated with higher levels of GIP and glucagon during a meal test.[90] An unexplored factor is the possibility of interactions between *PCSK1* polymorphisms with common substrate polymorphisms, for example, proglucagon.[93]

4.3 Rare Human Biallelic PCSK1 Mutations Cause Enteric and Systemic Endocrinopathies and Obesity

The first human case of *PCSK1* deficiency, identified over 18 years ago, was a patient with a loss-of-function G593R missense mutation on one allele, and a premature stop codon which resulted in an inactive truncated form on the other allele.[6] Several subsequent studies have confirmed that the total or near-total loss of PC1/3 activity observed in patients with rare inactivating mutations in both *PCSK1* alleles results in a complex age-dependent clinical phenotype that includes severe diarrhea, many other endocrine dysfunctions, and obesity.[6,8,9,46] These rare mutations are shown on the bottom portion of Fig. 3. Most of the severe human mutations shown in Fig. 3 were identified in infants via *PCSK1* exon sequencing[8,9,46,57,90] or total exome sequencing,[55] although the first case was identified in an adult.[6] This list of rare mutations is clearly not complete; as *PCSK1* insufficiency is increasingly recognized as a genetic cause of pediatric intestinal failure, the number of children being diagnosed with mutations in this gene is growing.

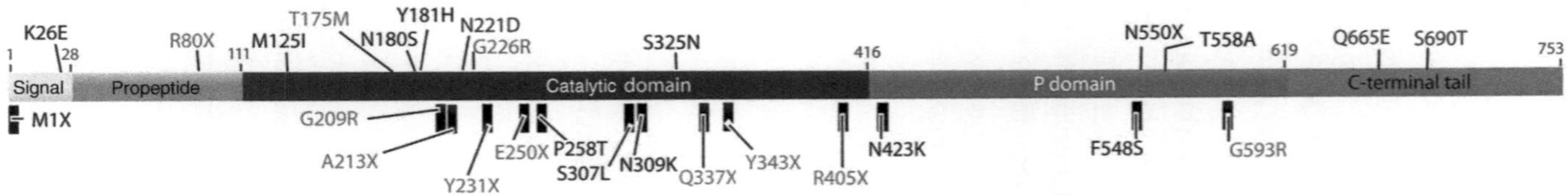

Figure 3 Human PC1/3 mutations and variations identified to date. Over two dozen different missense and nonsense mutations in *PCSK1* have been identified in infants with diarrhea and endocrine patients (shown below panel)[6,8,9,57] or in obese populations[81] (shown above panel). The three most frequent polymorphisms (N221D, Q665E/S690T) are shown in dark gray (blue in the web version); note that these may coexist with other mutations. Mutant proteins that are not secreted and likely are ER-retained are shown in light gray (red in the web version) *Diagram modified from Ref. [8].*

The earliest clinical phenotype in humans with loss-of-function mutations in *PCSK1* is severe generalized malabsorptive diarrhea that always begins in the immediate postnatal period.[8] All forms of nutrients induce the diarrhea in a dose-dependent manner, and infants present with severe weight loss and diarrhea that ceases only with fasting or the consumption of water. Endoscopic biopsies are generally normal and do not reveal any of the other classical findings seen in children with other forms of generalized malabsorptive diarrhea.[94] More importantly, the early clinical phenotype mirrors what is seen in children with loss-of-function mutations of *NEUROG3*, a transcription factor required for enteroendocrine and islet cell development.[95] Interestingly, children with both forms of enteric endocrinopathies require parenteral nutrition in addition to reduced enteral feeds during their first 18 months of life in order to thrive. Curiously, the requirement for parenteral nutrition is significantly diminished beyond 2 years of age, and the children can thrive despite the presence of persistent diarrhea that is lifelong. Unlike children with *NEUROG3* deficiency, patients with *PCSK1* mutations develop a profound appetite that results in significant weight gain and eventually obesity beyond 2 years of age.[8] Overall, our experience suggests that children with loss of PC1/3 activity will experience persistent diarrhea and malabsorption throughout life, and early in life will require intravenous support that may be tapered off as the child ages.

While children with *PCSK1* deficiency develop obesity as they age, there are important differences that distinguish them from children with other forms of monogenic obesity.[8] Children with mutations in the leptin–-melanocortin pathway experience profound obesity that becomes apparent within the first several months of life.[96] These disorders include mutations of leptin and its receptor; POMC; and MC4R, in which all patients exhibit obesity that persists throughout adulthood, with class III morbid obesity.[96,97] In contrast, as outlined earlier, patients with defects in PC1/3 have significant malnutrition and failure to thrive that requires intravenous support to manage.[8] Given PC1/3's extensive role in processing many of the peptidergic components of pathways regulating energy balance and appetite, we might have anticipated more profound obesity in probands with *PCSK1* deficiency. However, unlike other disorders that are associated with central forms of hyperphagia, *PCSK1* is also expressed in enteroendocrine cells, where it processes prohormones that presumably have an essential role in facilitating nutrient assimilation. Despite their early phase of poor growth—which is limited to children with *PCSK1* deficiency—these children eventually develop hyperphagia that results in significant weight gain despite their persistent diarrhea.[56] This diarrhea certainly results in malabsorption of nutrients that, if fully absorbed, would have likely resulted in profound weight gain and morbid obesity. Therefore, unlike other forms of monogenic obesity, significant alterations of PC1/3 action have both a central and enteric endocrinopathy that balances the weight gain as children age.

The malabsorption that occurs in subjects with both *PCSK1* and *NEUROG3* deficiency suggests that enteroendocrine cells must secrete peptide hormone(s) that normally facilitate nutrient assimilation. The murine counterparts of these disorders mimic some of the changes associated with the human disease, including early postnatal mortality.[4,98] Interestingly, none of the murine models that exhibit selective depletion of an enteric hormone or its receptor—including several of the more well-studied enteric hormones such as GLP-1 and 2, ghrelin and PYY—are associated with immediate postnatal complications.[77] These findings suggest either the existence of an uncharacterized gut hormone that requires PC1/3 processing; or that several of the currently established hormones have a redundant role in augmenting nutrient absorption. Several PC1/3-synthesized gut hormone agonists are used in clinical practice, including GLP-1 and 2 analogs that are used to treat diabetes mellitus and short bowel syndrome.[99,100] It is certainly plausible that one of these or other agonists can be used off-label to manage patients with PC1/3 deficiency; however, therapy should likely be limited to the first several years of life, since prolonged therapy could result in severe weight gain due to reversal of malabsorption. Overall, these various disorders highlight the important role of enteroendocrine cells and PC1/3 processing in the nutrient absorptive capacity of the gut.

Several *PCSK1*-deficient children exhibited linear growth abnormalities and received therapy for growth hormone (GH) deficiency. Very few of the other nonsyndromic monogenetic disorders that are associated with obesity also exhibit evidence of poor linear growth. For instance, patients with leptin deficiency exhibit normal height,[9] while those with haploinsufficient *MC4R* mutations actually show accelerated linear growth.[97,101,102] While GH deficiency was in fact reported in the few surviving *Pcsk1* null mice, as discussed earlier, the presence of an active furin consensus site in human proGHRH was thought to negate the requirement for PC1/3 processing.[4] In agreement, impaired linear growth has not been previously reported in prior reports of human *PCSK1* deficiency.[6,9,57] Given our recent findings of lowered GH in a few rare cases of biallelic PC1/3 deficiency,[8,55,103] we speculate that reduced linear growth might also contribute to elevated BMI in subjects carrying one of the common *PCSK1* polymorphisms known to be associated with obesity risk. Indeed, while the N221D polymorphism is not associated with decreased height, the more common Q665E/S690T variants are strongly associated with impaired growth in large cohorts of obese patients.[11] The exceedingly common frequency of this paired variant in the general population (heterozygous ~39%, and homozygous ~7%), suggests that it could contribute to impaired growth velocity, a highly heritable trait[104]; however, *PCSK1* has not thus far been identified as a gene contributing to height in the general population.[105] Collectively, these recent findings suggest that abnormalities in PC1/3 function might lead to reduced linear growth that can contribute to the degree of obesity as measured by BMI.

The rare missense human mutations fall into three general classes. The first class, exemplified by G593R, includes PC1/3 proteins that are retained within the endoplasmic reticulum (see red-colored mutants in Fig 3.) and always result in a severe clinical phenotype; these PC1/3 proteins are poorly folded and are likely all subject to endoplasmic reticulum-associated degradation. In the middle are mutant proteins which are somewhat impaired in secretory pathway trafficking, but are still able to traffic to the secretory granule; the mouse N222D PC1/3 mutant exemplifies this category,[79] although it is likely that certain rare human variants also fall into this class. In the third class are mutant proteins which are clearly well folded and well secreted, but exhibit impaired catalytic activity, such as N309K, an oxyanion hole mutant lacking all *in-trans* enzymatic activity.[103] However, many catalytic domain mutants do exhibit some degree of misfolding, as judged by severe impairments in secretion.[6,8,9] This is important because ER-retained mutant proteins both negatively impact the trafficking of wild-type PC1/3 and cause ER stress,[80] which is increasingly recognized as a contributing factor to obesity.[106,107] Interestingly, a dominant-negative effect on proPC1/3 processing was found in an obese patient heterozygous for a propeptide mutation resulting in a truncated protein,[108] further supporting the idea that dominant-negative interactions of PC1/3 mutations play an important role in human obesity. However, in a study of the four heterozygous children of the first *PCSK1* null patient, the proinsulin to des31,32 proinsulin ratios were found to vary between 0.75 and 5.25 (with a parental ratio of 2.8),[46] indicating a profound influence of unknown modifier genes on the proinsulin processing phenotype.

In obese patients undergoing bariatric surgery, levels of PC1/3 mRNA in jejunal mucosal samples were significantly reduced in patients who had type 2 diabetes as compared to those who did not.[109] PC1/3 expression is regulated by the transcription factor Pax6, and a GWAS study has shown that a polymorphism in this gene is associated with lower islet expression of PC1/3.[110]

Patients with rare inactivating *PCSK1* mutations share certain similarities with patients bearing loss-of-function POMC mutations, including early-onset obesity and adrenal insufficiency.[111–114] A similar phenotype is also observed in mouse POMC knockouts,[115] highlighting the critical importance of the POMC system in adrenal development and feeding behavior. Patients with *PCSK1*-inactivating mutations possess both similarities and differences with a recently described homozygous patient bearing a truncated *CPE* gene (encoding carboxypeptidase E) who not only is morbidly obese and has hypogonadotropic hypogonadism, but also exhibits diabetes mellitus and developmental delay.[116] In agreement, male *Cpe*$^{fat/fat}$ mice, who contain an inactivating *Cpe* mutation, are not only obese,[117] but also have lower testicular weights.[118] Interestingly, both these mice[119] and leptin-deficient mice[51] exhibit reduced PC1/3 levels in several brain regions, including the hypothalamus, where the lack of certain PC1/3-processed neuropeptides might plausibly result in obesity and hypogonadotropic hypogonadism. No individuals have yet been identified with inactivating mutations of *PCSK2*; based on mouse data,[5] these patients would be expected to be hypoglycemic, with little circulating glucagon, and also hyperproinsulinemic.

4.4 ProSAAS and Obesity

ProSAAS is a 30 kDa PC1/3 binding protein that was originally identified during a mass spectrometry search for novel brain peptides.[120] Homologous sequences (which are at best 30% conserved) have been identified only in vertebrates,[121] though PC1/3 has been found in invertebrates.[122] The proSAAS protein, abundantly distributed within neural and endocrine tissues,[123,124] consists of two domains separated by a furin cleavage site; both the carboxy- and the amino-terminal domains can be cleaved off to generate bioactive peptides, while the interior domain remains intact.[125–127] The C-terminal 41-residue peptide represents a nanomolar inhibitor of active PC1/3[128,129] and contains the inhibitory hexapeptide sequence Leu-Leu-Arg-Val-Lys-Arg, originally discovered as a PC1/3 inhibitor during a peptide library screen.[130] Interestingly, proSAAS transgenic mice exhibit an obesity phenotype,[131] while male proSAAS knockout mice are lean.[132] It is unclear at this point if either phenotype is related to the differential production of PC1/3-cleaved peptides, as the levels of most PC1/3-generated peptide products do not differ in proSAAS knockout mouse brains.[132]

Recent data obtained in mice show that proSAAS expression is downregulated by the transcription factor Pax6, while PC1/3 expression is upregulated.[133] *Pax6* heterozygote mice exhibit increased proSAAS expression, reduced PC1/3 expression, reduced proteolytic conversion to the 66 kDa form, and lower islet PC1/3 activity.[134,135] These effects clearly translate into differences in proinsulin conversion and glucose handling[134]; however, weight differences were not examined in these studies. No human proSAAS polymorphisms or mutations have been reported to date.

ProSAAS has often been identified in proteomics screens of cerebrospinal fluid in patients with various neurodegenerative diseases.[136] The internal unprocessed domain of the proSAAS protein potently blocks the aggregation of various fibrillating proteins, such as beta amyloid,[136] islet amyloid polypeptide[137] and synuclein (T. Jarvela and I. Lindberg, unpublished results). Obesity effects seen in proSAAS transgenic and knockout mice may be mediated in part through its antiaggregation bioactivity, rather than directly through PC1/3 inhibition.

5. CONCLUSIONS

The discovery that *PCSK1* mutations and polymorphisms represent an underlying cause of human obesity has provided considerable insight into a genetic cause of metabolic differences between humans. At the same time, many questions remain as to the physiologic mechanisms which underlie the powerful effects of certain *PCSK1* mutations on endocrine physiology. While human studies clearly demonstrate that rare inactivating mutations in *PCSK1* strongly impact PC1/3-mediated peptide hormone processing, heterozygotes show a variety of phenotypes that range from normal to obese.[46,108] Indeed, heterozygote *Pcsk1* null mice show no obesity phenotype,[4] suggesting that haploinsufficiency cannot explain the profound metabolic effects seen in certain *PCSK1*-deficient patients. Dominant-negative effects of mutant proteins on the disposition of wild-type PC1/3 protein, as well as a possible negative impact on proteostasis within various peptide hormone-producing cells, must be considered as possible contributing reasons to their metabolic phenotype.

An enduring puzzle remains the biochemical basis for the pediatric malabsorptive diarrhea typical of homozygote loss-of-function *PCSK1* patients. Whether this is due directly to the lack of specific PC1/3-synthesized peptide products arising from the lack of enzyme activity, or is secondary to general enteroendocrine cell dysfunction evolving from proteotoxicity of *PCSK1* mutants, is not yet clear. A possible role for proSAAS, the known PC1/3 binding protein, should also be examined in models of intestinal dysfunction. Another interesting future area for study is to investigate the developmental factors that enable infants with *PCSK1* mutations to overcome the severe malabsorptive syndrome between 2 and 3 years of age.

The contribution of the highly common *PCSK1* polymorphisms to human obesity risk is also puzzling, given their relatively benign effects on enzyme activity when expressed in cell culture. It is likely that a better grasp of their contribution to metabolism would come from studies of animal models of these polymorphisms, in which possible alterations in peptide hormone processing would take place in the same types of endocrine cells as in humans (rather than in the constitutive and endocrine tumor cell models traditionally used). Thus, the creation of an N221D mutated mouse, or of a mouse knock-in expressing a humanized *PCSK1* gene containing the Q665E/S690T paired polymorphisms, would permit a detailed analysis of peptide hormone processing in endocrine tissues and plasma, and enable studies of the interaction of these polymorphisms with diet. Analysis of the status of ER stress markers in the hypothalami of these mice could potentially explain the reduction in hypothalamic α-MSH in mice bearing *Pcsk1* mutations. Alternatively, the creation of human iPSC lines could enable the investigation of the physiological consequences of rare and common variants in an *in vitro* setting.

In summary, while many questions remain regarding the physiological mechanisms contributing to *PCSK1*-mediated obesity, the work of the last decade has clearly increased our understanding of the role of *PCSK1* in metabolism. With the recent identification of over 2 dozen human missense and nonsense variants that provide a range of enzyme inactivation from apparently nondetectable (Q665E/S690T) to total loss (ER-retained and truncated mutants), we are well poised to elucidate the precise relationship between enzyme activity and level of clinical impairment. The discoveries of dominant-negative impairment of wild-type enzyme function, as well as of the clear ER stress evoked by ER-retained mutant PC1/3 proteins, provide additional avenues for future investigation of the cell biology of peptide hormone processing in various peptidergic tissues. Accumulating data on pediatric cases of *PCSK1* deficiency will continue to provide clinicians with improved therapeutic regimens, while work in culture and in other animal models will deliver answers to many of the questions posed previously.

ACKNOWLEDGMENTS

This review was supported by NIH grant DA05087-27 to IL and NIH DK083762 and a grant from the California Institute of Regenerative Medicine (CIRM), RT2-01985, to MGM. We thank Malcolm Low for comments on the manuscript.

REFERENCES

1. Seidah NG. The proprotein convertases, 20 years later. *Methods Mol Biol.* 2011;768: 23–57.
2. Hoshino A, Lindberg I. In: Fricker LD, Devi L, eds. In: *Peptide Biosynthesis: Prohormone Convertases1/3and2. Princeton, NJ: Morgan & Claypool Life Sciences Publishers; 2012.*
3. Thomas G. Furin at the cutting edge: from protein traffic to embryogenesis and disease. *Nat Rev Mol Cell Biol.* 2002;3(10):753–766.
4. Zhu X, Zhou A, Dey A, et al. Disruption of PC1/3 expression in mice causes dwarfism and multiple neuroendocrine peptide processing defects. *Proc Natl Acad Sci USA.* 2002;99(16):10293–10298.
5. Furuta M, Yano H, Zhou A, et al. Defective prohormone processing and altered pancreatic islet morphology in mice lacking active SPC2. *Proc Natl Acad Sci USA.* 1997;94:6646–6651.
6. Jackson RS, Creemers JW, Ohagi S, et al. Obesity and impaired prohormone processing associated with mutations in the human prohormone convertase 1 gene. *Nat Genet.* 1997;16:303–306.
7. Farooqi S, O'Rahilly S. Genetics of obesity in humans. *Endocr Rev.* 2006;27(7):710–718.
8. Martin MG, Lindberg I, Solorzano-Vargas RS, et al. Congenital proprotein convertase 1/3 deficiency causes malabsorptive diarrhea and other endocrinopathies in a pediatric cohort. *Gastroenterology.* 2013;145(1):138–148.
9. Farooqi IS, Volders K, Stanhope R, et al. Hyperphagia and early-onset obesity due to a novel homozygous missense mutation in prohormone convertase 1/3. *J Clin Endocrinol Metab.* 2007;92(9):3369–3373.
10. Stijnen P, Tuand K, Varga TV, Franks PW, Aertgeerts B, Creemers JW. The association of common variants in PCSK1 with obesity: a HuGE review and meta-analysis. *Am J Epidemiol.* 2014;180(11):1051–1065.
11. Nead KT, Li A, Wehner MR, et al. Contribution of common non-synonymous variants in PCSK1 to body mass index variation and risk of obesity: a systematic review and meta-analysis with evidence from up to 331 175 individuals. *Hum Mol Genet.* 2015;24 (12):3582–3594.
12. Seidah NG, Marcinkiewicz M, Benjannet S, et al. Cloning and primary sequence of a mouse candidate prohormone convertase PC1 homologous to PC2, furin, and Kex2: distinct chromosomal localization and messenger RNA distribution in brain and pituitary compared to PC2. *Mol Endocrinol.* 1991;5(1):111–122.
13. Smeekens SP, Avruch AS, LaMendola J, Chan SJ, Steiner DF. Identification of a cDNA encoding a second putative prohormone convertase related to PC2 in AtT20 cells and islets of Langerhans. *Proc Natl Acad Sci USA.* 1991;88:340–344.
14. Seidah NG, Mattei MG, Gaspar L, Benjannet S, Mbikay M, Chretien M. Chromosomal assignments of the genes for neuroendocrine convertase PC1 (NEC1) to human 5q15-21, neuroendocrine convertase PC2 (NEC2) to human 20p11.1-11.2, and furin (mouse 7[D1-E2] region). *Genomics.* 1991;11(1):103–107.
15. Creemers JW, Vey M, Schafer W, et al. Endoproteolytic cleavage of its propeptide is a prerequisite for efficient transport of furin out of the endoplasmic reticulum. *J Biol Chem.* 1995;270(6):2695–2702.
16. Zhou A, Martin S, Lipkind G, LaMendola J, Steiner DF. Regulatory role of the P domain of subtilisin-like prohormone convertases. *J Biol Chem.* 1998;273:11107–11114.

17. Dikeakos JD, Di Lello P, Lacombe MJ, et al. Functional and structural characterization of a dense core secretory granule sorting domain from the PC1/3 protease. *Proc Natl Acad Sci USA.* 2009;106(18):7408–7413.
18. Hoshino A, Kowalska D, Jean F, Lazure C, Lindberg I. Modulation of PC1/3 activity by self-interaction and substrate binding. *Endocrinology.* 2011;152(4):1402–1411.
19. Benjannet S, Rondeau N, Paquet L, et al. Comparative biosynthesis, covalent post-translational modifications and efficiency of prosegment cleavage of the prohormone convertases PC1 and PC2: glycosylation, sulphation and identification of the intracellular site of prosegment cleavage of PC1 and PC2. *Biochem J.* 1993;294(Pt 3):735–743.
20. Lindberg I. Evidence for cleavage of the PC1/PC3 pro-segment in the endoplasmic reticulum. *Mol Cell Neurosci.* 1994;5(3):263–268.
21. Anderson ED, VanSlyke JK, Thulin CD, Jean F, Thomas G. Activation of the furin endoprotease is a multiple-step process: requirements for acidification and internal propeptide cleavage. *EMBO J.* 1997;16(7):1508–1518.
22. Feliciangeli SF, Thomas L, Scott GK, et al. Identification of a pH sensor in the furin propeptide that regulates enzyme activation. *J Biol Chem.* 2006;281(23):16108–16116.
23. Dillon SL, Williamson DM, Elferich J, et al. Propeptides are sufficient to regulate organelle-specific pH-dependent activation of furin and proprotein convertase 1/3. *J Mol Biol.* 2012;423(1):47–62.
24. Paquet L, Zhou A, Chang EY, Mains RE. Peptide biosynthetic processing: distinguishing prohormone convertases PC1 and PC2. *Mol Cell Endocrinol.* 1996;120(2):161–168.
25. Schnabel E, Mains RE, Farquahar MG. Proteolytic processing of proACTH-endorphin begins in the Golgi complex of pituitary corticotropes and AtT-20 cells. *Mol Endocrinol.* 1989;3(8):1223–1234.
26. Zandberg WF, Benjannet S, Hamelin J, Pinto BM, Seidah NG. N-Glycosylation controls trafficking, zymogen activation, and substrate processing of proprotein convertases PC1/3 and SKI-1. *Glycobiology.* 2011;21(10):1290–1300.
27. Zhou Y, Lindberg I. Enzymatic properties of carboxyl-terminally truncated prohormone convertase 1 (PC1/SPC3) and evidence for autocatalytic conversion. *J Biol Chem.* 1994;269(28):18408–18413.
28. Boudreault A, Gauthier D, Rondeau N, et al. Molecular characterization, enzymatic analysis, and purification of murine proprotein convertase-1/3 (PC1/PC3) secreted from recombinant baculovirus-infected insect cells. *Protein Expr Purif.* 1998;14(3):353–366.
29. Lloyd DJ, Bohan S, Gekakis N. Obesity, hyperphagia and increased metabolic efficiency in Pc1 mutant mice. *Hum Mol Genet.* 2006;15(11):1884–1893.
30. Dong W, Seidel B, Marcinkiewicz M, Chretien M, Seidah NG, Day R. Cellular localization of the prohormone convertases in the hypothalamic paraventricular and supraoptic nuclei: selective regulation of PC1 in corticotrophin-releasing hormone parvocellular neurons mediated by glucocorticoids. *J Neurosci.* 1997;17(2):563–575.
31. Damholt AB, Buchan AM, Holst JJ, Kofod H. Proglucagon processing profile in canine L cells expressing endogenous prohormone convertase 1/3 and prohormone convertase 2. *Endocrinology.* 1999;140(10):4800–4808.
32. Itoh Y, Tanaka S, Takekoshi S, Itoh J, Osamura RY. Prohormone convertases (PC1/3 and PC2) in rat and human pancreas and islet cell tumors: subcellular immunohistochemical analysis. *Pathol Int.* 1996;46(10):726–737.
33. Kurabuchi S, Tanaka S. Immunocytochemical localization of prohormone convertases PC1 and PC2 in the mouse thyroid gland and respiratory tract. *J Histochem Cytochem.* 2002;50(7):903–909.
34. Takumi I, Steiner DF, Sanno N, Teramoto A, Osamura RY. Localization of prohormone convertases 1/3 and 2 in the human pituitary gland and pituitary adenomas: analysis by immunohistochemistry, immunoelectron microscopy, and laser scanning microscopy. *Mod Pathol.* 1998;11(3):232–238.

35. Tanaka S, Kurabuchi S, Mochida H, et al. Immunocytochemical localization of prohormone convertases PC1/PC3 and PC2 in rat pancreatic islets. *Arch Histol Cytol.* 1996;59(3):261–271.
36. Zhou A, Bloomquist BT, Mains RE. The prohormone convertases PC1 and PC2 mediate distinct endoproteolytic cleavages in a strict temporal order during proopiomelanocortin biosynthetic processing. *J Biol Chem.* 1993;268(3):1763–1769.
37. Benjannet S, Rondeau N, Day R, Chretien M, Seidah NG. PC1 and PC2 are proprotein convertases capable of cleaving proopiomelanocortin at distinct pairs of basic residues. *Proc Natl Acad Sci USA.* 1991;88(9):3564–3568.
38. Thomas L, Leduc R, Thorne BA, Smeekens SP, Steiner D, Thomas G. Kex2-like endoproteases PC2 and PC3 accurately cleave a model prohormone in mammalian cells: evidence for a common core of neuroendocrine processing enzymes. *Proc Natl Acad Sci USA.* 1991;88:5297–5301.
39. Wardman JH, Zhang X, Gagnon S, et al. Analysis of peptides in prohormone convertase 1/3 null mouse brain using quantitative peptidomics. *J Neurochem.* 2010;114(1): 215–225.
40. Cowley MA, Smart JL, Rubinstein M, et al. Leptin activates anorexigenic POMC neurons through a neural network in the arcuate nucleus. *Nature.* 2001;411(6836): 480–484.
41. Belgardt BF, Okamura T, Bruning JC. Hormone and glucose signalling in POMC and AgRP neurons. *J Physiol.* 2009;587(Pt 22):5305–5314.
42. Xu Y, Elmquist JK, Fukuda M. Central nervous control of energy and glucose balance: focus on the central melanocortin system. *Ann NY Acad Sci.* 2011;1243:1–14.
43. Ollmann MM, Wilson BD, Yang YK, et al. Antagonism of central melanocortin receptors in vitro and in vivo by agouti-related protein. *Science.* 1997;278(5335): 135–138.
44. D'Agostino G, Diano S. Alpha-melanocyte stimulating hormone: production and degradation. *J Mol Med.* 2010;88(12):1195–1201.
45. Pan H, Nanno D, Che FY, et al. Neuropeptide processing profile in mice lacking prohormone convertase-1. *Biochemistry.* 2005;44(12):4939–4948.
46. Jackson RS, Creemers JW, Farooqi IS, et al. Small-intestinal dysfunction accompanies the complex endocrinopathy of human proprotein convertase 1 deficiency. *J Clin Invest.* 2003;112(10):1550–1560.
47. Dey A, Xhu X, Carroll R, Turck CW, Stein J, Steiner DF. Biological processing of the cocaine and amphetamine-regulated transcript precursors by prohormone convertases PC2 and PC1/3. *J Biol Chem.* 2003;278(17):15007–15014.
48. Creemers JW, Pritchard LE, Gyte A, et al. Agouti-related protein is posttranslationally cleaved by proprotein convertase 1 to generate agouti-related protein (AGRP)83-132: interaction between AGRP83-132 and melanocortin receptors cannot be influenced by syndecan-3. *Endocrinology.* 2006;147(4):1621–1631.
49. Kalra SP, Dube MG, Pu S, Xu B, Horvath TL, Kalra PS. Interacting appetite-regulating pathways in the hypothalamic regulation of body weight. *Endocr Rev.* 1999;20 (1):68–100.
50. Paquet L, Massie B, Mains RE. Proneuropeptide Y processing in large dense-core vesicles: manipulation of prohormone convertase expression in sympathetic neurons using adenoviruses. *J Neurosci.* 1996;16(3):964–973.
51. Nilaweera KN, Barrett P, Mercer JG, Morgan PJ. Precursor-protein convertase 1 gene expression in the mouse hypothalamus: differential regulation by ob gene mutation, energy deficit and administration of leptin, and coexpression with prepro-orexin. *Neuroscience.* 2003;119(3):713–720.
52. Zhu X, Orci L, Carroll R, Norrbom C, Ravazzola M, Steiner DF. Severe block in processing of proinsulin to insulin accompanied by elevation of des-64,65 proinsulin

intermediates in islets of mice lacking prohormone convertase 1/3. *Proc Natl Acad Sci USA*. 2002;99(16):10299–10304.
53. Smeekens S, Montag AG, Thomas G, et al. Proinsulin processing by the subtilisin-related proprotein convertases furin PC2, and PC3. *Proc Natl Acad Sci USA*. 1992;89:8822–8826.
54. Furuta M, Carroll R, Martin S, et al. Incomplete processing of proinsulin to insulin accompanied by elevation of Des-31,32 proinsulin intermediates in islets of mice lacking active PC2. *J Biol Chem*. 1998;273:1–7.
55. Yourshaw M, Solorzano-Vargas RS, Pickett LA, et al. Exome sequencing finds a novel PCSK1 mutation in a child with generalized malabsorptive diarrhea and diabetes insipidus. *J Pediatr Gastroenterol Nutr*. 2013;57(6):759–767.
56. Bandsma RH, Sokollik C, Chami R, et al. From diarrhea to obesity in prohormone convertase 1/3 deficiency: age-dependent clinical pathologic, and enteroendocrine characteristics. *J Clin Gastroenterol*. 2013;47(10):834–843.
57. Frank GR, Fox J, Candela N, et al. Severe obesity and diabetes insipidus in a patient with PCSK1 deficiency. *Mol Genet Metab*. 2013;110(1–2):191–194.
58. O'Rahilly S, Gray H, Humphreys PJ, et al. Brief report: impaired processing of prohormones associated with abnormalities of glucose homeostasis and adenal function. *New England J Med*. 1995;333:1386–1390.
59. Rouille Y, Westermark G, M. SK, Steiner DF. Proglucagon is processed to glucagon by prohormone convertase PC2 in alphaTC1-6 cells. *Proc Natl Acad Sci USA*. 1994;91:3242–3246.
60. Furuta M, Zhou A, Webb G, Ravazzola M, Orci L, Steiner DF. Severe defect in proglucagon processing in islet A-cells of prohormone convertase 2 null mice. *J Biol Chem*. 2001;276:27197–27202.
61. Rouille Y, Bianchi M, Irminger JC, Halban PA. Role of prohormone convertase PC2 in the processing of proglucagon to glucagon. *FEBS Lett*. 1997;413:119–123.
62. Rouille Y, Martin S, Steiner DF. Differential processing of proglucagon by the subtilisin-like prohormone convertases PC2 and PC3 to generate either glucagon or glucagon-like peptide. *J Biol Chem*. 1995;270(44):26488–26496.
63. Dhanvantari S, Seidah NG, Brubaker PL. Role of prohormone convertases in the tissue-specific processing of proglucagon. *Mol Endocrinol*. 1996;10:342–355.
64. Rouille Y, Kantengwa S, Irminger JC, Halban PA. Role of the prohormone convertases in the processing of proglucagon to glucagon-like peptides. *J Biol Chem*. 1997;72: 32810–32816.
65. Drucker DJ, Yusta B. Physiology and pharmacology of the enteroendocrine hormone glucagon-like peptide-2. *Ann Rev Physiol*. 2014;76:561–583.
66. Dailey MJ, Moran TH. Glucagon-like peptide 1 and appetite. *Trends Endocrinol Metab*. 2013;24(2):85–91.
67. Ugleholdt R, Poulsen ML, Holst PJ, et al. Prohormone convertase 1/3 is essential for processing of the glucose-dependent insulinotropic polypeptide precursor. *J Biol Chem*. 2006;281(16):11050–11057.
68. Usdin TB, Mezey E, Button DC, Brownstein MJ, Bonner TI. Gastric inhibitory polypeptide receptor, a member of the secretin-vasoactive intestinal peptide receptor family, is widely distributed in peripheral organs and the brain. *Endocrinology*. 1993;133 (6):2861–2870.
69. Paschetta E, Hvalryg M, Musso G. Glucose-dependent insulinotropic polypeptide: from pathophysiology to therapeutic opportunities in obesity-associated disorders. *Obes Rev*. 2011;12(10):813–828.
70. Zhu X, Cao Y, Voogd K, Steiner DF. On the processing of proghrelin to ghrelin. *J Biol Chem*. 2006;281(50):38867–38870.
71. Ozawa A, Cai Y, Lindberg I. Production of bioactive peptides in an in vitro system. *Anal Biochem*. 2007;366(2):182–189.

72. Shintani M, Ogawa Y, Ebihara K, et al. Ghrelin, an endogenous growth hormone secretagogue, is a novel orexigenic peptide that antagonizes leptin action through the activation of hypothalamic neuropeptide Y/Y1 receptor pathway. *Diabetes.* 2001;50: 227–232.
73. Gagnon J, Baggio LL, Drucker DJ, Brubaker PL. Ghrelin is a novel regulator of GLP-1 secretion. *Diabetes.* 2015;64:1513–1521.
74. Wang W, Birch NP, Beinfeld MC. Prohormone convertase 1 (PC1) when expressed with pro cholecystokinin (pro CCK) in L cells performs three endoproteolytic cleavages which are observed in rat brain and in CCK-expressing endocrine cells in culture, including the production of glycine and arginine extended CCK8. *Biochem Biophys Res Commun.* 1998;248(3):538–541.
75. Rehfeld JF, Zhu X, Norrbom C, et al. Prohormone convertases 1/3 and 2 together orchestrate the site-specific cleavages of progastrin to release gastrin-34 and gastrin-17. *Biochem J.* 2008;415(1):35–43.
76. Svendsen B, Pedersen J, Albrechtsen NJW, et al. An analysis of cosecretion and coexpression of gut hormones from male rat proximal and distal small intestine. *Endocrinology.* 2015;156:847–857.
77. Rehfeld JF. Gastrointestinal hormones and their targets. *Adv Exp Med Biol.* 2014;817: 157–175.
78. Mbikay M, Croissandeau G, Sirois F, et al. A targeted deletion/insertion in the mouse Pcsk1 locus is associated with homozygous embryo preimplantation lethality, mutant allele preferential transmission and heterozygous female susceptibility to dietary fat. *Dev Biol.* 2007;306(2):584–598.
79. Prabhu Y, Blanco EH, Liu M, et al. Defective transport of the obesity mutant PC1/3 N222D contributes to loss of function. *Endocrinology.* 2014;155(7):2391–2401.
80. Blanco EH, Ramos-Molina B, Lindberg I. Revisiting PC1/3 mutants: dominant-negative effect of endoplasmic reticulum-retained mutants. *Endocrinology.* 2015;156 (10):3625–3637. en20151068.
81. Creemers JW, Choquet H, Stijnen P, et al. Heterozygous mutations causing partial prohormone convertase 1 deficiency contribute to human obesity. *Diabetes.* 2012;61 (2):383–390.
82. Choquet H, Kasberger J, Hamidovic A, Jorgenson E. Contribution of common PCSK1 genetic variants to obesity in 8,359 subjects from multi-ethnic American population. *PLoS One.* 2013;8(2):e57857.
83. Villalobos-Comparan M, Villamil-Ramirez H, Villarreal-Molina T, et al. PCSK1 rs6232 is associated with childhood and adult class III obesity in the Mexican population. *PLoS One.* 2012;7(6):e39037.
84. Turpeinen H, Ortutay Z, Pesu M. Genetics of the first seven proprotein convertase enzymes in health and disease. *Curr Genomics.* 2013;14(7):453–467.
85. Benzinou M, Creemers JW, Choquet H, et al. Common nonsynonymous variants in PCSK1 confer risk of obesity. *Nat Genet.* 2008;40(8):943–945.
86. Heni M, Haupt A, Schafer SA, et al. Association of obesity risk SNPs in PCSK1 with insulin sensitivity and proinsulin conversion. *BMC Med Genet.* 2010;11:86.
87. Strawbridge RJ, Dupuis J, Prokopenko I, et al. Genome-wide association identifies nine common variants associated with fasting proinsulin levels and provides new insights into the pathophysiology of type 2 diabetes. *Diabetes.* 2011;60(10): 2624–2634.
88. Kilpelainen TO, Bingham SA, Khaw KT, Wareham NJ, Loos RJ. Association of variants in the PCSK1 gene with obesity in the EPIC-Norfolk study. *Hum Mol Genet.* 2009;18 (18):3496–3501.
89. Qi Q, Li H, Loos RJ, et al. Association of PCSK1 rs6234 with obesity and related traits in a Chinese Han population. *PLoS One.* 2010;5(5):e10590.

90. Gjesing AP, Vestmar MA, Jorgensen T, et al. The effect of PCSK1 variants on waist, waist-hip ratio and glucose metabolism is modified by sex and glucose tolerance status. *PLoS One.* 2011;6(9):e23907.
91. Mbikay M, Sirois F, Nkongolo KK, Basak A, Chretien M. Effects of rs6234/rs6235 and rs6232/rs6234/rs6235 PCSK1 single-nucleotide polymorphism clusters on proprotein convertase 1/3 biosynthesis and activity. *Mol Genet Metab.* 2011;104 (4):682–687.
92. Blanco EH, Peinado JR, Martin MG, Lindberg I. Biochemical and cell biological properties of the human prohormone convertase 1/3 Ser357Gly mutation: a PC1/3 hypermorph. *Endocrinology.* 2014;155(9):3434–3447.
93. Li L, Gao K, Zhao J, et al. Glucagon gene polymorphism modifies the effects of smoking and physical activity on risk of type 2 diabetes mellitus in Han Chinese. *Gene.* 2014;534:352–355.
94. Canani RB, Castaldo G, Bacchetta R, Martín MG, Goulet O. Congenital diarrhoeal disorders: advances in this evolving web of inherited enteropathies. *Nat Rev Gastroenterol Hepatol.* 2015;12:293–302.
95. Wang J, Cortina G, Wu SV, et al. Mutant neurogenin-3 in congenital malabsorptive diarrhea. *New England J Med.* 2006;355:270–280.
96. Walley AJ, Asher JE, Froguel P. The genetic contribution to non-syndromic human obesity. *Nat Rev Genet.* 2009;10:431–442.
97. Farooqi IS, Keogh JM, Yeo GSH, Lank EJ, Cheetham T, O'Rahilly S. Clinical spectrum of obesity and mutations in the melanocortin 4 receptor gene. *New England J Med.* 2003;348:1085–1095.
98. Gradwohl G, Dierich A, LeMeur M, Guillemot F. Neurogenin3 is required for the development of the four endocrine cell lineages of the pancreas. *Proc Natl Acad Sci USA.* 2000;97:1607–1611.
99. Jeppesen PB. Gut hormones in the treatment of short-bowel syndrome and intestinal failure. *Curr Diffid238/Opin Endocrinol Diab Obes.* 2015;22:14–20.
100. Syed YY, McCormack PL. Exenatide extended-release: an updated review of its use in type 2 diabetes mellitus. *Drugs.* 2015;75:1141–1152.
101. Martinelli CE, Keogh JM, Greenfield JR, et al. Obesity due to melanocortin 4 receptor (MC4R) deficiency is associated with increased linear growth and final height, fasting hyperinsulinemia, and incompletely suppressed growth hormone secretion. *J Clin Endocrinol Metabol.* 2011;96:E181–E188.
102. Clément K, Vaisse C, Lahlou N, et al. A mutation in the human leptin receptor gene causes obesity and pituitary dysfunction. *Nature.* 1998;392:398–401.
103. Wilschanski M, Abbasi M, Blanco E, et al. A novel familial mutation in the PCSK1 gene that alters the oxyanion hole residue of proprotein convertase 1/3 and impairs its enzymatic activity. *PLoS One.* 2014;9(10):e108878.
104. Durand C, Rappold GA. Height matters-from monogenic disorders to normal variation. *Nat Rev Endocrinol.* 2013;9:171–177.
105. Lango Allen H, Estrada K, Lettre G, et al. Hundreds of variants clustered in genomic loci and biological pathways affect human height. *Nature.* 2010;467:832–838.
106. Hotamisligil GS. Endoplasmic reticulum stress and the inflammatory basis of metabolic disease. *Cell.* 2010;140(6):900–917.
107. Ozcan U, Cao Q, Yilmaz E, et al. Endoplasmic reticulum stress links obesity, insulin action, and type 2 diabetes. *Science.* 2004;306(5695):457–461.
108. Philippe J, Stijnen P, Meyre D, et al. A nonsense loss-of-function mutation in PCSK1 contributes to dominantly inherited human obesity. *Int J Obes.* 2015;39(2):295–302.
109. Rohden F, Costa CS, Hammes TO, et al. Obesity associated with type 2 diabetes mellitus is linked to decreased PC1/3 mRNA expression in the Jejunum. *Obes Surg.* 2014;24:2075–2081.

110. Ahlqvist E, Turrini F, Lang ST, et al. A common variant upstream of the PAX6 gene influences islet function in man. *Diabetologia.* 2012;55:94–104.
111. Challis BG, Pritchard LE, Creemers JW, et al. A missense mutation disrupting a dibasic prohormone processing site in pro-opiomelanocortin (POMC) increases susceptibility to early-onset obesity through a novel molecular mechanism. *Hum Mol Genet.* 2002;11 (17):1997–2004.
112. Creemers JW, Lee YS, Oliver RL, et al. Mutations in the amino-terminal region of proopiomelanocortin (POMC) in patients with early-onset obesity impair POMC sorting to the regulated secretory pathway. *J Clin Endocrinol Metab.* 2008;93(11): 4494–4499.
113. Farooqi IS, Drop S, Clements A, et al. Heterozygosity for a POMC-null mutation and increased obesity risk in humans. *Diabetes.* 2006;55(9):2549–2553.
114. Krude H, Biebermann H, Luck W, Horn R, Brabant G, Gruters A. Severe early-onset obesity, adrenal insufficiency and red hair pigmentation caused by POMC mutations in humans. *Nat Genet.* 1998;19(2):155–157.
115. Yaswen L, Diehl N, Brennan M, Hochgeschwender U. Obesity in the mouse model of pro-opimelanocortin deficiency responds to peripheral melanocortin. *Nat Med.* 1999;5 (9):1066–1070.
116. Alsters SIM, Goldstone AP, Buxton JL, et al. Truncating homozygous mutation of carboxypeptidase E (CPE) in a morbidly obese female with type 2 diabetes mellitus intellectual disability and hypogonadotrophic hypogonadism. *PloS One.* 2015;10: e0131417.
117. Naggert JK, Fricker LD, Varlamov O, et al. Hyperproinsulinemia in obese fat/fat mice associated with a point mutation in the carboxypeptidase R gene and reduced enzyme activity. *Nat Genet.* 1995;1995:135–142.
118. Srinivasan S, Bunch DO, Feng Y, et al. Deficits in reproduction and pro-gonadotropin-releasing hormone processing in male Cpefat mice. *Endocrinology.* 2004;145 (4):2023–2034.
119. Berman Y, Mzhavia N, Polonskaia A, Devi LA. Impaired prohormone convertases in cpefat/cpefat mice. *J Biol Chem.* 2001;276:1466–1473.
120. Fricker LD, McKinzie AA, Sun J, et al. Identification and characterization of proSAAS, a granin-like neuroendocrine peptide precursor that inhibits prohormone processing. *J Neurosci.* 2000;20(2):639–648.
121. Kudo H, Liu J, Jansen EJ, et al. Identification of proSAAS homologs in lower vertebrates: conservation of hydrophobic helices and convertase-inhibiting sequences. *Endocrinology.* 2009;150(3):1393–1399.
122. Chun JY, Korner J, Kreiner T, Scheller RH, Axel R. The function and differential sorting of a family of aplysia prohormone processing enzymes. *Neuron.* 1994;12 (4):831–844.
123. Feng Y, Reznik SE, Fricker LD. Distribution of proSAAS-derived peptides in rat neuroendocrine tissues. *Neuroscience.* 2001;105(2):469–478.
124. Feng Y, Reznik SE, Fricker LD. ProSAAS and prohormone convertase 1 are broadly expressed during mouse development. *Brain Res Gene Expr Patterns.* 2002;1(2):135–140.
125. Atkins Jr N, Mitchell JW, Romanova EV, et al. Circadian integration of glutamatergic signals by little SAAS in novel suprachiasmatic circuits. *PLoS One.* 2010;5(9):e12612.
126. Mzhavia N, Berman Y, Che FY, Fricker LD, Devi LA. ProSAAS processing in mouse brain and pituitary. *J Biol Chem.* 2001;276(9):6207–6213.
127. Sayah M, Fortenberry Y, Cameron A, Lindberg I. Tissue distribution and processing of proSAAS by proprotein convertases. *J Neurochem.* 2001;76(6):1833–1841.
128. Qian Y, Devi LA, Mzhavia N, Munzer S, Seidah NG, Fricker LD. The C-terminal region of proSAAS is a potent inhibitor of prohormone convertase 1. *J Biol Chem.* 2000;275(31):23596–23601.

129. Cameron A, Fortenberry Y, Lindberg I. The SAAS granin exhibits structural and functional homology to 7B2 and contains a highly potent hexapeptide inhibitor of PC1. *FEBS Lett.* 2000;473(2):135–138.
130. Apletalina E, Appel J, Lamango NS, Houghten RA, Lindberg I. Identification of inhibitors of prohormone convertases 1 and 2 using a peptide combinatorial library. *J Biol Chem.* 1998;273(41):26589–26595.
131. Wei S, Feng Y, Che FY, et al. Obesity and diabetes in transgenic mice expressing proSAAS. *J Endocrinol.* 2004;180(3):357–368.
132. Morgan DJ, Wei S, Gomes I, et al. The propeptide precursor proSAAS is involved in fetal neuropeptide processing and body weight regulation. *J Neurochem.* 2010;113(5):1275–1284.
133. Gosmain Y, Katz LS, Masson MH, Cheyssac C, Poisson C, Philippe J. Pax6 is crucial for β-cell function, insulin biosynthesis, and glucose-induced insulin secretion. *Mol Endocrinol.* 2012;26:696–709.
134. Wen JH, Chen YY, Song SJ, et al. Paired box 6 (PAX6) regulates glucose metabolism via proinsulin processing mediated by prohormone convertase 1/3 (PC1/3). *Diabetologia.* 2009;52(3):504–513.
135. Liu T, Zhao Y, Tang N, et al. Pax6 directly down-regulates Pcsk1n expression thereby regulating PC1/3 dependent proinsulin processing. *PloS One.* 2012;7:e46934.
136. Hoshino A, Helwig M, Razaei S, Berridge C, Eriksen J, Lindberg I. A novel function for proSAAS as an amyloid anti-aggregant in Alzheimer's disease. *J Neurochem.* 2014;128(3):419–430.
137. Peinado JR, Sami F, Rajpurohit N, Lindberg I. Blockade of islet amyloid polypeptide fibrillation and cytotoxicity by the secretory chaperones 7B2 and proSAAS. *FEBS Lett.* 2013;587(21):3404–3411.

CHAPTER THREE

Rare Syndromes and Common Variants of the Brain-Derived Neurotrophic Factor Gene in Human Obesity

J.C. Han[1]

Department of Pediatrics, University of Tennessee Health Science Center and Children's Foundation Research Institute, Le Bonheur Children's Hospital, Memphis, TN, United States of America

[1] Corresponding author: e-mail address: jhan14@uthsc.edu.

Contents

Abstract

Rare genetic disorders that cause *BDNF* haploinsufficiency, such as WAGR syndrome, 11p deletion, and 11p inversion, serve as models for understanding the role of BDNF in human energy balance and neurocognition. Patients with *BDNF* haploinsufficiency or inactivating mutations of the BDNF receptor exhibit hyperphagia, childhood-onset obesity, intellectual disability, and impaired nociception. Prader–Willi, Smith–Magenis, and ROHHAD syndromes are separate genetic disorders that do not directly affect the *BDNF* locus but share many similar clinical features with *BDNF* haploinsufficiency, and

Progress in Molecular Biology and Translational Science, Volume 140
ISSN 1877-1173
http://dx.doi.org/10.1016/bs.pmbts.2015.12.002

BDNF insufficiency is believed to possibly contribute to the pathophysiology of each of these conditions. In the general population, common variants of *BDNF* that affect *BDNF* gene expression or BDNF protein processing have also been associated with modest alterations in energy balance and cognitive functioning. Thus, variable degrees of BDNF insufficiency appear to contribute to a spectrum of excess weight gain and cognitive impairment that ranges in phenotypic severity. In this modern era of precision medicine, genotype-specific therapies aimed at increasing BDNF signaling in patients with rare and common disorders associated with BDNF insufficiency could serve as useful approaches for treating obesity and neurodevelopmental disorders.

1. INTRODUCTION

Brain-derived neurotrophic factor (BDNF) is a protein that is widely expressed in the nervous system and plays an important role in neuronal growth and survival. Human BDNF is encoded by the *BDNF* gene, located at chromosome 11p14.1, spanning 67 kb, and comprised of 11 exons with 9 transcriptional promoters. The combination of alternate splicing involving the 5' untranslated exons and two different 3' polyadenylation sites results in multiple transcript variants that all encode the prohormone precursor peptide, pro-BDNF, which undergoes amino-terminal cleavage to yield a common 119 amino acid mature BDNF protein.[1] Pro-BDNF exerts synaptic long-term depression, acting primarily through the $p75^{NTR}$ receptor, while mature BDNF preferentially binds the tropomyosin-related kinase B receptor (TrkB), a tyrosine kinase receptor, activation of which promotes neuronal differentiation and synaptic potentiation.[2–4]

In 1989, BDNF was the second protein, after nerve growth factor, to be sequenced from the neurotrophin family.[5] BDNF has since been extensively investigated, with over 16,000 publications in the medical literature to date, elucidating its role in memory, learning, neuropsychiatric and neurodegenerative disorders, and energy homeostasis.[6–10] BDNF's role in energy homeostasis emerged in the 1990s with the observation that intracerebroventricular (ICV) BDNF administration suppresses appetite and induces weight loss in rodents.[11–13] *Bdnf* heterozygous knockout mice exhibit hyperphagia and obesity,[14] which is more pronounced with high-fat diet,[15] and can be reversed with ICV BDNF administration[16] or prevented by pair feeding that restricts food intake to that of wild-type littermates.[17] Complete lack of BDNF during embryologic development is perinatally lethal, but postnatal brain-specific *Bdnf* deletion permits normal survival and leads to marked hyperphagia and severe obesity.[18] Furthermore, mice with an *Ntrk2* mutation that leads to reduced expression of the BDNF receptor, TrkB, also display hyperphagia and obesity.[19]

BDNF appears to regulate both energy intake and expenditure. BDNF administration to either the ventromedial hypothalamus (VMH) or the paraventricular hypothalamus (PVH) in wild-type rats suppresses food intake and increases energy expenditure.[20–23] Brain region specific deletion of *Bdnf* results in hyperphagia and modest obesity after deletion of *Bdnf* in the VMH and dorsomedial hypothalamus (DMH), and like the whole body BDNF-deficient mice, weight gain is completely prevented by pair feeding.[24] On the other hand, *Bdnf* deletion in the PVH causes severe hyperphagia, reduced locomotor activity, impaired thermogenesis, and severe obesity that is only partially ameliorated with pair feeding.[25] More specifically, the anterior PVH appears to regulate appetite and spontaneous physical activity, while the medial and posterior PVH appears to regulate thermogenesis.[25] Heterozygous *Bdnf* knockout mice and total-brain *Bdnf*-deficient mice have increased locomotor activity, which could attenuate the impact of BDNF on energy expenditure and thereby explain the differences observed in pair-feeding experiments of these mouse models compared to PVH-specific *Bdnf* knockouts.[16,18]

Several studies suggest that BDNF controls appetite through a combination of central and peripheral pathways.[15,26] In the VMH, BDNF is abundantly expressed in the fed state but suppressed after food deprivation.[19] Peripherally, vagal sensory innervation in the gastrointestinal track and vagal afferents to the dorsal vagal motor nucleus are altered in BDNF-deficient mice.[26,27] With respect to BDNF's regulation of metabolic rate, this function appears to be mediated by BDNF neurons in the medial and posterior VMH that project to the spinal cord and form polysynaptic connections with brown adipose, where sympathetic outflow induces thermogenesis.[25] The mechanism by which BDNF regulates locomotor activity, however, remains to be elucidated.

Regulation of *BDNF* expression is mediated by multiple promoters controlled by varied mechanisms, including calcium ion influx, electrophysiological activity, binding of activators and repressors, DNA methylation, histone acylation, and chromatin remodeling.[28] Promoters I and IV are activity dependent and calcium responsive, while promoters II and VI display tonic basal activity and are controlled by transcriptional regulators.[28] Different regions of the brain express different patterns of *BDNF* transcripts, and these individual transcripts appear to play specific roles in neurologic function.[1] Loss of BDNF protein translation from just one transcript appears to be sufficient to cause significant phenotypic deficits.[29] Furthermore, alternate polyadenylation sites produce short- and long-3'UTR transcripts with separate subcellular localizations; dendritic targeting of the long-3'UTR transcript appears to be essential for body-weight regulation in mice.[30]

Mechanistically, BDNF is believed to function as a downstream regulator of the leptin–proopiomelanocortin pathway.[19] Delivery of a *Bdnf*-overexpressing gene vector to the hypothalamus of leptin receptor deficient *db/db* mice effectively reverses their hyperphagia and obesity.[31] Similarly, *Mc4r* homozygous knockout mice have reduced VMH *Bdnf* expression, and ICV administration of BDNF reduces food intake and body weight in these mice.[19] Moreover, administration of MC4R agonist to wild-type mice attenuates food deprivation induced suppression of *Bdnf* expression.[19] From a clinical translational perspective, BDNF and its receptor, TrkB, may serve as important therapeutic targets for both rare and common forms of obesity that share etiologies converging on the leptin signaling pathway (Fig. 1).[32]

2. HUMAN *BDNF* HAPLOINSUFFICIENCY

2.1 WAGR Syndrome

The natural occurrence of functional loss of one of the two alleles for the *BDNF* gene in patients has permitted the study of human *BDNF* haploinsufficiency. Wilms tumor, aniridia, genitourinary abnormalities, and mental retardation (WAGR) syndrome is a rare genetic disorder, with a prevalence of approximately 1 in 1,000,000, and is caused by contiguous gene deletions in the chromosome 11p13 region.[33] The acronym, WAGR, identifies the key clinical features of the syndrome: Wilms tumor, aniridia, genitourinary abnormalities, and mental retardation. The core features of the syndrome are caused by haploinsufficiency of *WT1* and *PAX6*, which are involved in genitourinary and eye development, respectively. *BDNF* resides at 11p14.1, which is 4 Mb from the WAGR critical region (Fig. 2), and is included in the deleted segment of approximately half of the patients with WAGR syndrome.[33] In a cohort of 33 patients with WAGR syndrome, *BDNF* haploinsufficiency was associated with higher body mass index (BMI, standardized for age and sex, and adjusted for parental BMI), fivefold higher frequency of developing childhood obesity (100 vs. 20% for patients

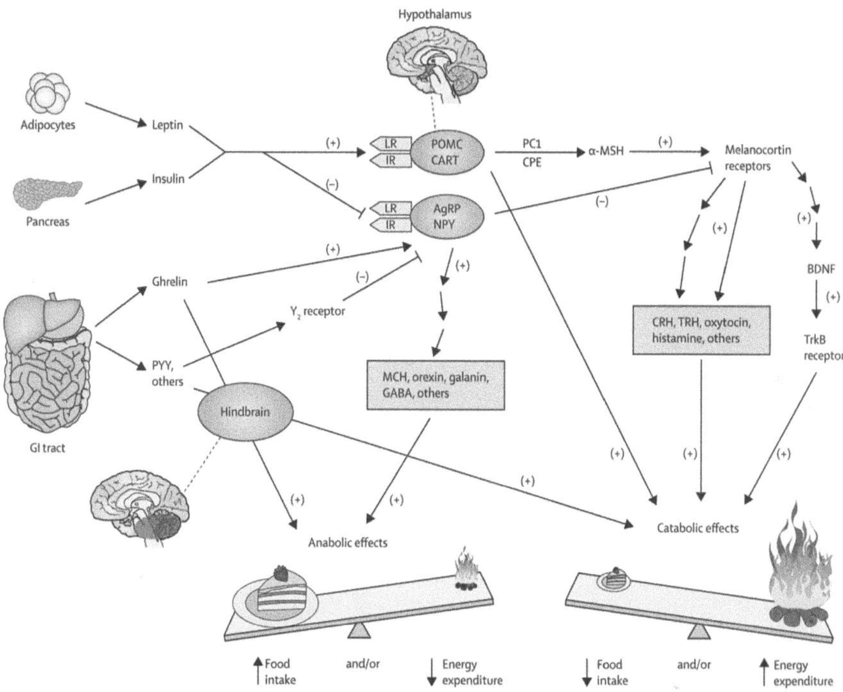

Figure 1 Simplified overview of the leptin pathway. *Lines with arrowheads* show stimulatory action. *Lines with perpendicular endblocks* show inhibitory action. *AgRP,* agouti-related protein; *BDNF,* brain-derived neurotrophic factor; *CART,* cocaine–amphetamine related transcript; *CPE,* carboxypeptidase E; *CRH,* corticotropin-releasing hormone; *GABA,* gamma amino butyric acid; *GI,* gastrointestinal; *IR,* insulin receptor; *LR,* leptin receptor; *MCH,*melanin-concentrating hormone; *MSH,* melanocyte-stimulating hormone; *NPY,* neuropeptide Y; *PC1,* prohormone convertase 1; *POMC,* proopio-melanocortin; *PYY,* peptide YY; *TRH,* thyrotropin-releasing hormone; *TrkB,* tropomyosin-related kinase B. *Reprinted with permission from Ref. [32].*

with and without *BDNF* deletion, respectively), reduced serum BDNF concentrations, and higher scores on a parent-reported hyperphagia questionnaire[34] that assessed symptoms such as persistent thoughts about food, stealing of food, and foraging for food in the trash.[33] Representative anthropometrics and photographs of patients with WAGR syndrome demonstrating the association of obesity with *BDNF* haploinsufficiency are shown in Table 1 and Fig. 3.

Interestingly, one of the patients with obesity in this cohort had heterozygous deletion of the first three exons of *BDNF*, but sparing of the remaining eight exons, including the single mature BDNF protein-coding exon.[33]

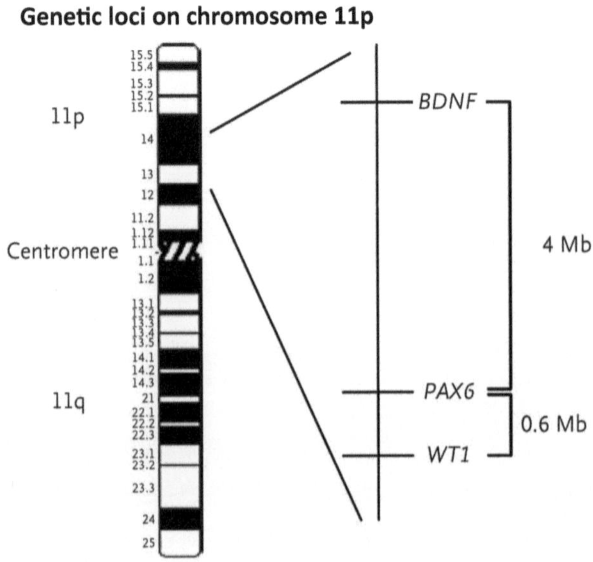

Figure 2 WAGR syndrome critical region on chromosome 11p13. WAGR syndrome is caused by heterozygous contiguous gene deletions involving *WT1* and *PAX6*. Approximately half of the patients will also have deletion of *BDNF*, which is 4 Mb telomeric of *PAX6*. *Reprinted with permission from Ref. [33].*

BDNF transcript splice variants that contain exons I, II, or III, comprise 64% of the total mRNA transcripts expressed in human hypothalamic tissue derived from normal adult subjects, suggesting that these early 5' exons are important in energy homeostasis.[33] Furthermore, another patient with a deletion boundary that ended 72.5-kb upstream of *BDNF* was quite thin with a BMI at the 12th percentile (patient E in Table 1 and Fig. 3). This intronic 72.5-kb region lacks protein-coding sequences, which supports the hypothesis that *BDNF* is a specific obesity-risk locus in WAGR syndrome.[33]

In addition to obesity, patients with *BDNF* haploinsufficiency exhibit neurocognitive impairments similar to the deficits in memory,[35] social behavior,[14] and nociception[36] reported in *Bdnf* heterozygous knockout mice. In a cohort of 28 patients with WAGR syndrome who underwent neurocognitive assessments, patients with *BDNF* haploinsufficiency had an average IQ that was 20-points lower than patients with intact *BDNF*.[37] *BDNF* haploinsufficiency was also associated with greater social impairment and higher rate of meeting criteria for autism on the Autism-Diagnostic Interview-Revised.[37] Parent-reported behavioral responses[38] to typically painful stimuli were also lower for patients with *BDNF* haploinsufficiency.[33]

Table 1 Association of Obesity with *BDNF* Haploinsufficiency in Patients with WAGR Syndrome.

			11p Deletion Boundaries							
Patient	Age	Sex	Start	Stop	*BDNF* Deletion	Height (cm)	Weight (kg)	BMI (kg/m^2)	BMI (%)	Weight Status
A	2	M	28,761,503	41,388,352	No	82.5	11.3	16.6	48	Normal
B	6	M	28,200,894	32,820,697	No	112.1	22.6	18.0	92	Overweight
C	7	F	31,459,056	33,682,710	No	127.4	28.3	17.4	83	Normal
D	11	M	32,380,218	39,188,328	No	137.1	36.6	19.5	80	Normal
E	12	M	27,773,897	33,998,994	No	154.3	37.3	15.7	12	Normal
F	27	F	30,035,387	33,602,848	No	158.7	54.7	21.7	50	Normal
G	3	M	27,204,073	36,512,238	Yes	93.5	16.9	19.3	98	Obesity
H	6	F	24,744,722	32,740,440	Yes	124.0	34.5	22.4	99	Obesity
I	9	M	22,962,873	36,138,961	Yes	124.1	36.2	23.5	98	Obesity
J	12	F	23,887,358	42,707,993	Yes	143.1	61.7	30.1	99	Obesity
K	14	M	24,748,424	36,203,063	Yes	155.8	81.6	33.6	≥99	Obesity
L	23	F	17,181,487	36,424,081	Yes	157.2	119.5	48.3	>99	Obesity

Deletion boundaries were determined by high-resolution oligonucleotide comparative genomic hybridization (reference sequence: human genome 18). Individual patients are shown in Fig. 3.

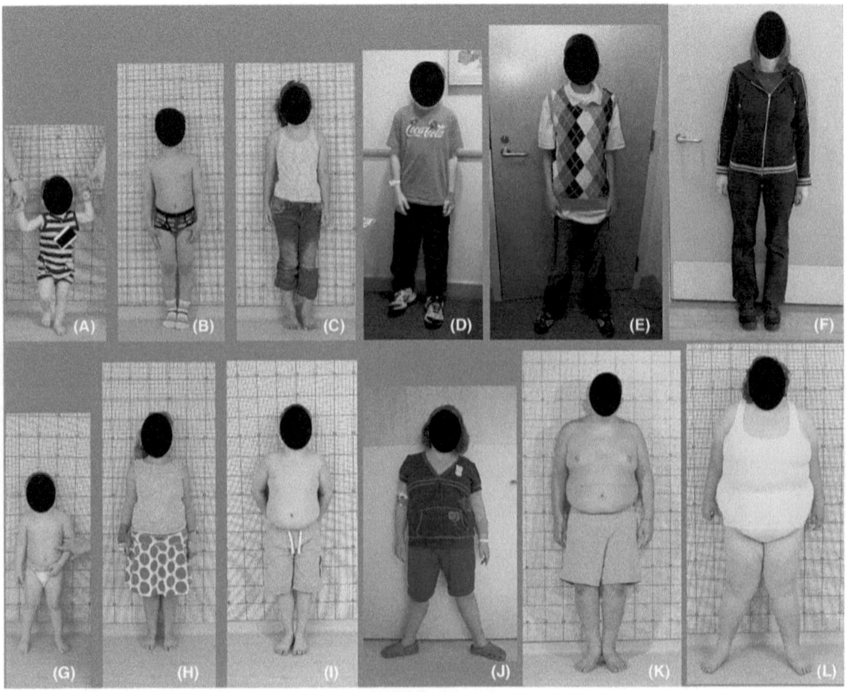

Figure 3 Children and adults with WAGR syndrome. Patient characteristics are shown in Table 1. Patients (A–F), whose BMIs range from normal to overweight, have both alleles of *BDNF* intact. Patients (G–L) have heterozygous deletion of *BDNF*, and all have obesity.

2.2 11p14 Deletion

Shinawi et al.[39] and Ernst et al.[40] together have identified nine individuals with *BDNF* haploinsufficiency due to heterozygous 11p14 deletions sparing the WAGR region critical genes. All subjects had neurodevelopmental abnormalities, including three reporting impaired nociception.[40] Obesity was observed in eight subjects, while the ninth subject was overweight (89th percentile BMI) at age 45 months, but had a rapid weight-gain trajectory that would be anticipated to progress to obesity.[39] Combining both reports, the 2 Mb common region shared by the deletions of all nine individuals includes four genes (*BDNF, LIN7C, LGR4*, and *CCDC34*), of which, only *BDNF* has a strong molecular basis for being involved in energy balance.

2.3 11p Inversion

Another form of *BDNF* haploinsufficiency can be due to loss of expression from one *BDNF* allele even if the allele is still present. Gray et al.[41] described an 8-year-old girl with a heterozygous paracentric 11p13p15.3 inversion that encompassed *BDNF* without disrupting the sequence of the gene itself, but interfered with expression of the inverted *BDNF* allele, which was distinguishable from the unaffected *BDNF* allele due to heterozygosity for the common G196A (Val66Met) polymorphism. Her serum BDNF concentration was significantly lower compared to obese and lean controls, and she displayed hyperphagia, obesity, impaired nociception, and intellectual disability.[41]

While the sum of human *BDNF* haploinsufficiency cases remains only in the dozens, this rare condition has shed insight into the critical role of BDNF in human energy balance and neurocognitive development.

3. BDNF AND OBESITY IN OTHER DISORDERS

3.1 *NTRK2* (TrkB) Mutation

Yeo et al.[42] described an 8-year-old male with a heterozygous missense mutation of *NTRK2*, the gene which encodes the BDNF receptor, TrkB, resulting in a Y722C substitution that interferes with receptor autophosphorylation.[43] This patient also exhibited hyperphagia, obesity, impaired nociception, and intellectual disability.[42] Thus, functional loss of one allele of *BDNF* or *NTRK2* appears to be sufficient to cause a highly penetrant phenotype of obesity and other neurodevelopmental abnormalities.

3.2 Prader–Willi Syndrome

Prader–Willi syndrome (PWS) is a hyperphagic obesity disorder with an estimated prevalence of 1 in 15,000–30,000. PWS is caused by lack of expression of paternally derived genes on chromosome 15q11-13, due to heterozygous deletions of paternal alleles (70%), uniparental disomy in which two copies of maternal alleles are inherited (20–30%), or imprinting defects in which paternal alleles are silenced due to inappropriate methylation (2–5%).[44] In classical PWS, major diagnostic features include neonatal hypotonia, feeding problems in infancy, rapid weight gain after infancy, hyperphagia, developmental delay, hypogonadism, and characteristic facial features (narrow face, almond-shaped eyes, small mouth, thin upper lip, and downturned corners of mouth). Minor diagnostic features include decreased fetal movement, weak cry and lethargy in infancy, behavior problems, sleep disturbance, short stature, hypopigmentation, small hands and feet, and skin picking. Other common features include high pain threshold, decreased vomiting, temperature instability, scoliosis, early adrenarche, and unusual skill with jigsaw puzzles.[45]

Several features of PWS can be observed in patients with inactivating mutations of *MAGEL2*, a gene that resides at 15q11.2 within the PWS critical region, and screening for *MAGEL2* mutations is recommended for patients who present with atypical PWS.[46] *Magel2*-null mice recapitulate many of the features of PWS and are used as models for studying the pathophysiology of PWS. *Magel2*-null mice display diminished physical activity, reduced food consumption, and lower body weight before weaning, but they have higher adiposity after weaning as well as disrupted circadian rhythms and reduced fertility.

The cause of the disordered satiety in PWS is not well understood. Hyperghrelinemia is observed in patients with PWS,[47–49] but pharmacologic suppression of ghrelin has not been therapeutically successful in reducing hyperphagia or body weight, calling into question the pathophysiologic role of ghrelin in PWS.[50,51] Adipose tissue leptin RNA and serum leptin concentrations in patients with PWS are appropriately increased in proportion with their higher fat mass and at levels similar to patients with nonsyndromic forms of obesity.[52] Therefore, leptin production is preserved in PWS, but disruption in downstream mediators of leptin signaling continue to be investigated as etiologies for PWS.

Leptin acts within the hypothalamic arcuate nucleus to suppress orexigenic signaling from neurons that express neuropeptide Y (NPY) and Agouti-related protein (AgRP) and promotes anorexigenic signaling from neurons that express proopiomelanocortin (POMC) and cocaine- and amphetamine-regulated transcript (CART) (Fig. 1). NPY and AgRP appear to be appropriately suppressed in postmortem hypothalamic tissue from patients with PWS, indicating that overexpression of these appetite-stimulating peptides is not likely to be the cause of hyperphagia in PWS.[53] However, several lines of evidence point to defective signaling along the POMC/CART branch of the leptin pathway. *Magel2* knockout mice are resistant to the appetite-suppressing effect of leptin administration,[54–56] with development of leptin insensitivity occurring between 4 and 6 weeks of age, concomitant with a decline in the number of arcuate POMC neurons, suggesting that the transition from failure to thrive in infancy to hyperphagia in early childhood for patients with PWS may be attributable to a neurodegenerative process.

Consistent with the hypothesis that disruption of POMC signaling could be the etiology of hyperphagia in PWS is the observation that patients with PWS have reduced serum and plasma BDNF concentrations compared to obese and lean controls.[57] Considering BDNF's role as a downstream mediator of melanocortin-receptor signaling (Fig. 1) and its function in neurocognition and pain perception,[58] insufficiency of BDNF could potentially account for the hyperphagia, intellectual disability, behavior abnormalities, and high pain tolerance associated with PWS.

3.3 *RAI1* Haploinsufficiency in Smith–Magenis and ROHHAD Syndromes

RAI1 is located at 17p11.2 and encodes a protein that is believed to be a transcriptional regulator involved in craniofacial and nervous system development. In frogs, knockdown of *Rai1* using antisense morpholinos results in lower *Bdnf* mRNA expression and abnormal brain and face development.[59] Heterozygous *Rai1* knockout mice have diminished hypothalamic *Bdnf* expression and display hyperphagia and obesity after age 20 weeks on normal chow,[60] and by 16 weeks on high-fat or high-carbohydrate diets.[61]

In human patients, heterozygous deletions of the 17p11.2 region and point mutations of *RAI1* cause Smith–Magenis syndrome, which has an estimated prevalence of 1 in 15,000–25,000, and is characterized by intellectual disability, maladaptive and self-injurious behaviors, sleep disturbance, and dysmorphic facial features (brachycephaly, broad face, frontal bossing, synophrys, hypertelorism, upslanting eyes, midface hypoplasia with a depressed nasal bridge, a tented upper lip, prognathism, and low-set or abnormally shaped ears). Hyperphagia and obesity are typically not observed until later childhood or adolescence, mimicking the older onset of these symptoms in mice, and are more pronounced in patients with *Rai1* mutations compared to those with deletions.[62] Examination of functional effects of mutations found in patients has revealed that mutated RA1 protein fails to localize to the nucleus and does not activate expression of a reporter gene driven by an endogenous *BDNF* promoter.[63,64] Presence of the mutated protein may have a dominant negative effect on the remaining normal protein, thus leading to a more severe obesity phenotype in patients with mutations.

Recently, Thaker et al.[65] described an 11-year-old boy with a BMI of 62 kg/m^2 who had been clinically diagnosed with rapid-onset obesity with hypothalamic dysfunction, hypoventilation, and autonomic dysfunction (ROHHAD) syndrome and was found to have a truncating mutation of *RAI1*. In addition to severe obesity, this patient also exhibited intellectual disability, autism spectrum disorder, dysmorphic facial features (macrocephaly, hypertelorism, flat nasal bridge, prominent forehead, and anteverted nares), high pain tolerance, excessive sweating, lack of fever with infections, obstructive and central sleep apnea, and hypoventilation leading to tracheostomy placement. ROHHAD is currently a clinically diagnosed condition without a known genetic cause. *BDNF* and *NTRK2* have been examined as candidate genes but mutations have not been identified to date among these patients.[66] Pituitary hormone abnormalities and neural crest tumors are typically associated with ROHHAD but were lacking in this patient. Thus, he could be categorized as atypical ROHHAD or possibly atypical SMS, and screening for *RAI1* haploinsufficiency should be considered in the differential for patients presenting with features of ROHHAD.

The overlap in symptoms among patients with *BDNF* haploinsufficiency, Prader–Willi, Smith–Magenis, and ROHHAD syndromes suggest that BDNF insufficiency may be a common etiology that could serve as a useful therapeutic target for each of these conditions.

4. COMMON VARIANTS OF *BDNF*

4.1 Val66Met Polymorphism

Multiple genome-wide association studies of obesity in children and adults of different racial and ethnic populations have found associations for single-nucleotide polymorphisms (SNPs) at the *BDNF* locus and BMI.[67–74] The most extensively studied of these SNPs is the G196A variant (rs6265), which leads to a valine to methionine substitution at the 66th amino acid position (Val66Met) of the N-terminal prodomain of pro-BDNF (Fig. 4). The ancestral allele is G, but there is wide variability in the frequency of the G allele across populations (>0.99, 0.81, 0.79, and 0.38 in African, European Caucasian, Hispanic, and Asian cohorts, respectively).[75] Mechanistically, the amino acid change does not alter the mature BDNF protein, but rather appears to disrupt intracellular trafficking of pro-BDNF and decrease depolarization-induced secretion of BDNF without altering its baseline constitutive secretion.[76,77] Initial reports of neurocognitive dysfunction in Met carriers have since been followed by other studies showing no association or contrary findings, and a metaanalysis of >7000 subjects found no significant associations of Val66Met with general cognitive ability, memory, executive function, visual processing skills, or cognitive fluency.[78] However, the small

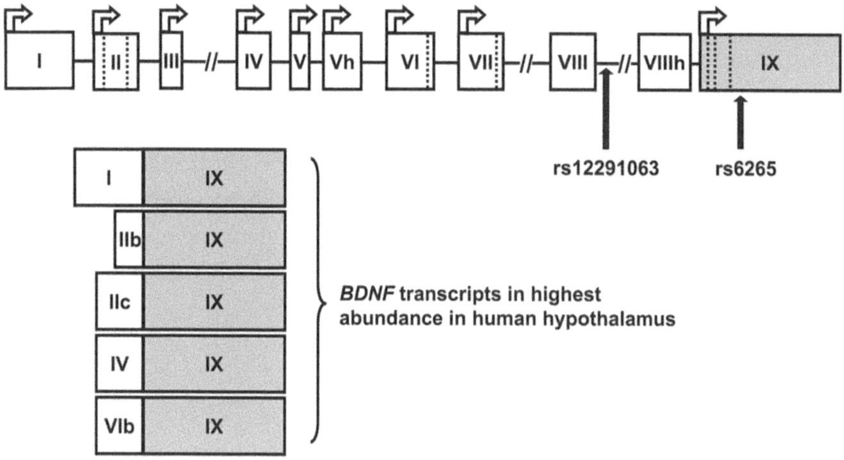

Figure 4 Human *BDNF* gene. *BDNF* spans approximately 67 kb and contains 11 exons. Alternative splicing of noncoding exons with a single coding exon IX *(gray rectangle)* produces 17 transcript variants with expression driven by nine different promoters *(white arrows)*. Exons II, VI, and VII contain multiple splice sites *(dashed lines)*, which create additional transcript variants IIa, IIb, IIc, VIa, VIb, VIIa, and VIIb.[1] Transcripts I, IIb, IIc, IV, and VIb are the most abundant variants in human hypothalamus.[33] *BDNF* SNP rs12291063 is located within the intron between exon VIII and VIIIh, approximately 1.4-kb downstream of exon VIII and 12.3-kb upstream of VIIIh. *BDNF* SNP rs6265 is located within the N-terminal cleaved portion of pro-BDNF. *Reprinted with permission from Ref. [100].*

sample sizes and between study heterogeneity limit drawing firm conclusions from the available extant data, and more studies are needed to understand the role of Val66Met in neurocognitive function.

With regard to obesity, the preponderance of studies, particularly those with larger sample sizes, have consistently identified the G allele of rs6265 (Val66 variant) as associated with higher BMI in both children and adults across diverse racial and ethnic populations, including European Caucasian, Asian, and Hispanic individuals (data are lacking for individuals of African descent due to the near absent, low minor allele frequency).[67,68,72–74,79–83] In addition to increased risk for obesity, the G allele also appears to predict weight regain in treatment studies. In patients with impaired glucose tolerance in the Diabetes Prevention Program (DPP) and patients with diabetes in the Look AHEAD study, the G allele was associated with greater weight regain across interventions, reaching significance in the DPP[84] and borderline significance in Look AHEAD.[85]

However, some studies have shown lack of association[86–90] or even an opposite direction of association,[91,92] with the A genotype (Met66 variant) being identified as the obesity risk allele. Many of these discrepant findings came from smaller case-control comparisons[86–88] or longitudinal cohort studies[89,90] with fewer than 1000 subjects per group, which may have been underpowered to detect relatively small effect sizes. Furthermore, modifying factors—particularly sex, lifestyle behaviors, and psychotropic medication use—appear to be important confounders for the association between rs6265 and BMI. In a Puerto Rican adult cohort, the GG genotype was associated with higher BMI in men but lower BMI in women,[93] suggesting that there may be sexual dimorphism for the effect of rs6265 on BMI. In a cohort of Chinese children, physical activity and sedentary behavior were found to modify the association between rs6265 and obesity risk.[94] In children with high physical activity and limited sedentary behavior, obesity risk did not differ by rs6265 genotype, whereas in children with low physical activity and high sedentary behavior, the A allele was associated with obesity. On the other hand, the G allele in inactive, sedentary children was associated with greater obesity risk compared to the A allele in active, nonsedentary children. Therefore, differences in physical activity and sedentary behavior can greatly influence the interpretation of the observed associations. Smoking is an additional lifestyle behavior that may be important. In a Korean adult cohort, the G allele was associated with higher BMI only among smokers.[95] Lastly, the A allele has been observed in several studies to predict greater weight gain with psychotropic medication use in patients with schizophrenia,[96–99] and other medications may be important to consider in patients with other neuropsychiatric conditions as well.

Taken together, the extant literature points toward a protective effect against obesity for the Met66 variant in the general population—perhaps more strongly in inactive, sedentary males who smoke and are not taking psychotropic medications. The mechanism for this protective effect is unclear, and is somewhat counterintuitive because the Met66 variant is thought to impair intracellular trafficking and depolarization-induced release of BDNF.[76,77] One possible explanation is that diminished activity-dependent secretion of BDNF may lead to chronically higher constitutive basal secretion of BDNF, which in turn suppresses appetite. Further studies are needed to delineate the role of lifestyle factors and potential therapeutic targets in modifying the relationship between rs6265 and BMI.

4.2 Intronic rs12291063 Single-Nucleotide Polymorphism

Besides rs6265 (Val66Met), all the other *BDNF* locus SNPs that have been identified in obesity-related GWAS are located within intronic, untranslated, and intergenic regions. Whether these noncoding region variants are pathogenic and how they might affect *BDNF* expression are not well understood. To explore these questions, Mou et al.[100] examined the association of *BDNF* expression in cadaveric VMH brain tissue with 44 *BDNF* locus SNPs, and observed that the minor C allele of rs12291063 was associated with lower *BDNF* expression and higher BMI in a postmortem cohort of 84 adults. The minor C allele was confirmed to be associated with BMI and adiposity in living subjects from racially diverse adult and pediatric cohorts, totaling >31,000 subjects.

BDNF SNP rs12291063 is a T to C substitution located within the intron between exon VIII and VIIIh, approximately 1.4-kb downstream of exon VIII and 12.3-kb upstream of VIIIh (Fig. 4). The proposed mechanism (Fig. 5) for the increased obesity risk with the C allele is diminished binding of the transcriptional enhancer, heterogeneous nuclear ribonucleoprotein

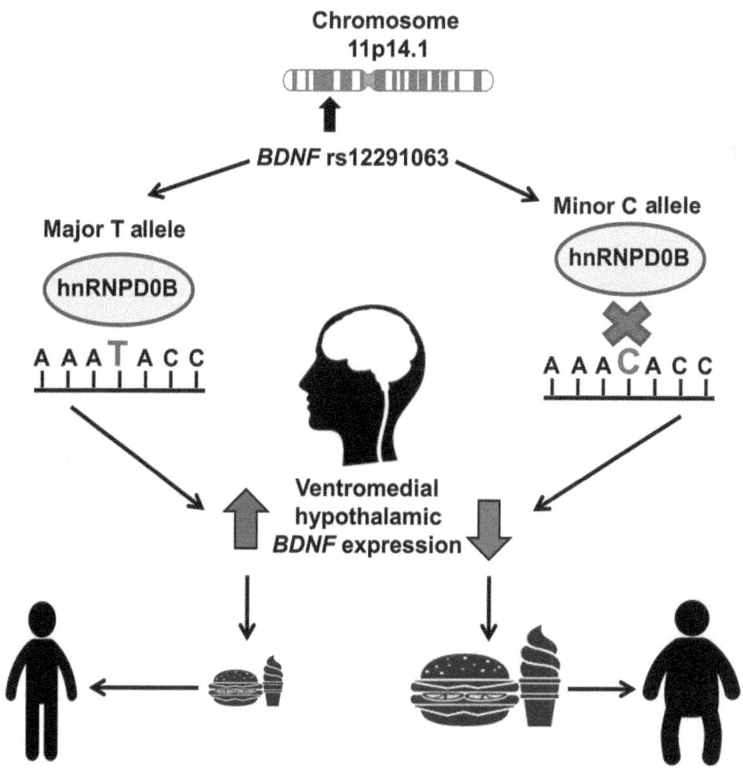

Figure 5 Proposed mechanism for obesity caused by intronic *BDNF* SNP rs12291063. *Reprinted with permission from Ref. [100].*

D0B, to the C allele sequence, leading to loss of transactivation function, decreased transcription of *BDNF*, diminished hypothalamic BDNF protein, overeating, and weight gain.[100] The minor allele frequency for rs12291063 varies with race and ethnicity (<0.01, 0.12, 0.26, and 0.38 for Caucasian European, Asian, Hispanic, and African, respectively).[75] The higher frequencies of the obesity risk allele in individuals with Hispanic or African ancestry may be a contributing factor for the higher obesity rates in these populations. Therapies aimed at increasing BDNF signaling in individuals with the rs12291063 CC genotype would provide a targeted approach for weight management.

As rs6265 and rs12291063 are only two of the many SNPs in the *BDNF* locus, further investigations are needed for better understanding of the regulation of *BDNF* expression and the relationship of these SNPs with obesity and possibly cognitive functioning.

5. CONCLUSIONS

BDNF is an important regulator of energy balance. Rare mutations, affecting the *BDNF* gene directly and indirectly, and common variants of *BDNF* lead to a spectrum of disordered energy balance and cognitive phenotypes. Complete loss of function of one *BDNF* allele is exceedingly rare and can lead to hyperphagia, childhood obesity, intellectual disability, impaired pain perception, and behavioral abnormalities, while minor variants of *BDNF* are widespread, and patients may only exhibit subtle changes in executive cognitive functioning or slightly higher BMI in adulthood. Identifying the genetic etiologies at both ends of this spectrum could be beneficial for the application of genotype-specific approaches for both rare and common forms of BDNF insufficiency. Approximately 10% of individuals of African–American race or Hispanic ethnicity have the obesity risk rs12291063 CC genotype. Therefore, CC-specific therapies could potentially impact a sizeable portion of the population and would advance the field of personalized precision medicine. Therapies identified and tested in patients with rare forms of *BDNF* loss may be translatable into the general population on a genotype-specific basis.

ACKNOWLEDGMENTS

This work was supported by the Memphis Research Consortium, the Intramural Research Program of the *Eunice Kennedy Shriver* National Institute of Child Health and Human Development (ZIAHD008898), and the National Institutes of Health Bench to Bedside Award.

REFERENCES

1. Pruunsild P, Kazantseva A, Aid T, Palm K, Timmusk T. Dissecting the human BDNF locus: bidirectional transcription, complex splicing, and multiple promoters. *Genomics.* 2007;90:397–406.
2. Martinowich K, Manji H, Lu B. New insights into BDNF function in depression and anxiety. *Nat Neurosci.* 2007;10:1089–1093.
3. Greenberg ME, Xu B, Lu B, Hempstead BL. New insights in the biology of BDNF synthesis and release: implications in CNS function. *J Neurosci.* 2009;29:12764–12767.
4. Deinhardt K, Chao MV. Shaping neurons: long and short range effects of mature and proBDNF signalling upon neuronal structure. *Neuropharmacology.* 2014;76(Pt C): 603–609.
5. Leibrock J, Lottspeich F, Hohn A, et al. Molecular cloning and expression of brain-derived neurotrophic factor. *Nature.* 1989;341:149–152.
6. Zuccato C, Cattaneo E. Brain-derived neurotrophic factor in neurodegenerative diseases. *Nat Rev Neurol.* 2009;5:311–322.
7. Noble EE, Billington CJ, Kotz CM, Wang C. The lighter side of BDNF. *Am J Physiol Regul Integr Comp Physiol.* 2011;300:R1053–R1069.
8. Autry AE, Monteggia LM. Brain-derived neurotrophic factor and neuropsychiatric disorders. *Pharmacol Rev.* 2012;64:238–258.
9. Waterhouse EG, Xu B. The skinny on brain-derived neurotrophic factor: evidence from animal models to GWAS. *J Mol Med.* 2013;91:1241–1247.
10. Andero R, Choi DC, Ressler KJ. BDNF-TrkB receptor regulation of distributed adult neural plasticity, memory formation, and psychiatric disorders. *Prog Mol Biol Transl Sci.* 2014;122:169–192.
11. Lapchak PA, Hefti F. BDNF and NGF treatment in lesioned rats: effects on cholinergic function and weight gain. *Neuroreport.* 1992;3:405–408.
12. Martin-Iverson MT, Todd KG, Altar CA. Brain-derived neurotrophic factor and neurotrophin-3 activate striatal dopamine and serotonin metabolism and related behaviors: interactions with amphetamine. *J Neurosci.* 1994;14:1262–1270.
13. Pelleymounter MA, Cullen MJ, Wellman CL. Characteristics of BDNF-induced weight loss. *Exp Neurol.* 1995;131:229–238.
14. Lyons WE, Mamounas LA, Ricaurte GA, et al. Brain-derived neurotrophic factor-deficient mice develop aggressiveness and hyperphagia in conjunction with brain serotonergic abnormalities. *Proc Natl Acad Sci USA.* 1999;96:15239–15244.
15. Fox EA, Byerly MS. A mechanism underlying mature-onset obesity: evidence from the hyperphagic phenotype of brain-derived neurotrophic factor mutants. *Am J Physiol Regul Integr Comp Physiol.* 2004;286:R994–R1004.
16. Kernie SG, Liebl DJ, Parada LF. BDNF regulates eating behavior and locomotor activity in mice. *EMBO J.* 2000;19:1290–1300.
17. Coppola V, Tessarollo L. Control of hyperphagia prevents obesity in BDNF heterozygous mice. *Neuroreport.* 2004;15:2665–2668.
18. Rios M, Fan G, Fekete C, et al. Conditional deletion of brain-derived neurotrophic factor in the postnatal brain leads to obesity and hyperactivity. *Mol Endocrinol.* 2001;15:1748–1757.
19. Xu B, Goulding EH, Zang K, et al. Brain-derived neurotrophic factor regulates energy balance downstream of melanocortin-4 receptor. *Nat Neurosci.* 2003;6:736–742.
20. Wang C, Bomberg E, Levine A, Billington C, Kotz CM. Brain-derived neurotrophic factor in the ventromedial nucleus of the hypothalamus reduces energy intake. *Am J Physiol Regul Integr Comp Physiol.* 2007;293:R1037–R1045.
21. Wang C, Bomberg E, Billington CJ, Levine AS, Kotz CM. Brain-derived neurotrophic factor (BDNF) in the hypothalamic ventromedial nucleus increases energy expenditure. *Brain Res.* 2010;1336:66–77.

22. Wang C, Bomberg E, Billington C, Levine A, Kotz CM. Brain-derived neurotrophic factor in the hypothalamic paraventricular nucleus reduces energy intake. *Am J Physiol Regul Integr Comp Physiol.* 2007;293:R1003–R1012.
23. Wang C, Bomberg E, Billington C, Levine A, Kotz CM. Brain-derived neurotrophic factor in the hypothalamic paraventricular nucleus increases energy expenditure by elevating metabolic rate. *Am J Physiol Regul Integr Comp Physiol.* 2007;293:R992–R1002.
24. Unger TJ, Calderon GA, Bradley LC, Sena-Esteves M, Rios M. Selective deletion of Bdnf in the ventromedial and dorsomedial hypothalamus of adult mice results in hyperphagic behavior and obesity. *J Neurosci.* 2007;27:14265–14274.
25. An JJ, Liao GY, Kinney CE, Sahibzada N, Xu B. Discrete BDNF neurons in the paraventricular hypothalamus control feeding and energy expenditure. *Cell Metab.* 2015;22:175–188.
26. Fox EA, Biddinger JE, Jones KR, McAdams J, Worman A. Mechanism of hyperphagia contributing to obesity in brain-derived neurotrophic factor knockout mice. *Neuroscience.* 2013;229:176–199.
27. Murphy MC, Fox EA. Mice deficient in brain-derived neurotrophic factor have altered development of gastric vagal sensory innervation. *J Comp Neurol.* 2010;518:2934–2951.
28. Zheng F, Zhou X, Moon C, Wang H. Regulation of brain-derived neurotrophic factor expression in neurons. *Int J Physiol Pathophysiol Pharmacol.* 2012;4:188–200.
29. Sakata K, Duke SM. Lack of BDNF expression through promoter IV disturbs expression of monoamine genes in the frontal cortex and hippocampus. *Neuroscience.* 2014;260:265–275.
30. Liao GY, An JJ, Gharami K, et al. Dendritically targeted Bdnf mRNA is essential for energy balance and response to leptin. *Nat Med.* 2012;18:564–571.
31. Cao L, Lin EJ, Cahill MC, Wang C, Liu X, During MJ. Molecular therapy of obesity and diabetes by a physiological autoregulatory approach. *Nat Med.* 2009;15:447–454.
32. Han JC, Lawlor DA, Kimm SY. Childhood obesity. *Lancet.* 2010;375:1737–1748.
33. Han JC, Liu QR, Jones M, et al. Brain-derived neurotrophic factor and obesity in the WAGR syndrome. *N Engl J Med.* 2008;359:918–927.
34. Dykens EM, Maxwell MA, Pantino E, Kossler R, Roof E. Assessment of hyperphagia in Prader–Willi syndrome. *Obesity.* 2007;15:1816–1826.
35. Linnarsson S, Bjorklund A, Ernfors P. Learning deficit in BDNF mutant mice. *Eur J Neurosci.* 1997;9:2581–2587.
36. MacQueen GM, Ramakrishnan K, Croll SD, et al. Performance of heterozygous brain-derived neurotrophic factor knockout mice on behavioral analogues of anxiety, nociception, and depression. *Behav Neurosci.* 2001;115:1145–1153.
37. Han JC, Thurm A, Golden Williams C, et al. Association of brain-derived neurotrophic factor (BDNF) haploinsufficiency with lower adaptive behaviour and reduced cognitive functioning in WAGR/11p13 deletion syndrome. *Cortex.* 2013;49:2700–2710.
38. Breau LM, Finley GA, McGrath PJ, Camfield CS. Validation of the non-communicating children's pain checklist-postoperative version. *Anesthesiology.* 2002;96:528–535.
39. Shinawi M, Sahoo T, Maranda B, et al. 11p14.1 microdeletions associated with ADHD, autism, developmental delay, and obesity. *Am J Med Genet A.* 2011;155A:1272–1280.
40. Ernst C, Marshall CR, Shen Y, et al. Highly penetrant alterations of a critical region including BDNF in human psychopathology and obesity. *Arch Gen Psychiatry.* 2012;69:1238–1246.
41. Gray J, Yeo GS, Cox JJ, et al. Hyperphagia, severe obesity, impaired cognitive function, and hyperactivity associated with functional loss of one copy of the brain-derived neurotrophic factor (BDNF) gene. *Diabetes.* 2006;55:3366–3371.
42. Yeo GS, Connie Hung CC, Rochford J, et al. A de novo mutation affecting human TrkB associated with severe obesity and developmental delay. *Nat Neurosci.* 2004;7:1187–1189.

43. Gray J, Yeo G, Hung C, et al. Functional characterization of human NTRK2 mutations identified in patients with severe early-onset obesity. *Int J Obes.* 2007;31:359–364.
44. Cassidy SB, Driscoll DJ. Prader–Willi syndrome. *Eur J Hum Genet.* 2009;17:3–13.
45. Holm VA, Cassidy SB, Butler MG, et al. Prader–Willi syndrome: consensus diagnostic criteria. *Pediatrics.* 1993;91:398–402.
46. Schaaf CP, Gonzalez-Garay ML, Xia F, et al. Truncating mutations of MAGEL2 cause Prader–Willi phenotypes and autism. *Nat Genet.* 2013;45:1405–1408.
47. Cummings DE, Clement K, Purnell JQ, et al. Elevated plasma ghrelin levels in Prader–Willi syndrome. *Nat Med.* 2002;8:643–644.
48. Haqq AM, Farooqi IS, O'Rahilly S, et al. Serum ghrelin levels are inversely correlated with body mass index, age, and insulin concentrations in normal children and are markedly increased in Prader–Willi syndrome. *J Clin Endocrinol Metab.* 2003;88: 174–178.
49. Kweh FA, Miller JL, Sulsona CR, et al. Hyperghrelinemia in Prader–Willi syndrome begins in early infancy long before the onset of hyperphagia. *Am J Med Genet A.* 2015;167A:69–79.
50. De Waele K, Ishkanian SL, Bogarin R, et al. Long-acting octreotide treatment causes a sustained decrease in ghrelin concentrations but does not affect weight, behaviour and appetite in subjects with Prader–Willi syndrome. *Eur J Endocrinol.* 2008;159:381–388.
51. Haqq AM, Stadler DD, Rosenfeld RG, et al. Circulating ghrelin levels are suppressed by meals and octreotide therapy in children with Prader–Willi syndrome. *J Clin Endocrinol Metab.* 2003;88:3573–3576.
52. Lindgren AC, Marcus C, Skwirut C, et al. Increased leptin messenger RNA and serum leptin levels in children with Prader–Willi syndrome and nonsyndromal obesity. *Pediatr Res.* 1997;42:593–596.
53. Goldstone AP, Unmehopa UA, Bloom SR, Swaab DF. Hypothalamic NPY and agouti-related protein are increased in human illness but not in Prader–Willi syndrome and other obese subjects. *J Clin Endocrinol Metab.* 2002;87:927–937.
54. Mercer RE, Michaelson SD, Chee MJ, Atallah TA, Wevrick R, Colmers WF. Magel2 is required for leptin-mediated depolarization of POMC neurons in the hypothalamic arcuate nucleus in mice. *PLoS Genet.* 2013;9:e1003207.
55. Merighi A, Salio C, Ghirri A, et al. BDNF as a pain modulator. *Prog Neurobiol.* 2008;85:297–317.
56. Groth R, Aanonsen L. Spinal brain-derived neurotrophic factor (BDNF) produces hyperalgesia in normal mice while antisense directed against either BDNF or trkB, prevent inflammation-induced hyperalgesia. *Pain.* 2002;100:171–181.
57. Han JC, Muehlbauer MJ, Cui HN, Newgard CB, Haqq AM. Lower brain-derived neurotrophic factor in patients with Prader–Willi syndrome compared to obese and lean control subjects. *J Clin Endocrinol Metab.* 2010;95:3532–3536.
58. Allen SJ, Dawbarn D. Clinical relevance of the neurotrophins and their receptors. *Clin Sci.* 2006;110:175–191.
59. Tahir R, Kennedy A, Elsea SH, Dickinson AJ. Retinoic acid induced-1 (Rai1) regulates craniofacial and brain development in *Xenopus*. *Mech Dev.* 2014;133:91–104.
60. Burns B, Schmidt K, Williams SR, Kim S, Girirajan S, Elsea SH. Rai1 haploinsufficiency causes reduced Bdnf expression resulting in hyperphagia, obesity and altered fat distribution in mice and humans with no evidence of metabolic syndrome. *Hum Mol Genet.* 2010;19:4026–4042.
61. Alaimo JT, Hahn NH, Mullegama SV, Elsea SH. Dietary regimens modify early onset of obesity in mice haploinsufficient for Rai1. *PLoS One.* 2014;9:e105077.
62. Edelman EA, Girirajan S, Finucane B, et al. Gender, genotype, and phenotype differences in Smith–Magenis syndrome: a meta-analysis of 105 cases. *Clin Genet.* 2007;71: 540–550.

63. Carmona-Mora P, Encina CA, Canales CP, et al. Functional and cellular characterization of human Retinoic Acid Induced 1 (RAI1) mutations associated with Smith–Magenis syndrome. *BMC Mol Biol.* 2010;11:63.
64. Carmona-Mora P, Canales CP, Cao L, et al. RAI1 transcription factor activity is impaired in mutants associated with Smith–Magenis syndrome. *PLoS One.* 2012;7: e45155.
65. Thaker VV, Esteves KM, Towne MC, et al. Whole exome sequencing identifies RAI1 mutation in a morbidly obese child diagnosed with ROHHAD syndrome. *J Clin Endocrinol Metab.* 2015;100:1723–1730.
66. Weese-Mayer DE, Bolk S, Silvestri JM, Chakravarti A. Idiopathic congenital central hypoventilation syndrome: evaluation of brain-derived neurotrophic factor genomic DNA sequence variation. *Am J Med Genet.* 2002;107:306–310.
67. Thorleifsson G, Walters GB, Gudbjartsson DF, et al. Genome-wide association yields new sequence variants at seven loci that associate with measures of obesity. *Nat Genet.* 2009;41:18–24.
68. Zhao J, Bradfield JP, Li M, et al. The role of obesity-associated loci identified in genome-wide association studies in the determination of pediatric BMI. *Obesity.* 2009;17:2254–2257.
69. Speliotes EK, Willer CJ, Berndt SI, et al. Association analyses of 249,796 individuals reveal 18 new loci associated with body mass index. *Nat Genet.* 2010;42:937–948.
70. Croteau-Chonka DC, Marvelle AF, Lange EM, et al. Genome-wide association study of anthropometric traits and evidence of interactions with age and study year in Filipino women. *Obesity.* 2011;19:1019–1027.
71. Okada Y, Kubo M, Ohmiya H, et al. Common variants at CDKAL1 and KLF9 are associated with body mass index in east Asian populations. *Nat Genet.* 2012;44:302–306.
72. Wen W, Cho YS, Zheng W, et al. Meta-analysis identifies common variants associated with body mass index in east Asians. *Nat Genet.* 2012;44:307–311.
73. Gong J, Schumacher F, Lim U, et al. Fine mapping and identification of BMI loci in African Americans. *Am J Hum Genet.* 2013;93:661–671.
74. Leon-Mimila P, Villamil-Ramirez H, Villalobos-Comparan M, et al. Contribution of common genetic variants to obesity and obesity-related traits in Mexican children and adults. *PLoS One.* 2013;8:e70640.
75. Sherry ST, Ward MH, Kholodov M, et al. dbSNP: the NCBI database of genetic variation. *Nucleic Acids Res.* 2001;29:308–311.
76. Egan MF, Kojima M, Callicott JH, et al. The BDNF val66met polymorphism affects activity-dependent secretion of BDNF and human memory and hippocampal function. *Cell.* 2003;112:257–269.
77. Chen ZY, Patel PD, Sant G, et al. Variant brain-derived neurotrophic factor (BDNF) (Met66) alters the intracellular trafficking and activity-dependent secretion of wild-type BDNF in neurosecretory cells and cortical neurons. *J Neurosci.* 2004;24:4401–4410.
78. Mandelman SD, Grigorenko EL. BDNF Val66Met and cognition: all, none, or some? A meta-analysis of the genetic association. *Genes Brain Behav.* 2012;11:127–136.
79. Gunstad J, Schofield P, Paul RH, et al. BDNF Val66Met polymorphism is associated with body mass index in healthy adults. *Neuropsychobiology.* 2006;53:153–156.
80. Wu L, Xi B, Zhang M, et al. Associations of six single nucleotide polymorphisms in obesity-related genes with BMI and risk of obesity in Chinese children. *Diabetes.* 2010;59:3085–3089.
81. Mitchell JA, Hakonarson H, Rebbeck TR, Grant SF. Obesity-susceptibility loci and the tails of the pediatric BMI distribution. *Obesity.* 2013;21:1256–1260.
82. Xi B, Cheng H, Shen Y, et al. Study of 11 BMI-associated loci identified in GWAS for associations with central obesity in the Chinese children. *PLoS One.* 2013;8: e56472.

83. Shugart YY, Chen L, Day IN, et al. Two British women studies replicated the association between the Val66Met polymorphism in the brain-derived neurotrophic factor (BDNF) and BMI. *Eur J Hum Genet.* 2009;17:1050–1055.
84. Delahanty LM, Pan Q, Jablonski KA, et al. Genetic predictors of weight loss and weight regain after intensive lifestyle modification, metformin treatment, or standard care in the Diabetes Prevention Program. *Diabetes Care.* 2012;35:363–366.
85. McCaffery JM, Papandonatos GD, Huggins GS, et al. FTO predicts weight regain in the Look AHEAD clinical trial. *Int J Obes.* 2013;37:1545–1552.
86. Friedel S, Horro FF, Wermter AK, et al. Mutation screen of the brain derived neurotrophic factor gene (BDNF): identification of several genetic variants and association studies in patients with obesity, eating disorders, and attention-deficit/hyperactivity disorder. *Am J Med Genet B Neuropsychiatr Genet.* 2005;132B:96–99.
87. Krabbe KS, Nielsen AR, Krogh-Madsen R, et al. Brain-derived neurotrophic factor (BDNF) and type 2 diabetes. *Diabetologia.* 2007;50:431–438.
88. Marti A, Santos JL, Gratacos M, et al. Association between leptin receptor (LEPR) and brain-derived neurotrophic factor (BDNF) gene variants and obesity: a case-control study. *Nutr Neurosci.* 2009;12:183–188.
89. Nikolac Perkovic M, Mustapic M, Pavlovic M, et al. Lack of association between brain-derived neurotrophic factor Val66Met polymorphism and body mass index change over time in healthy adults. *Neurosci Lett.* 2013;545:127–131.
90. Zhao XY, Zhang MX, Cheng H, et al. Risk of obesity-related gene polymorphism on the incidence and durative of childhood obesity. *Zhonghua Liu Xing Bing Xue Za Zhi.* 2013;34:560–565.
91. Skledar M, Nikolac M, Dodig-Curkovic K, Curkovic M, Borovecki F, Pivac N. Association between brain-derived neurotrophic factor Val66Met and obesity in children and adolescents. *Prog Neuropsychopharmacol Biol Psychiatry.* 2012;36:136–140.
92. Beckers S, Peeters A, Zegers D, Mertens I, Van Gaal L, Van Hul W. Association of the BDNF Val66Met variation with obesity in women. *Mol Genet Metab.* 2008;95:110–112.
93. Ma XY, Qiu WQ, Smith CE, et al. Association between BDNF rs6265 and obesity in the Boston Puerto Rican Health Study. *J Obes.* 2012;2012:102942.
94. Xi B, Wang C, Wu L, et al. Influence of physical inactivity on associations between single nucleotide polymorphisms and genetic predisposition to childhood obesity. *Am J Epidemiol.* 2011;173:1256–1262.
95. Hong KW, Lim JE, Go MJ, et al. Recapitulation of the association of the Val66Met polymorphism of BDNF gene with BMI in Koreans. *Obesity.* 2012;20:1871–1875.
96. Bonaccorso S, Sodhi M, Li J, et al. The brain-derived neurotrophic factor (BDNF) Val66Met polymorphism is associated with increased body mass index and insulin resistance measures in bipolar disorder and schizophrenia. *Bipolar Disord.* 2015;17:528–535.
97. Fonseka TM, Tiwari AK, Goncalves VF, et al. The role of genetic variation across IL-1beta, IL-2, IL-6, and BDNF in antipsychotic-induced weight gain. *World J Biol Psychiatry.* 2015;16:45–56.
98. Zhang Y, Chen M, Wu Z, et al. Association study of Val66Met polymorphism in brain-derived neurotrophic factor gene with clozapine-induced metabolic syndrome: preliminary results. *PLoS One.* 2013;8:e72652.
99. Zhang XY, Zhou DF, Wu GY, et al. BDNF levels and genotype are associated with antipsychotic-induced weight gain in patients with chronic schizophrenia. *Neuropsychopharmacology.* 2008;33:2200–2205.
100. Mou Z, Hyde TM, Lipska BK, et al. Human obesity associated with an intronic SNP in the brain-derived neurotrophic factor locus. *Cell Rep.* 2015;13:1073–1080.

CHAPTER FOUR

Mutations in Melanocortin-3 Receptor Gene and Human Obesity

Z. Yang, Y.-X. Tao[1]
Department of Anatomy, Physiology and Pharmacology, College of Veterinary Medicine, Auburn University, Auburn, AL, United States of America
[1] Corresponding author. E-mail address: taoyaxi@auburn.edu

Contents

Abstract

The prevalence of obesity calls for novel therapeutic targets. The melanocortin-3 receptor (MC3R) has been increasingly recognized as an important regulator of energy homeostasis and *MC3R* has been intensively analyzed in molecular genetic studies for obesity-related traits. Twenty-seven *MC3R* mutations and two common polymorphic variants have been identified so far in different cohorts. The mutant MC3Rs demonstrate multiple defects in functional analysis and can be cataloged into different classes according to receptor life cycle based classification system. Although the

Progress in Molecular Biology and Translational Science, Volume 140
ISSN 1877-1173
http://dx.doi.org/10.1016/bs.pmbts.2016.01.002

pathogenic role of *MC3R* in human obesity remains controversial, recent findings in the noncanonical signaling pathway of MC3R mutants have provided new insights. Potential therapeutic strategies for obesity related to *MC3R* mutations are highlighted.

ABBREVIATIONS

ACTH	Adrenocorticotropin
AgRP	Agouti-related peptide
BMI	Body mass index
ERK1/2	Extracellular signal-regulated kinase 1 and 2
GPCR	G protein-coupled receptor
JAK	Janus kinase
LOF	Loss of function
MAPK	Mitogen-activated protein kinase
MC3R	Melanocortin-3 receptor
MCR	Melanocortin receptor
MSH	Melanocyte-stimulating hormone
ObR	Leptin receptor
PC	Prohormone convertases
PI3K	Phosphoinositide 3-kinase
POMC	Proopiomelanocortin
SNP	Single nucleotide polymorphism
STAT	Signal transducers and activators of transcription
TM	Transmembrane
WHO	World Health Organization
WT	Wild type

1. OBESITY AND CURRENT TREATMENT

1.1 The Prevalence of Obesity

Obesity is characterized as a medical condition in which excessive fat is accumulated in the adipose tissue and other organs. The National Heart, Lung, and Blood Institute and the World Health Organization (WHO) define obesity as a body mass index (BMI, body weight in kilograms divided by height in meters squared.) equal to or greater than 30 kg/m^2 and overweight as a BMI of 25–29.9 kg/m^2.[1]

Obesity has become a global epidemic. According to WHO, 1.46 billion adults are overweight and over 200 million men and 300 million women are obese globally.[2] In the United States, the most recent National Health and Nutrition Examination Survey indicates that 68.5% of the adults are presently classified as overweight or obese with 34.9% as obese.[3] The obesity epidemic is not restricted to developed countries. Sharp increase of obesity has also been observed in developing countries.

Obesity in children and adolescents is also increasing at an alarming rate. About 170 million children globally are classified as overweight or obese, a quarter of which are under 5 years of age.[4] Although the prevalence of obesity in certain age groups of childhood from some developed countries appears to be flattening, the overall prevalence is still high.[3,5] Obese children are more likely to be afflicted by obesity-related diseases such as type 2 diabetes mellitus and heart diseases before or during adulthood.[4,6]

Obesity is becoming a significant public health burden since it is associated with a number of comorbidities including type 2 diabetes mellitus, cardiovascular diseases, hypertension, arteriosclerosis, sleep apnea, and certain types of cancer.[7,8] According to the new guidelines released on November 2013 by the American Heart Association, American College of Cardiology, and The Obesity Society, obesity itself is currently recognized as a disease.[9]

In addition to the health crisis, obesity is also associated with enormous economic costs. The annual medical expenditures attributed to obesity and obesity-related diseases in the United States increased from $75 billion in 2003 to $147 billion in 2008, and the number was estimated to reach $344 billion by 2018.[10] Obesity is now accounting for up to 6% of the total healthcare cost in the United States while this number is estimated to rise to roughly 21% by 2018.[11] Indeed, obesity has now overtaken tobacco as the largest preventable cause of disease burden in some countries. Moreover, the negative psychological effect of obesity, such as discrimination and exclusion, cannot be ignored since these negative emotions can lead to social problems.[12]

1.2 Current Treatments for Obesity

Although considerable attention has been paid to obesity research during the past few decades, there are very limited effective therapeutic options currently available. Current treatments for obesity include lifestyle intervention, pharmacological therapy, and bariatric surgery.

Lifestyle interventions or modifications, such as exercise and dieting, have conventionally been used for treating obesity and can achieve short-term success for the majority of obese patients.[13,14] However, strong physiological and behavioral responses to caloric restriction together with the "obesogenic environment" make the lifestyle modification extremely difficult for obese individuals to maintain the weight loss. The human body's defense of body weight prevents the obese patients from long-term success in weight loss.[15,16]

Pharmacotherapy will be considered if lifestyle interventions are ineffective for obese individuals. Many antiobesity drugs have been developed to manage obesity over the past several decades.[17] However, most of the drugs, such as phentermine and rimonabant, have now been withdrawn due to serious side effects.[18,19] The most recently withdrawn drug for obesity control is sibutramine, a selective noradrenaline/serotonin reuptake inhibitor,[20] which has a strong association with increased cardiovascular events and stroke.[21] The only currently available drug for the long-term management of obesity is orlistat, a potent and reversible gastrointestinal lipase inhibitor that prevents dietary fat absorption by inhibiting pancreatic and gastric lipases. Although the efficacy of orlistat has been proven in clinic,[22,23] it is associated with side effects such as diarrhea, flatulence, and dyspepsia.[24]

Bariatric surgery, such as laparoscopic-adjustable gastric banding, Roux-en-Y gastric bypass, and sleeve gastrectomy, can restrict the flow of food through the gastrointestinal tract and is currently considered the most effective treatment for morbid obesity. The average percentage of excess weight loss by bariatric surgery is 61.2%[25] and it can also reduce obesity-related comorbidities including diabetes, hypertension, heart diseases, and certain types of cancers.[25–28] Moreover, both the short- and long-term efficacies of bariatric surgery have been established.[29] However, the significant cost, risks, and complication rates prevent majority of the obese patients from undergoing the surgeries.[30,31]

Obesity is a metabolic disorder caused by the imbalance of energy intake and expenditure. Both environmental and genetic factors contribute to human obesity. Studies of BMI correlations of twins and adopted siblings indicated that heredity plays a significant role in body weight regulation.[32,33] Therefore, there is an urgent need for understanding the regulation of energy homeostasis from a genetic perspective. This chapter summarizes the study of a receptor molecule that is involved in regulating energy homeostasis and potentially contributes to obesity pathogenesis, the melanocortin-3 receptor (MC3R).

2. THE LEPTIN-REGULATED MELANOCORTIN CIRCUIT

2.1 The Melanocortin System

The melanocortin system is comprised of six major known endogenous ligands (four agonists and two antagonists) and five receptors that were named according to the order of their cloning.[34–36]

The four endogenous agonists, α-melanocyte-stimulating hormone (MSH), β-MSH, γ-MSH, and adrenocorticotropin (ACTH), are produced from a preprohormone proopiomelanocortin (POMC), which is posttranslationally processed by prohormone convertases (PC1/PC2). POMC neurons are mainly located in the arcuate nucleus of the hypothalamus with a small population found in the nucleus tractus solitaries in the brainstem. The melanocortin system is unique to have two endogenous antagonists, Agouti and Agouti-related peptide (AgRP).

The diverse actions of melanocortins are mediated by five melanocortin receptors (MCRs). The melanocortin-1 receptor (MC1R) regulates skin and hair pigmentation while MC2R is the regulator of adrenal steroidogenesis.[37,38] Two neural MCRs, MC3R and MC4R, have been shown to play important roles in energy homeostasis.[39] MC5R is mainly involved in the regulation of exocrine gland secretion.[40]

The endogenous ligands exhibit variable degrees of specificity for different MCRs. α-MSH and β-MSH display agonist activities on MC1R, MC3R, MC4R, and MC5R. γ-MSH exhibits modest selectivity for MC3R while MC2R only binds to ACTH.[41] Agouti acts as a selective antagonist on MC1R and MC4R,[42,43] whereas AgRP functions specifically on MC3R and MC4R.[44,45] Recent studies also indicated that AgRP could inhibit the basal activity of MC3R and MC4R, suggesting inverse agonist property.[46–48]

Among the five MCRs, MC3R and MC4R are primarily expressed in the central nervous system and are involved in the physiological regulation of body weight and metabolic activity. The hypothalamic melanocortin system serves as the intersection connecting the neural pathways governing satiety and metabolism with signals of metabolic status to maintain energy homeostasis through the leptin-melanocortin pathway.

2.2 The Leptin-Melanocortin Pathway

There are two subsets of neurons expressing the endogenous ligands for the neural MCRs within the arcuate nucleus. One subset of neurons coexpresses POMC and cocaine and amphetamine-regulated transcript in the lateral part of the arcuate nucleus, while the other subset of neurons coexpresses AgRP and neuropeptide Y in the medial arcuate nucleus.[49,50] These two neuronal populations have distinct functional effects on the regulation of energy homeostasis. POMC neurons inhibit food intake and elevate energy expenditure by releasing α-MSH while AgRP neurons have opposite functions. Both POMC and AgRP neurons are described as "first order" neurons since they are the primary targets of signals of metabolic status including leptin,[51–53] insulin,[54] ghrelin,[55] serotonin,[56] orexin,[57] and glucose,[58] among which leptin is the most intensively studied and functionally highlighted.

Leptin is a 16-kDa peptide hormone secreted from adipocytes in proportion to the degree of adiposity.[59,60] It can be transported across the blood–brain barrier and acts directly in the hypothalamus by stimulating the leptin receptor (ObR) that is highly expressed in "first order" neurons.[61,62] The ObR is a member of Class I cytokine receptor family with a single transmembrane (TM) domain. Activation of ObR initiates a cascade of signal transduction pathways, including Janus kinase/signal transducers and activators of transcription (JAK/STAT) and phosphoinositide 3-kinase (PI3K) pathways, which consequently activate the POMC neurons but inhibit the AgRP neurons.[63,64] The released α-MSH or AgRP affects the "second order" neurons where MC3R or MC4R are highly expressed and finally leads to the maintenance of energy homeostasis.

The importance of this leptin-regulated melanocortin circuit has been highlighted in both rodent and human since mutations of any component in this circuit can lead to obesity and other metabolic disorders. Typical models in rodent, such as *ob/ob* mice, *db/db* mice, and *POMC* knockout (KO) mice, and mutations in human including *leptin*, *ObR*, *POMC*, and *MC4R* mutations, can all result in severe obesity.[59,65–73]

Although MC4R shows a much wider expression than MC3R in the central nervous system and functional evidence detailing the role of MC4R in the regulation of energy homeostasis is more substantial, the study of MC3R is still of significant interest since it is the only MCR expressed by both POMC and AgRP neurons in the arcuate nucleus[74,75] and the role it plays in metabolism is now emerging in several interesting aspects.

3. MELANOCORTIN-3 RECEPTOR AND ENERGY HOMEOSTASIS

The human *MC3R* was first cloned by Gantz and coworkers and predicted to encode a 360 amino acid protein as a single exon gene located on chromosome 20q13.2.[76,77] Recent studies suggest that the translational starting site of human MC3R is at the evolutionarily conserved second ATG codon instead of the originally assumed nonconserved first ATG, resulting in a 37 amino acid shorter protein.[78,79] Park et al. further indicated that human

MC3R has an additional upstream exon that directs the utilization of the second ATG codon as the translational start site.[79]

The MC3R is primarily expressed in the hypothalamus, especially in the arcuate nucleus, the ventromedial nucleus, and the posterior hypothalamic region.[74] It is also expressed in several peripheral tissues, including the placenta, gut, heart, kidney, and peritoneal macrophages.[76,80–82] Based on its wide distribution, the MC3R has been shown to be involved in the regulation of hypothalamic–pituitary–thyroid axis,[83] cardiovascular function,[84,85] natriuresis,[82,86] and inflammation.[87–89]

The functional role of MC3R in energy homeostasis is less clear compared to that of MC4R that has been shown to be crucial for both food intake and energy expenditure. However, the more severe obesity phenotype observed in mice lacking both *Mc3r and Mc4r* suggested that these two receptors serve nonredundant roles in energy homeostasis.[90] This nonredundancy has been further confirmed by the study of the anorectic response of mice to the mixed MC3/4R agonist melanotan II, showing that the double knockout mice are completely unresponsive while both $Mc3r^{-/-}$ and $Mc4r^{-/-}$ mice show partial anorectic response.[91]

Unlike the hyperphagic and obese $Mc4r^{-/-}$ mice, $Mc3r^{-/-}$ mice display a moderate obesity syndrome characterized with no hyperphagia but increased fat mass and reduced lean mass.[90,92,93] The increase in adiposity is exacerbated by feeding a high-fat diet.[94] Reduced energy expenditure is also observed in $Mc3r^{-/-}$ mice with decreasing locomotor behavior.[90,95] Therefore, MC3R is primarily involved in the mediation of feed efficiency and nutrient partitioning instead of direct regulation of food intake.

One possible mechanism for the regulation of feed efficiency by MC3R is suggested to be the subtle imbalance between fat intake and oxidation, since the fatty acid oxidation was significantly reduced in the skeletal muscle of $Mc3r^{-/-}$ mice.[96] Renquist et al. suggested that loss of MC3R in mice might impair the metabolic adaptation to fasting with defective white adipose tissue lipolysis, reduced liver triglyceride accumulation, and blunted hypothalamic-pituitary-adrenal axis activation.[97]

Recently, studies from Butler's laboratory have revealed that MC3R is also involved in the regulation of inputs of feeding-related signals into systems expressing rhythms of food anticipatory activity. Their studies showed that $Mc3r^{-/-}$ mice under restricted feeding exhibit reduced wakefulness before food presentation indicating impaired behavioral adaptation.[95,98,99] Abnormal rhythmic expression of clock genes such as *Bmal1* and *Rev-erba* in *Mc3r*$^{-/-}$ mice is also observed.[100] In addition to the impaired behavioral adaptation, *Mc3r*$^{-/-}$ mice also display abnormal metabolic adaptation to the restricted feeding protocol. Although previous studies suggested that *Mc3r*$^{-/-}$ mice are protected from the development of fatty liver disease, severe insulin resistance, and obesity-induced inflammation in white adipose tissue, different from the *Mc4r*$^{-/-}$ mice,[96,101,102] *Mc3r*$^{-/-}$ mice under restricted feeding was found to show a sophisticated phenotype with decreased body weight but deteriorating metabolism including hyperinsulinemia, glucose intolerance, enhanced expression of lipogenic genes, and increased ketogenesis.[103]

Several recent intriguing studies revisited the potential direct participation of MC3R in food intake regulation. It is reported that both peripheral and central administration of selective MC3R agonist stimulates feeding in rats,[104,105] and another study demonstrated that treatment of a mixed MC3R antagonist and MC4R agonist tetrapeptide reduces food intake of *Mc4r*$^{-/-}$ rats.[106] These studies suggested that MC3R might act as an inhibitory autoreceptor on POMC neurons but the conclusion could not be easily drawn since there was also some inconsistency within these studies. For example, the same researchers of the *Mc4r*$^{-/-}$ rat model study also reported that treatment of a mixed MC3R and MC4R full agonist terapeptide results in reduced food intake in both WT and *Mc4r*$^{-/-}$ rats.[106]

Collectively, it is generally accepted that MC3R participates in the regulation of energy homeostasis in a different way from MC4R, including regulating feed efficiency, maintaining circadian rhythm, and adapting to fasting.

4. NATURALLY OCCURRING MUTATIONS IN *MC3R* AND HUMAN OBESITY

4.1 Identification of *MC3R* Variants and Mutations

Prior to the revelation of the role of MC3R in energy homeostasis using *Mc3r*$^{-/-}$ animal models, human genetic studies have demonstrated a positive linkage between obesity and the chromosomal region 20q13 that harbors *MC3R*, making the *MC3R* a plausible candidate gene for human obesity and type 2 diabetes.[107–111] Several groups conducted large-scale screening studies in obese and type 2 diabetic individuals in order to identify potential *MC3R* mutation. The first two variants identified in *MC3R* are T6K and V81I, which are polymorphic variants in complete linkage disequilibrium.[112,113] Since then, 27 *MC3R* mutations have been identified so far in different cohorts from Asia, Europe, and North America.[114–121] These mutations are all missense mutations and scattered throughout the MC3R as shown in Fig. 1. TM6 of MC3R is the most variable area harboring six mutations while TM5 is the only TM without any mutation identified so far.

Since the *MC3R* locus and expression have a positive linkage to obesity, several studies were performed to identify potential mutations in the regulatory sequences of MC3R, especially the promoter region. Two single nucleotide polymorphisms (SNPs) in the promoter region of *MC3R*, rs11575886 and rs6127698, have been studied and shown to be not associated with obesity,[122,123] although the rs11575886 ($-239A > G$) reduces the binding of transcription factor GATA-4 to the *MC3R* promoter region.[124] These data suggest that defect in *MC3R* gene transcription is unlikely to be a major cause of severe early-onset obesity.

As for the other variants in regulatory sequences, +2138InsCAGACC, which is a 6 base-pair insertion located 2138 base pairs downstream of the original *MC3R* start codon, has been suggested to be related to body composition and eating disorder.[125,126] However, no association was identified between the +2138InsCAGACC polymorphism and differences in energy expenditure or glucose homeostasis.[127] Although an association has been shown between D20S32e, which lies in the 5'-flanking region of the *MC3R*, and fasting insulin and insulin resistance index, the association was independent of BMI.[128] Therefore, these studies of variations in the regulatory regions suggest potential participation of *MC3R* in energy homeostasis but the contribution or involvement of other genes near *MC3R* could not be completely excluded.

4.2 Prevalence of *MC3R* Mutations

The human *MC3R* has two common nonsynonymous variants, T6K and V81I, in complete linkage disequilibrium. Therefore, genotype analysis is usually categorized based on the combination of these two variants. However, large-scale case control studies indicate that the frequencies of these two variants have always been similar in obese and nonobese cohorts.[115,118,124] (Phenotype and functional studies will be discussed later in the chapter.) As shown in Table 1, a subanalysis of the data from different cohorts reveals some diversity of the frequencies of the T6K/V81I variants in obese individuals from different countries or ethnic groups. In the

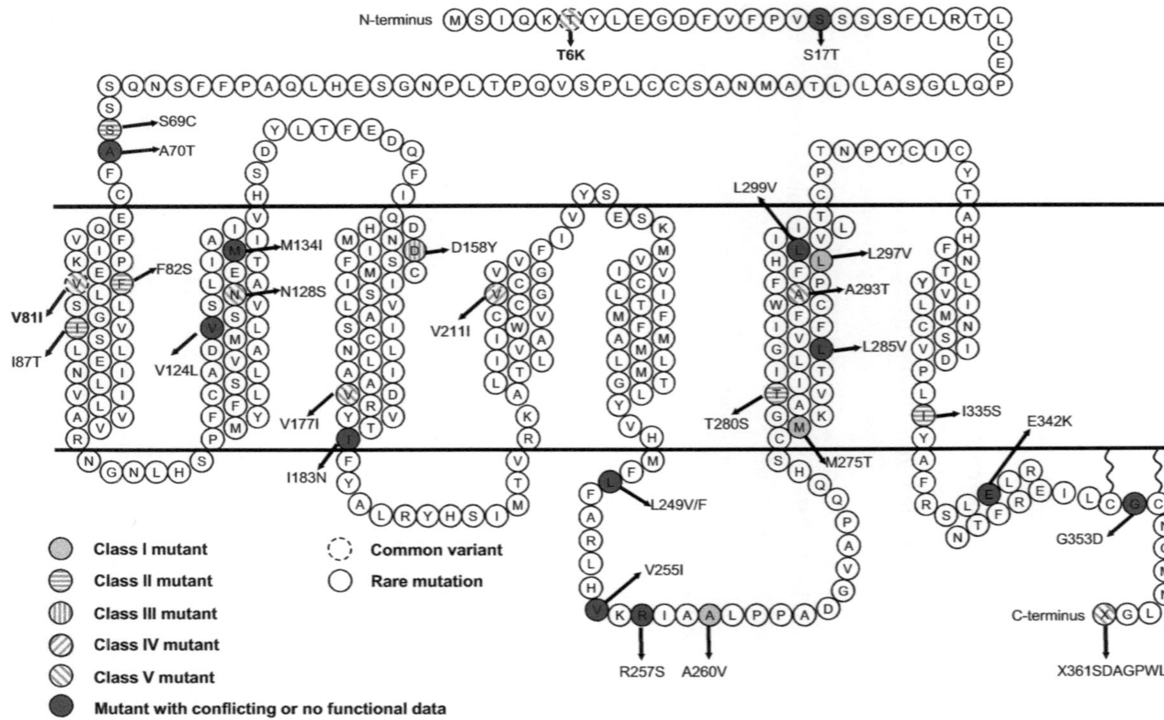

Figure 1 Naturally occurring variants and mutations in the *MC3R*. Two common variants in the *MC3R*, Thr6Lys, and Val81Ile, are indicated. Mutants of different Classes are also indicated separately.

Table 1 Distribution of *MC3R* Common Variants (T6K/V81I) in Obese Individuals From Different Cohorts.

Groups (*n*)		Wt/Wt[a]	Het[b]	Hom/Hom[c]	References
United States	African-American (152)	31.6	52.6	15.8	[134]
	Caucasian (176)	80.1	18.2	1.7	
Singapore	Chinese (105)	60.9	36.2	2.8	[115]
	Malays (68)	64.7	35.5	0	
Polish (257)		77.4	21.4	1.2	[137]
Italian (184)		89.1	10.9	0	[135]
Chilean (229)		88.8	10.7	0.44	[133]
Belgian (1008)		85.9	13.5	0.6	[122]

Data are given in percentage.
[a]Wt/Wt, wide type for both variant alleles (Thr/Thr and Val/Val).
[b]Het, heterozygous for at least one allele (Thr/Thr and Val/Ile, Thr/Lys and Val/Ile, Lys/Lys and Val/Ile, or Thr/Lys and Ile/Ile).
[c]Hom/Hom, homozygous for both variant alleles (Lys/Lys and Ile/Ile).

United States, the variants are more prevalent in African American than in Caucasian obese individuals, and homozygosity frequency of African American obese subjects for both variants is higher compared to all the other cohorts (15.8%). This is supported by another study of 416 American children (obese and nonobese) indicating that the frequency of K6 alleles in African American is 64.7% but 16.7% in Caucasian.[127] The prevalence of these two variants in Asians is intermediate and the lowest prevalence has been identified in Italy and Chile.

Except for the two common variants, the prevalence of the other *MC3R* mutations is generally low. In a study of 201 obese children and adolescents from Singapore, three rare *MC3R* mutations were identified with a prevalence of 1.49%.[115] Screening of 290 obese individuals in Italy revealed three rare *MC3R* mutations with a prevalence of 1.03%.[117] Another study of 448 severely obese children and adolescents from Belgium identified three rare mutations of *MC3R*, with a prevalence of 0.67%.[120] No rare mutations of *MC3R* were found in the control groups of all the above three studies.

In a cohort of North American adults consisting of 889 obese and 932 nonobese controls, the prevalence of rare *MC3R* mutations is 0.67% in the obese group and 0.32% in the control group.[118] In another subsequent study of Belgians, 1.6% of obese individuals have *MC3R* mutations compared to 1.7% in nonobese subjects.[121] Another large-scale study of 839 obese and 967 control subjects performed in Italy revealed that the frequency of rare *MC3R* mutations is 1.55% in the obese group compared to 0.83% in the nonobese group.[119] In these three studies, there was no significant difference between the frequencies of rare *MC3R* mutations in obese and nonobese subjects.

The similar prevalence of *MC3R* mutations in obese and nonobese individuals might raise doubt about the role of MC3R in obesity pathogenesis. But it needs to be mentioned that functional studies of the mutations were not performed in all of these studies and therefore, the mutation might not have functional significance. For example, in the large-scale study in Italy with 839 obese and 967 control subjects, the prevalence of *MC3R* mutations with functional alterations was actually significantly higher in the obese group.[119] Nonetheless, in contrast to *MC4R* that has been characterized as the most common monogenic form of obesity in humans,[129,130] the role of *MC3R* in obesity pathogenesis remains controversial.

4.3 Clinical Phenotypes of *MC3R* Mutation Carriers

In the past decade, a number of epidemiological studies have focused on the effects of *MC3R* mutations on obesity and obesity-related phenotypes. Among the 29 identified *MC3R* mutations and variants, the two common variants (T6K and V81I) have been most extensively studied. The original studies did not identify an association between the 6K/81I haplotype and obesity or diabetes, and these two variants were characterized as benign polymorphisms,[124,128] which were confirmed in another study.[123] In one study of school pupils aged 11–16 years in South Africa, the T6K and V81I variants were even suggested to be negatively associated with body weight and blood pressure.[131] However, several studies obtained different results. Lee et al. reported that obese subjects harboring the 6K/81I haplotype have higher leptin levels, percentage of body fat, and lower insulin resistance index compared to noncarriers.[115] The 6K/81I alleles carriers were shown to have significantly lower rates of lipid oxidation, higher rates of glucose oxidation, higher insulin levels, and increased respiratory quotient compared to noncarriers.[132,133] Yanovski and coworkers further demonstrated that homozygosity for the two variants is associated with higher BMI, more body fat (both fat mass and percentage fat mass), and increased insulin resistance compared to wild type (WT) or heterozygous subjects.[134] This was supported by the study conducted by Savastano et al. indicating that the carriers of double homozygosity have greater BMI and fat mass with promoted energy intake.[127] Effects of these two variants in weight loss and eating behavior regulation have also been investigated.[126,135] These results provided evidence for the contribution of common *MC3R* variants in human obesity.

Of the 27 naturally occurring mutations of *MC3R*, 16 mutations (S17T, A70T, N128S, M134I, D158Y, V177I, I183N, V211I, L249 V, A260 V, M275T, T280S, A293T, L297 V, L299 V, and I335S) were identified only from obese individuals and 5 mutations (F82S, I87T, L249F, R257S, and X361S) were identified from both obese and nonobese subjects, and 6 mutations (S69C, V124L, V255I, L285 V, E342K, and G353D) were found only in individuals with normal body weight. Therefore, *MC3R* mutations are not associated with obesity in all cases.

For the mutations identified only from obese individuals, the epidemiologic data was limited. Probands of A70T, M134I, and I183N have higher leptin levels and body fat mass but lower hunger score, reminiscent of the $Mc3r^{-/-}$ mice.[115] Probands of N128S, V211I, and L299 V have higher BMI compared to the control group.[120] In order to establish one mutation as a cause of monogenic obesity, one line of evidence needed to be provided should be the cosegregation of this mutation with obesity in the family.[136] In this regard, only two *MC3R* mutations, I183N and I335S, could be regarded as potential pathogenic mutations.

I183N was first detected in a 13-year-old Indian girl (proband) and her father in Singapore with excessive percentage body fat which, after 4 years of follow-up, had risen from 49 to 71% (daughter) and from 30 to 48% (father), respectively.[114,115] The proband's brother was found to be heterozygous of I183N, and he was overweight as a child but was able to lose weight in his teenage years by maintaining active lifestyle and careful dieting.[115] I335S was first identified in an Italian family and the carriers had early-onset obesity.[117] Another study of a three-generation Polish family showed that all carriers of I335S *MC3R* were obese or overweight.[137] These studies suggested that these two *MC3R* mutations, I183N and I335S, might be at least predisposing genetic factors conferring susceptibility to excessive weight gain.

It is worth noting that recent studies of *MC3R* promoter sequences indicated that SNPs rs6127698 and rs11575886 are significantly associated with pulmonary tuberculosis susceptibility.[138–140] These results suggested that the genetic effect of *MC3R* mutations might not be confined to obesity pathogenesis and provided new rationale into *MC3R* investigation.

4.4 Functional Characterizations of MC3R Mutants

4.4.1 *Molecular Classification of MC3R Mutations*

The cosegregation of one mutation with obesity in the patient's family and absence of the mutation in ethnically matched controls would provide solid support for the causative role of this mutation in obesity. Functional characterization of the mutant receptor in vitro is also indispensable for establishing the role of the mutation in obesity pathogenesis.[141,142]

MC3R is a typical member of rhodopsin-like Family A of the G protein-coupled receptors (GPCRs) superfamily, consisting of seven TM helices with an extracellular N-terminus and an intracellular C-terminus. It primarily couples to G_s protein. Activation of MC3R by agonists results in the GDP/GTP exchange in G_s protein, activating adenylyl cyclase to increase the production of cAMP, the second messenger in the system. Signaling assays, either directly measuring the cAMP levels or indirectly determining reporter gene expression, have been employed to characterize the signaling properties of WT and mutant MC3Rs. For signaling-impaired mutants, ligand binding assays would be used to study whether the mutant receptors are defective in ligand binding or behave normally in ligand binding but fail to initiate signaling. For binding-defective mutants, cell surface expression would be measured to determine whether the binding defect of the mutants is due to direct impairment in ligand recognition or is caused by reduced cell surface expression. For the latter case, total receptor expression would be measured by permeabilization to detect whether the receptors could be normally synthesized or stabilized. Based on this life cycle of receptors from biosynthesis to intracellular signaling, we proposed a classification scheme for MC4R mutants,[143] which can also be applied in other GPCRs including the MC3R.[144]

Class I. Null mutants that have decreased receptor levels either due to defective protein synthesis and/or increased protein degradation.

Class II. Intracellularly retained mutants that, despite normal synthesis of the mutant protein, are impaired in cell surface expression due to misfolding of the mutant receptor being detected by cellular quality control system.

Class III. Binding defective mutants that lead to normal cell surface expression but defective ligand binding due to either decreased binding capacity and/or affinity.

Class IV. Signaling defective mutants that have normal cell surface expression and ligand binding but impaired signaling with either decreased efficacy and/or potency.

Class V. Mutants with unknown defects that behave like WT receptors with normal cell surface expression, ligand binding, and cAMP production. Whether and how these variants cause energy imbalance is unclear.

As shown in Table 2, the 29 naturally occurring *MC3R* mutations or variants have been cataloged into different classes according to this classification system. Apparently, the Class II comprises the largest set of *MC3R*

Table 2 Functional Analysis for the MC3R Variants and Mutants.

Domain	Mutation	Mutation Identification References	Function	Classification According to Tao[141]	Functional Analyses References
Extracellular N-terminus	Thr6Lys	[113]	Like WT	Class V	[128,146]
	Ser17Thr	[119]	Like WT	Class V	[149]
			RF, reduced cAMP production	Class IV	[119]
	Ser69Cys	[118]	RF, reduced cell surface expression, binding, and cAMP production	Class II/III	[161]
	Ala70Thr	[115]	RF, reduced cAMP production, normal binding	Class IV	[115]
			RF, reduced cell surface expression	Class II	[161]
Transmembrane domain 1	Val81Ile	[113]	Like WT	Class V	[128,146]
	Phe82Ser	[118]	RF, reduced cell surface expression, binding, and cAMP production	Class II	[149]
	Ile87Thr	[118]	RF, reduced cell surface expression	Class II	[161]
Transmembrane domain 2	Val124Leu	[121]	ND	ND	ND
	Asn128Ser	[120]	Like WT	Class V	[120]
	Met134Ile	[115]	RF, slightly higher EC_{50}	Class IV	[115]
			RF, reduced cell surface expression and binding, normal cAMP production	Class II	[161]
Transmembrane domain 3	Asp158Tyr	[119]	LOF, normal cell surface expression, no binding	Class III	[119,149]
	Val177Ile	[119]	Like WT	Class V	[119,149]
	Ile183Asn	[114]	LOF, normal cell surface expression and binding, no cAMP production	Class IV	[115,146]
			LOF, no cell surface expression	Class II	[147]

(*Continued*)

Table 2 Functional Analysis for the MC3R Variants and Mutants.—cont'd.

Domain	Mutation	Mutation Identification References	Function	Classification According to Tao[141]	Functional Analyses References
Transmembrane domain 4	Val211Ile	[120]	RF, reduced cAMP production	Class IV	[120]
Intracellular loop 3	Leu249Val	[116]	RF, reduced cell surface expression	Class II	[161]
	Leu249Phe	[119]	Like WT	Class V	[119]
			RF, reduced cAMP production	Class IV	[149]
	Val255Ile	[121]	ND	Class ND	ND
	Arg257Ser	[118]	Like WT	Class V	[119]
			RF, reduced cAMP production	Class IV	[149]
	Ala260Val	[118]	RF, reduced total and cell surface expression	Class I	[161]
Transmembrane domain 6	Met275Thr	[118]	RF, reduced total and cell surface expression and binding, normal cAMP production	Class I	[161]
	Thr280Ser	[118]	RF, reduced cell surface expression, impaired binding, and cAMP production	Class II	[161]
	Leu285Val	[118]	Like WT	Class V	[119]
			RF, reduced cAMP production	Class IV	[149]
	Ala293Thr	[117]	Like WT	Class V	[117,148]
	Leu297Val	[118]	RF, reduced total and cell surface expression	Class I	[161]
	Leu299Val	[120]	RF, reduced cell surface expression, and cAMP production	Class II	[120]

			RE, normal cell surface expression, impaired binding, and cAMP production	Class III	[149]
Transmembrane domain 7	Ile335Ser	[117]	LOF, intracellular retention	Class II	[117,148]
Intracellular C-terminus	Glu342Lys	[121]	ND	ND	ND
	Gly353Asp	[121]	ND	ND	ND
	X361SerAspAla; GlyProTryLys	[117]	Like WT	Class V	[117,148]

All *MC3R* mutations currently identified are sorted by amino acid position and domain, including functional data if available. If functional data of a mutation from different groups are conflicting, all the data and corresponding classification are presented. Abbreviations: *WT*, wide type; *RF*, reduced function; *LOF*, loss of function; *ND*, not determined.

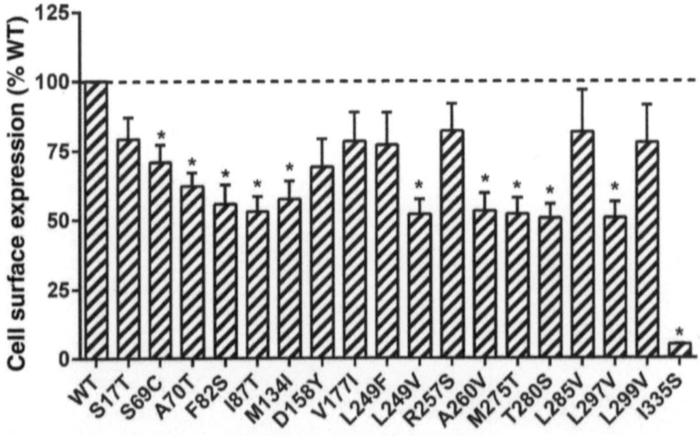

Figure 2 Cell surface expression of the WT and selected mutant MC3Rs. Flow cytometry was used to determine the cell surface expression of the MC3Rs using the HA epitope tag inserted at the N-terminus of receptor cDNA construct. The results are expressed as percentage of cell surface expression level of WT MC3R after correction of the nonspecific staining in cells transiently transfected with the empty vector. * Indicates significantly different from WT MC3R ($P < 0.05$). *Data were originally published in Refs. [149,161,163].*

mutations with 10 mutants (Table 2, Fig. 2). This is in accordance with *MC4R* mutations[141] and moreover, receptor retention or mislocalization has been described as the major defect for human disease caused by GPCR mutations.[144,145] The common variants T6K and V81I belong to Class V since in vitro studies demonstrated that they behave similarly as the WT MC3R.[146] Studies from Feng et al. suggested that the double variant (T6K/V81I) MC3R is partially inactive, with significantly decreased protein expression and ligand binding and reduced signal transduction.[134] However, Lee et al. reported that the double variant only displays reduced receptor activity without any abnormal cell surface expression or ligand binding.[115] Considering the disputes about the phenotypes of these two variants, further investigation is needed.

I183N and I335S, which were suggested to be potential pathogenic mutations, have been shown to completely abolish cAMP response to agonist stimulation by different groups.[117,146–148] But the mechanisms for the loss of function (LOF) of these two mutants are different. I335S results in a complete loss of ligand binding and signaling due to intracellular retention, which was characterized as a Class II mutation (Fig. 2).[117,148] Discordant results were reported for I183N. We reported that I183N mutant receptor completely lacks signaling in response to agonist stimulation, although it could be expressed at cell surface and bind to ligand with normal affinity. However, the functional analysis of I183N MC3R by Penhoat and coworkers showed that the loss of signaling of the mutant receptor is due to intracellular retention with no cell surface expression.[147] A subsequent study from Lee et al. using endogenous agonist α-MSH obtained results similar to ours, demonstrating that the mutant receptor is expressed on cell surface with normal ligand binding affinity with only a slight response to high concentrations of α-MSH.[115]

Besides I183N and I335S, there are three other mutations which lead to severely impaired or even completely abolished cAMP response. D158Y is expressed at cell surface but is almost totally devoid of ligand binding and cAMP signaling.[149] We showed previously that D158 is directly involved in ligand binding.[150] T280S has reduced cell surface expression and dramatic functional defects in both ligand binding and intracellular cAMP generation.[149] L299 V was initially identified to result in impaired functions due to decreased cell surface expression,[120] but our subsequent study suggested that it has similar expression but significantly reduced ligand binding and cAMP signaling compared to the WT MC3R.[149]

Functional studies have highlighted the pathogenic roles of I183N and I335S, and also suggested the potential participation of other mutations, especially D158Y, T280S, and L299 V, in obesity. However, the data for the carriers of the latter three mutations is rather limited and the roles of these mutations in obesity pathogenesis require further study.

4.4.2 MAPK Signaling Properties of MC3R Mutants

It is important to note that several mutations, which were identified from obese individuals and were not present in nonobese controls, belong to Class V since they behave similarly as the WT MC3R in functional studies (Table 2). In addition, some mutations that were only identified from lean controls, such as F82S, have impaired responses to ligand stimulation. Therefore, current indices for obesity identification might not comprehensively characterize the real status of energy homeostasis. And "deep phenotyping," which is not only confined to BMI determination, has been suggested in human genetic studies of obesity.[151]

In addition to the G_s-cAMP pathway, MC3R also activates several other signaling pathways, such as the inositol phospholipid–Ca^{2+} pathway,[152] protein kinase B pathway,[153] and mitogen-activated protein kinases (MAPKs), especially extracellular signal-regulated kinase 1 and 2 (ERK1/2).[154] Therefore, whether MC3R mutants might behave differently from the WT MC3R in other signaling pathways is also of interest. Indeed, we suggested that defect in ERK1/2 signaling in Class V mutants might contribute to obesity pathogenesis in *MC4R* mutation-carrying patients.[155] Therefore we characterized the ERK1/2 signaling properties of 22 MC3R mutants stimulated by endogenous agonist α-MSH and the result indicated the prevalence of biased signaling in MC3R.[149] Four mutants (I87T, N128S, M134I, and L297 V) with normal cAMP production have diminished ERK1/2 activation, whereas three cAMP-defective mutants (F82S, I183N, and I335S) can initiate ERK1/2 phosphorylation upon α-MSH stimulation. Two mutants, D158Y and T280S, cannot activate any of the two signaling pathways.[149] The result provided evidence and support for the potential roles of the mutants, which have been cataloged into Class V due to normal cAMP response, in obesity pathogenesis. It also added a new layer of complexity to the signaling of MC3R mutants and shed light for the future study of pathogenic roles of *MC3R* mutations.

4.4.3 Other Perspectives of MC3R Mutant Functions

In MC4R, most mutations lead to a haploinsufficiency of the receptor and only for a few mutations, dominant negative effects have been reported.[130,141,142,156–160] Similarly, in MC3R, most of the mutations do not exert dominant negative activity on the WT receptor,[115,146,148,161] although L297 V was shown to have a slight effect on WT MC3R signaling.[161]

Constitutively active mutations of GPCRs have been linked to large amount of human diseases, while loss of constitutive activity also causes diseases.[162] The WT MC3R has low constitutive activity and there is only one lab-generated MC3R mutant that is constitutively active.[48,163] Of the naturally occurring *MC3R* mutations, several mutants, including S69C, T280S, D158Y, and L285 V, have moderately decreased basal cAMP activities. Although no MC3R mutant is constitutively active in cAMP activity, I183N has significantly increased basal phosphorylated ERK1/2 level.[149] The potential physiological function of the altered basal activities of MC3R mutants remains unclear.

4.5 Insight Into the Structure and Function of the *MC3R*

A better understanding of the structure–function relationship of the GPCRs could be obtained by analyzing the functional alteration induced by naturally occurring mutations.[141] Intracellular retention of the receptors has been described as the major defect for human disease caused by GPCR mutations.[144,145] The majority of the naturally occurring *MC4R* mutations associated with obesity are intracellularly retained.[130,141,142] Functional studies of *MC3R* mutations revealed that one potentially pathogenic variant, I335S, led to no signaling response due to intracellular retention.[148] I335 resides in the signature motif N/DPxxY (DPLIY in MC3R) at the cytoplasmic end of the TM7. This highly conserved motif has been characterized as a versatile regulator in GPCR expression, trafficking, ligand binding, and signal transduction.[163–165] A substitution of the corresponding residue I301 to Thr in MC4R has been described as a LOF mutation,[166] suggesting the critical role of this residue for maintaining the normal functions of both neural MCRs. We systematically studied I335 in MC3R by performing additional mutagenesis.[148] The results showed that mutations of I335 into charged residues such as Asp and Arg lead to intracellular retention while mutations into other residues, including Ala, Leu, and Asn, do not affect cell surface expression but result in severely impaired ligand binding. One mutation, I335 V, has normal binding but decreased cAMP response. These results suggested that I335 might interact with the residues in Helix 8 by hydrophobic interactions to maintain the receptor conformation necessary for normal trafficking, binding, and signaling properties.[148]

Compared to GPCR mutants leading to intracellular retention, naturally occurring mutations that specifically result in defects in either ligand binding or signaling are relatively uncommon. Examples have been identified in some other GPCRs, such as MC1R, MC4R, and arginine vasopressin receptor 2.[167–169] I183N MC3R is a classical Class IV mutant. I183 resides at the cytoplasmic end of TM3 right after the highly conserved D/ERY motif, which has been suggested to directly interact with G protein. This Ile is substituted by Val in some other GPCRs and the Leu/Val residue is highly conserved accounting for 94% occupancy of all GPCRs.[170,171] Additional mutagenesis experiments showed that I183D and I183R completely lack signaling, I183A and I183L retain partial signaling, and I183 V behaves similarly as WT MC3R.[146] Considering that all the above mutants have normal ligand binding, we suggest that I183 might interact with hydrophobic residues at the extreme C terminus of G_s subunit.[146]

In summary, these studies provide a better insight into the structure–function relationship of MC3R.

5. POTENTIAL THERAPEUTIC STRATEGIES

Obesity is a polygenic trait that is recognized to be the result of interactions between multiple genetic loci and the environment. In contrast to *MC4R*, mutations which represent the most frequent cause of monogenic obesity in humans, the *MC3R* is generally accepted as one of the predisposing genes contributing to the susceptibility to excessive weight gain. As mentioned previously, one of the I183N-harboring individuals was overweight as a child but was able to lose weight in his teenage years by maintaining active lifestyle and careful dieting, and the situation was similar for another A70T carrier.[115] These cases might suggest that regular treatment such as lifestyle intervention would be helpful enough for decreasing the severity of adiposity at a young age. However, in the study of three-generation carriers of I335S, one subject had to undergo intestinal surgery and keep restricted diet to maintain normal body weight.[137] Therefore, the severity of the obesity induced by *MC3R* mutations cannot be underestimated.

Since most of the GPCR mutations associated with human disease are misfolded and intracellularly retained, approaches leading to increased cell surface expression could potentially be of therapeutic value. Some small molecules could cross the cell membrane and act as pharmacoperones to increase the cell surface expression of the receptors. Pharmacoperones have been identified and applied to correct the trafficking and signaling of many GPCRs, such as rhodopsin, arginine vasopressin receptor 2, calcium-sensing receptor, gonadotrophin-releasing hormone receptor, and MC4R (reviewed in Ref. [145]). Recent studies showed that small molecules such as Ipsen 5i (MC4R inverse agonist) and THIQ (MC4R agonist) can rescue the cell surface expression of several intracellularly retained mutant MC4Rs but neither of them has any effect on the trafficking of I335S MC3R, suggesting specificity of pharmacoperones.[172,173] Therefore, it would be interesting to explore MC3R-specific pharmacoperones and test their effects in the functional rescue of Class II MC3R mutants.

6. CONCLUSIONS

Obesity is a multifactorial epidemic with enormous health, economic, and social burdens. Therapeutic options for obesity available now are far from ideal and novel avenues are urgently sought. The MC3R is increasingly recognized as an important regulator of energy homeostasis and *MC3R* mutations have been intensively analyzed for obesity-related traits. Although a number of significant studies have been published, the pathogenic role of the *MC3R* gene in human adiposity pathogenesis remains controversial. Recent studies of MC3R signaling pathways other than cAMP-protein kinase A pathway have shed light for future additional research of MC3R, and more detailed studies would be needed to unravel the role of MC3R in human body weight regulation and obesity pathogenesis.

ACKNOWLEDGMENTS

We received funding for our work on the melanocortin-3 receptor from National Institutes of Health (Grant R15-DK077213), American Diabetes Association (Grant 1-12-BS212), and Animal Health and Diseases Research Program of Auburn University College of Veterinary Medicine (to Y.-X.T.). Z.Y. received graduate fellowship from China Scholarship Council of the People's Republic of China. We thank Fan Yang and Hui Huang for their contribution to some of the original studies described in this review.

REFERENCES

1. Flegal KM, Carroll MD, Ogden CL, Curtin LR. Prevalence and trends in obesity among US adults, 1999–2008. *JAMA*. 2010;303:235–241.
2. Finucane MM, Stevens GA, Cowan MJ, Danaei G, Lin JK, Paciorek CJ, Singh GM, Gutierrez HR, Lu Y, Bahalim AN, Farzadfar F, Riley LM, Ezzati M. Global Burden of Metabolic Risk Factors of Chronic Diseases Collaborating Group.National, regional, and global trends in body-mass index since 1980: systematic analysis of health examination surveys and epidemiological studies with 960 country-years and 9.1 million participants. *Lancet*. 2011;377:557–567.
3. Ogden CL, Carroll MD, Kit BK, Flegal KM. Prevalence of childhood and adult obesity in the United States, 2011–2012. *JAMA*. 2014;311:806–814.
4. Lobstein T, Baur L, Uauy R. Obesity in children and young people: a crisis in public health. *Obes Rev*. 2004;5(suppl 1):4–85.
5. Ogden CL, Carroll MD, Kit BK, Flegal KM. Prevalence of obesity and trends in body mass index among US children and adolescents, 1999–2010. *JAMA*. 2012;307:483–490.
6. Baker JL, Olsen LW, Sorensen TI. Childhood body-mass index and the risk of coronary heart disease in adulthood. *N Engl J Med*. 2007;357:2329–2337.
7. Desvergne B, Michalik L, Wahli W. Be fit or be sick: peroxisome proliferator-activated receptors are down the road. *Mol Endocrinol*. 2004;18:1321–1332.
8. Tao YX, Yuan ZH, Xie J. G protein-coupled receptors as regulators of energy homeostasis. *Prog Mol Biol Transl Sci*. 2013;114:1–43.
9. Jensen MD, Ryan DH, Apovian CM, Ard JD, Comuzzie AG, Donato KA, Hu FB, Hubbard VS, Jakicic JM, Kushner RF, Loria CM, Millen BE, Nonas CA, Pi-Sunyer FX, Stevens J, Stevens VJ, Wadden TA, Wolfe BM, Yanovski SZ, Jordan HS, Kendall KA, Lux LJ, Mentor-Marcel R, Morgan LC, Trisolini MG, Wnek J, Anderson JL, Halperin JL, Albert NM, Bozkurt B, Brindis RG, Curtis LH, DeMets D, Hochman JS, Kovacs RJ, Ohman EM, Pressler SJ, Sellke FW, Shen WK, Smith Jr SC, Tomaselli GF. American College of Cardiology/American Heart Association Task Force on Practice Guidelines and The Obesity Society.2013 AHA/ACC/TOS guideline for the management of overweight and obesity in adults: a report of the American College of Cardiology/American Heart Association Task Force on Practice Guidelines and The Obesity Society. *Circulation*. 2014;129:S102–S138.
10. Trogdon JG, Finkelstein EA, Feagan CW, Cohen JW. State- and payer-specific estimates of annual medical expenditures attributable to obesity. *Obesity (Silver Spring)*. 2012;20:214–220.

11. Finkelstein EA, Fiebelkorn IC, Wang G. State-level estimates of annual medical expenditures attributable to obesity. *Obes Res.* 2004;12:18–24.
12. Puhl R, Peterson JL, Luedicke J. Fighting obesity or obese persons? Public perceptions of obesity-related health messages. *Int J Obes (Lond).* 2013;37:774–782.
13. Bray GA. Lifestyle and pharmacological approaches to weight loss: efficacy and safety. *J Clin Endocrinol Metab.* 2008;93:S81–S88.
14. Wadden TA, Webb VL, Moran CH, Bailer BA. Lifestyle modification for obesity: new developments in diet, physical activity, and behavior therapy. *Circulation.* 2012;125:1157–1170.
15. Woods SC, Schwartz MW, Baskin DG, Seeley RJ. Food intake and the regulation of body weight. *Annu Rev Psychol.* 2000;51:255–277.
16. Guyenet SJ, Schwartz MW. Clinical review: regulation of food intake, energy balance, and body fat mass: implications for the pathogenesis and treatment of obesity. *J Clin Endocrinol Metab.* 2012;97:745–755.
17. Hofbauer KG, Nicholson JR, Boss O. The obesity epidemic: current and future pharmacological treatments. *Annu Rev Pharmacol Toxicol.* 2007;47:565–592.
18. Connolly HM, Crary JL, McGoon MD, Hensrud DD, Edwards BS, Edwards WD, Schaff HV. Valvular heart disease associated with fenfluramine-phentermine. *N Engl J Med.* 1997;337:581–588.
19. Christensen R, Kristensen PK, Bartels EM, Bliddal H, Astrup A. Efficacy and safety of the weight-loss drug rimonabant: a meta-analysis of randomised trials. *Lancet.* 2007;370:1706–1713.
20. Nisoli E, Carruba MO. An assessment of the safety and efficacy of sibutramine, an anti-obesity drug with a novel mechanism of action. *Obes Rev.* 2000;1:127–139.
21. James WP, Caterson ID, Coutinho W, Finer N, Van Gaal LF, Maggioni AP, Torp-Pedersen C, Sharma AM, Shepherd GM, Rode RA, Renz CL. Effect of sibutramine on cardiovascular outcomes in overweight and obese subjects. *N Engl J Med.* 2010;363:905–917.
22. Davidson MH, Hauptman J, DiGirolamo M, Foreyt JP, Halsted CH, Heber D, Heimburger DC, Lucas CP, Robbins DC, Chung J, Heymsfield SB. Weight control and risk factor reduction in obese subjects treated for 2 years with orlistat: a randomized controlled trial. *JAMA.* 1999;281:235–242.
23. Torgerson JS, Hauptman J, Boldrin MN, Sjostrom L. XENical in the prevention of diabetes in obese subjects (XENDOS) study: a randomized study of orlistat as an adjunct to lifestyle changes for the prevention of type 2 diabetes in obese patients. *Diab Care.* 2004;27:155–161.
24. Filippatos TD, Derdemezis CS, Gazi IF, Nakou ES, Mikhailidis DP, Elisaf MS. Orlistat-associated adverse effects and drug interactions: a critical review. *Drug Saf.* 2008;31:53–65.
25. Buchwald H, Avidor Y, Braunwald E, Jensen MD, Pories W, Fahrbach K, Schoelles K. Bariatric surgery: a systematic review and meta-analysis. *JAMA.* 2004;292:1724–1737.
26. Pontiroli AE, Morabito A. Long-term prevention of mortality in morbid obesity through bariatric surgery. A systematic review and meta-analysis of trials performed with gastric banding and gastric bypass. *Ann Surg.* 2011;253:484–487.
27. Ashrafian H, Ahmed K, Rowland SP, Patel VM, Gooderham NJ, Holmes E, Darzi A, Athanasiou T. Metabolic surgery and cancer: protective effects of bariatric procedures. *Cancer.* 2011;117:1788–1799.
28. Stefater MA, Wilson-Perez HE, Chambers AP, Sandoval DA, Seeley RJ. All bariatric surgeries are not created equal: insights from mechanistic comparisons. *Endocr Rev.* 2012;33:595–622.
29. Sjostrom L, Lindroos AK, Peltonen M, Torgerson J, Bouchard C, Carlsson B, Dahlgren S, Larsson B, Narbro K, Sjostrom CD, Sullivan M, Wedel H, Swedish Obese Subjects

Study Scientific Group. Lifestyle, diabetes, and cardiovascular risk factors 10 years after bariatric surgery. *N Engl J Med.* 2004;351:2683–2693.
30. DeMaria EJ. Bariatric surgery for morbid obesity. *N Engl J Med.* 2007;356:2176–2183.
31. Terranova L, Busetto L, Vestri A, Zappa MA. Bariatric surgery: cost-effectiveness and budget impact. *Obes Surg.* 2012;22:646–653.
32. Stunkard AJ, Harris JR, Pedersen NL, McClearn GE. The body-mass index of twins who have been reared apart. *N Engl J Med.* 1990;322:1483–1487.
33. Rankinen T, Zuberi A, Chagnon YC, Weisnagel SJ, Argyropoulos G, Walts B, Perusse L, Bouchard C. The human obesity gene map: the 2005 update. *Obesity (Silver Spring).* 2006;14:529–644.
34. Cone RD, Lu D, Koppula S, Vage DI, Klungland H, Boston B, Chen W, Orth DN, Pouton C, Kesterson RA. The melanocortin receptors: agonists, antagonists, and the hormonal control of pigmentation. *Recent Prog Horm Res.* 1996;51:287–317.
35. Gantz I, Fong TM. The melanocortin system. *Am J Physiol.* 2003;284:E468–E474.
36. Cone RD. Anatomy and regulation of the central melanocortin system. *Nat Neurosci.* 2005;8:571–578.
37. Mountjoy KG, Robbins LS, Mortrud MT, Cone RD. The cloning of a family of genes that encode the melanocortin receptors. *Science.* 1992;257:1248–1251.
38. Valverde P, Healy E, Jackson I, Rees JL, Thody AJ. Variants of the melanocyte-stimulating hormone receptor gene are associated with red hair and fair skin in humans. *Nat Genet.* 1995;11:328–330.
39. Cone RD. Studies on the physiological functions of the melanocortin system. *Endocr Rev.* 2006;27:736–749.
40. Chen W, Kelly MA, Opitz-Araya X, Thomas RE, Low MJ, Cone RD. Exocrine gland dysfunction in MC5-R-deficient mice: evidence for coordinated regulation of exocrine gland function by melanocortin peptides. *Cell.* 1997;91:789–798.
41. Abdel-Malek ZA, Scott MC, Furumura M, Lamoreux ML, Ollmann M, Barsh GS, Hearing VJ. The melanocortin 1 receptor is the principal mediator of the effects of agouti signaling protein on mammalian melanocytes. *J Cell Sci.* 2001;114:1019–1024.
42. Lu D, Willard D, Patel IR, Kadwell S, Overton L, Kost T, Luther M, Chen W, Woychik RP, Wilkison WO, Cone RD. Agouti protein is an antagonist of the melanocyte-stimulating-hormone receptor. *Nature.* 1994;371:799–802.
43. Fan W, Boston BA, Kesterson RA, Hruby VJ, Cone RD. Role of melanocortinergic neurons in feeding and the agouti obesity syndrome. *Nature.* 1997;385:165–168.
44. Ollmann MM, Wilson BD, Yang YK, Kerns JA, Chen Y, Gantz I, Barsh GS. Antagonism of central melanocortin receptors in vitro and in vivo by agouti-related protein. *Science.* 1997;278:135–138.
45. Fong TM, Mao C, MacNeil T, Kalyani R, Smith T, Weinberg D, Tota MR, Van der Ploeg LH. ART (protein product of agouti-related transcript) as an antagonist of MC-3 and MC-4 receptors. *Biochem Biophys Res Commun.* 1997;237:629–631.
46. Nijenhuis WA, Oosterom J, Adan RA. AgRP(83-132) acts as an inverse agonist on the human melanocortin-4 receptor. *Mol Endocrinol.* 2001;15:164–171.
47. Haskell-Luevano C, Monck EK. Agouti-related protein functions as an inverse agonist at a constitutively active brain melanocortin-4 receptor. *Regul Pept.* 2001;99:1–7.
48. Tao YX, Huang H, Wang ZQ, Yang F, Williams JN, Nikiforovich GV. Constitutive activity of neural melanocortin receptors. *Methods Enzymol.* 2010;484:267–279.
49. Elias CF, Saper CB, Maratos-Flier E, Tritos NA, Lee C, Kelly J, Tatro JB, Hoffman GE, Ollmann MM, Barsh GS, Sakurai T, Yanagisawa M, Elmquist JK. Chemically defined projections linking the mediobasal hypothalamus and the lateral hypothalamic area. *J Comp Neurol.* 1998;402:442–459.
50. Hahn TM, Breininger JF, Baskin DG, Schwartz MW. Coexpression of Agrp and NPY in fasting-activated hypothalamic neurons. *Nat Neurosci.* 1998;1:271–272.

51. Cheung CC, Clifton DK, Steiner RA. Proopiomelanocortin neurons are direct targets for leptin in the hypothalamus. *Endocrinology.* 1997;138:4489–4492.
52. Schwartz MW, Seeley RJ, Woods SC, Weigle DS, Campfield LA, Burn P, Baskin DG. Leptin increases hypothalamic pro-opiomelanocortin mRNA expression in the rostral arcuate nucleus. *Diabetes.* 1997;46:2119–2123.
53. Cowley MA, Smart JL, Rubinstein M, Cerdan MG, Diano S, Horvath TL, Cone RD, Low MJ. Leptin activates anorexigenic POMC neurons through a neural network in the arcuate nucleus. *Nature.* 2001;411:480–484.
54. Konner AC, Janoschek R, Plum L, Jordan SD, Rother E, Ma X, Xu C, Enriori P, Hampel B, Barsh GS, Kahn CR, Cowley MA, Ashcroft FM, Bruning JC. Insulin action in AgRP-expressing neurons is required for suppression of hepatic glucose production. *Cell Metab.* 2007;5:438–449.
55. Cowley MA. Hypothalamic melanocortin neurons integrate signals of energy state. *Eur J Pharmacol.* 2003;480:3–11.
56. Xu Y, Jones JE, Kohno D, Williams KW, Lee CE, Choi MJ, Anderson JG, Heisler LK, Zigman JM, Lowell BB, Elmquist JK. 5-HT_{2C}Rs expressed by pro-opiomelanocortin neurons regulate energy homeostasis. *Neuron.* 2008;60:582–589.
57. van den Top M, Lee K, Whyment AD, Blanks AM, Spanswick D. Orexigen-sensitive NPY/AgRP pacemaker neurons in the hypothalamic arcuate nucleus. *Nat Neurosci.* 2004;7:493–494.
58. Parton LE, Ye CP, Coppari R, Enriori PJ, Choi B, Zhang CY, Xu C, Vianna CR, Balthasar N, Lee CE, Elmquist JK, Cowley MA, Lowell BB. Glucose sensing by POMC neurons regulates glucose homeostasis and is impaired in obesity. *Nature.* 2007;449:228–232.
59. Zhang Y, Proenca R, Maffei M, Barone M, Leopold L, Friedman JM. Positional cloning of the mouse obese gene and its human homologue. *Nature.* 1994;372:425–432.
60. Frederich RC, Hamann A, Anderson S, Lollmann B, Lowell BB, Flier JS. Leptin levels reflect body lipid content in mice: evidence for diet-induced resistance to leptin action. *Nat Med.* 1995;1:1311–1314.
61. Tartaglia LA, Dembski M, Weng X, Deng N, Culpepper J, Devos R, Richards GJ, Campfield LA, Clark FT, Deeds J, Muir C, Sanker S, Moriarty A, Moore KJ, Smutko JS, Mays GG, Wool EA, Monroe CA, Tepper RI. Identification and expression cloning of a leptin receptor, OB-R. *Cell.* 1995;83:1263–1271.
62. Mercer JG, Hoggard N, Williams LM, Lawrence CB, Hannah LT, Morgan PJ, Trayhurn P. Coexpression of leptin receptor and preproneuropeptide Y mRNA in arcuate nucleus of mouse hypothalamus. *J Neuroendocrinol.* 1996;8:733–735.
63. Ghilardi N, Skoda RC. The leptin receptor activates janus kinase 2 and signals for proliferation in a factor-dependent cell line. *Mol Endocrinol.* 1997;11:393–399.
64. Niswender KD, Morton GJ, Stearns WH, Rhodes CJ, Myers Jr MG, Schwartz MW. Intracellular signalling. Key enzyme in leptin-induced anorexia. *Nature.* 2001;413:794–795.
65. Chen H, Charlat O, Tartaglia LA, Woolf EA, Weng X, Ellis SJ, Lakey ND, Culpepper J, Moore KJ, Breitbart RE, Duyk GM, Tepper RI, Morgenstern JP. Evidence that the diabetes gene encodes the leptin receptor: identification of a mutation in the leptin receptor gene in db/db mice. *Cell.* 1996;84:491–495.
66. Yaswen L, Diehl N, Brennan MB, Hochgeschwender U. Obesity in the mouse model of pro-opiomelanocortin deficiency responds to peripheral melanocortin. *Nat Med.* 1999;5:1066–1070.
67. Graham M, Shutter JR, Sarmiento U, Sarosi I, Stark KL. Overexpression of Agrt leads to obesity in transgenic mice. *Nat Genet.* 1997;17:273–274.
68. Montague CT, Farooqi IS, Whitehead JP, Soos MA, Rau H, Wareham NJ, Sewter CP, Digby JE, Mohammed SN, Hurst JA, Cheetham CH, Earley AR, Barnett AH, Prins JB,

O'Rahilly S. Congenital leptin deficiency is associated with severe early-onset obesity in humans. *Nature.* 1997;387:903–908.
69. Clement K, Vaisse C, Lahlou N, Cabrol S, Pelloux V, Cassuto D, Gourmelen M, Dina C, Chambaz J, Lacorte JM, Basdevant A, Bougneres P, Lebouc Y, Froguel P, Guy-Grand B. A mutation in the human leptin receptor gene causes obesity and pituitary dysfunction. *Nature.* 1998;392:398–401.
70. Krude H, Biebermann H, Luck W, Horn R, Brabant G, Gruters A. Severe early-onset obesity, adrenal insufficiency and red hair pigmentation caused by POMC mutations in humans. *Nat Genet.* 1998;19:155–157.
71. Vaisse C, Clement K, Guy-Grand B, Froguel P. A frameshift mutation in human MC4R is associated with a dominant form of obesity. *Nat Genet.* 1998;20:113–114.
72. Yeo GS, Farooqi IS, Aminian S, Halsall DJ, Stanhope RG, O'Rahilly S. A frameshift mutation in MC4R associated with dominantly inherited human obesity. *Nat Genet.* 1998;20:111–112.
73. Hinney A, Schmidt A, Nottebom K, Heibult O, Becker I, Ziegler A, Gerber G, Sina M, Gorg T, Mayer H, Siegfried W, Fichter M, Remschmidt H, Hebebrand J. Several mutations in the melanocortin-4 receptor gene including a nonsense and a frameshift mutation associated with dominantly inherited obesity in humans. *J Clin Endocrinol Metab.* 1999;84:1483–1486.
74. Jegou S, Boutelet I, Vaudry H. Melanocortin-3 receptor mRNA expression in pro-opiomelanocortin neurones of the rat arcuate nucleus. *J Neuroendocrinol.* 2000;12: 501–505.
75. Mounien L, Bizet P, Boutelet I, Vaudry H, Jegou S. Expression of melanocortin MC3 and MC4 receptor mRNAs by neuropeptide Y neurons in the rat arcuate nucleus. *Neuroendocrinology.* 2005;82:164–170.
76. Gantz I, Konda Y, Tashiro T, Shimoto Y, Miwa H, Munzert G, Watson SJ, DelValle J, Yamada T. Molecular cloning of a novel melanocortin receptor. *J Biol Chem.* 1993;268:8246–8250.
77. Magenis RE, Smith L, Nadeau JH, Johnson KR, Mountjoy KG, Cone RD. Mapping of the ACTH, MSH, and neural (MC3 and MC4) melanocortin receptors in the mouse and human. *Mamm Genome.* 1994;5:503–508.
78. Tarnow P, Rediger A, Schulz A, Gruters A, Biebermann H. Identification of the translation start site of the human melanocortin 3 receptor. *Obes Facts.* 2012;5:45–51.
79. Park J, Sharma N, Cutting GR. Melanocortin 3 receptor has a 5' exon that directs translation of apically localized protein from the second in-frame ATG. *Mol Endocrinol.* 2014;28:1547–1557.
80. Chhajlani V. Distribution of cDNA for melanocortin receptor subtypes in human tissues. *Biochem Mol Biol Int.* 1996;38:73–80.
81. Getting SJ, Christian HC, Lam CW, Gavins FN, Flower RJ, Schioth HB, Perretti M. Redundancy of a functional melanocortin 1 receptor in the anti-inflammatory actions of melanocortin peptides: studies in the recessive yellow (e/e) mouse suggest an important role for melanocortin 3 receptor. *J Immunol.* 2003;170:3323–3330.
82. Ni XP, Butler AA, Cone RD, Humphreys MH. Central receptors mediating the cardiovascular actions of melanocyte stimulating hormones. *J Hypertens.* 2006;24:2239–2246.
83. Kim MS, Small CJ, Russell SH, Morgan DG, Abbott CR, alAhmed SH, Hay DL, Ghatei MA, Smith DM, Bloom SR. Effects of melanocortin receptor ligands on thyrotropin-releasing hormone release: evidence for the differential roles of melanocortin 3 and 4 receptors. *J Neuroendocrinol.* 2002;14:276–282.
84. Versteeg DH, Van Bergen P, Adan RA, De Wildt DJ. Melanocortins and cardiovascular regulation. *Eur J Pharmacol.* 1998;360:1–14.
85. Mioni C, Giuliani D, Cainazzo MM, Leone S, Bazzani C, Grieco P, Novellino E, Tomasi A, Bertolini A, Guarini S. Further evidence that melanocortins prevent

myocardial reperfusion injury by activating melanocortin MC3 receptors. *Eur J Pharmacol.* 2003;477:227–234.
86. Chandramohan G, Durham N, Sinha S, Norris K, Vaziri ND. Role of © melanocyte-stimulating hormone-renal melanocortin 3 receptor system in blood pressure regulation in salt-resistant and salt-sensitive rats. *Metabolism.* 2009;58:1424–1429.
87. Catania A, Gatti S, Colombo G, Lipton JM. Targeting melanocortin receptors as a novel strategy to control inflammation. *Pharmacol Rev.* 2004;56:1–29.
88. Getting SJ, Lam CW, Leoni G, Gavins FN, Grieco P, Perretti M. [D-Trp^{8}]-©-melanocyte-stimulating hormone exhibits anti-inflammatory efficacy in mice bearing a nonfunctional MC1R (recessive yellow e/e mouse). *Mol Pharmacol.* 2006;70: 1850–1855.
89. Getting SJ, Riffo-Vasquez Y, Pitchford S, Kaneva M, Grieco P, Page CP, Perretti M, Spina D. A role for MC3R in modulating lung inflammation. *Pulm Pharmacol Ther.* 2008;21:866–873.
90. Chen AS, Marsh DJ, Trumbauer ME, Frazier EG, Guan XM, Yu H, Rosenblum CI, Vongs A, Feng Y, Cao L, Metzger JM, Strack AM, Camacho RE, Mellin TN, Nunes CN, Min W, Fisher J, Gopal-Truter S, MacIntyre DE, Chen HY, Van der Ploeg LH. Inactivation of the mouse melanocortin-3 receptor results in increased fat mass and reduced lean body mass. *Nat Genet.* 2000;26:97–102.
91. Rowland NE, Schaub JW, Robertson KL, Andreasen A, Haskell-Luevano C. Effect of MTII on food intake and brain c-Fos in melanocortin-3, melanocortin-4, and double MC3 and MC4 receptor knockout mice. *Peptides.* 2010;31:2314–2317.
92. Butler AA, Kesterson RA, Khong K, Cullen MJ, Pelleymounter MA, Dekoning J, Baetscher M, Cone RD. A unique metabolic syndrome causes obesity in the melanocortin-3 receptor-deficient mouse. *Endocrinology.* 2000;141:3518–3521.
93. Zhang Y, Kilroy GE, Henagan TM, Prpic-Uhing V, Richards WG, Bannon AW, Mynatt RL, Gettys TW. Targeted deletion of melanocortin receptor subtypes 3 and 4, but not CART, alters nutrient partitioning and compromises behavioral and metabolic responses to leptin. *FASEB J.* 2005;19:1482–1491.
94. Trevaskis JL, Gawronska-Kozak B, Sutton GM, McNeil M, Stephens JM, Smith SR, Butler AA. Role of adiponectin and inflammation in insulin resistance of Mc3r and Mc4r knockout mice. *Obesity (Silver Spring).* 2007;15:2664–2672.
95. Sutton GM, Perez-Tilve D, Nogueiras R, Fang J, Kim JK, Cone RD, Gimble JM, Tschop MH, Butler AA. The melanocortin-3 receptor is required for entrainment to meal intake. *J Neurosci.* 2008;28:12946–12955.
96. Sutton GM, Trevaskis JL, Hulver MW, McMillan RP, Markward NJ, Babin MJ, Meyer EA, Butler AA. Diet-genotype interactions in the development of the obese, insulin-resistant phenotype of C57BL/6J mice lacking melanocortin-3 or -4 receptors. *Endocrinology.* 2006;147:2183–2196.
97. Renquist BJ, Murphy JG, Larson EA, Olsen D, Klein RF, Ellacott KL, Cone RD. Melanocortin-3 receptor regulates the normal fasting response. *Proc Natl Acad Sci USA.* 2012;109:E1489–E1498.
98. Begriche K, Girardet C, McDonald P, Butler AA. Melanocortin-3 receptors and metabolic homeostasis. *Prog Mol Biol Transl Sci.* 2012;114:109–146.
99. Begriche K, Marston OJ, Rossi J, Burke LK, McDonald P, Heisler LK, Butler AA. Melanocortin-3 receptors are involved in adaptation to restricted feeding. *Genes Brain Behav.* 2012;11:291–302.
100. Girardet C, Butler AA. Neural melanocortin receptors in obesity and related metabolic disorders. *Biochim Biophys Acta.* 2014;1842:482–494.
101. Albarado DC, McClaine J, Stephens JM, Mynatt RL, Ye J, Bannon AW, Richards WG, Butler AA. Impaired coordination of nutrient intake and substrate oxidation in melanocortin-4 receptor knockout mice. *Endocrinology.* 2004;145:243–252.

102. Ellacott KL, Murphy JG, Marks DL, Cone RD. Obesity-induced inflammation in white adipose tissue is attenuated by loss of melanocortin-3 receptor signaling. *Endocrinology.* 2007;148:6186–6194.
103. Sutton GM, Begriche K, Kumar KG, Gimble JM, Perez-Tilve D, Nogueiras R, McMillan RP, Hulver MW, Tschop MH, Butler AA. Central nervous system melanocortin-3 receptors are required for synchronizing metabolism during entrainment to restricted feeding during the light cycle. *FASEB J.* 2010;24:862–872.
104. Marks DL, Hruby V, Brookhart G, Cone RD. The regulation of food intake by selective stimulation of the type 3 melanocortin receptor (MC3R). *Peptides.* 2006;27:259–264.
105. Lee M, Kim A, Conwell IM, Hruby V, Mayorov A, Cai M, Wardlaw SL. Effects of selective modulation of the central melanocortin-3-receptor on food intake and hypothalamic POMC expression. *Peptides.* 2008;29:440–447.
106. Irani BG, Xiang Z, Yarandi HN, Holder JR, Moore MC, Bauzo RM, Proneth B, Shaw AM, Millard WJ, Chambers JB, Benoit SC, Clegg DJ, Haskell-Luevano C. Implication of the melanocortin-3 receptor in the regulation of food intake. *Eur J Pharmacol.* 2011;660:80–87.
107. Lembertas AV, Perusse L, Chagnon YC, Fisler JS, Warden CH, Purcell-Huynh DA, Dionne FT, Gagnon J, Nadeau A, Lusis AJ, Bouchard C. Identification of an obesity quantitative trait locus on mouse chromosome 2 and evidence of linkage to body fat and insulin on the human homologous region 20q. *J Clin Invest.* 1997;100: 1240–1247.
108. Zouali H, Hani EH, Philippi A, Vionnet N, Beckmann JS, Demenais F, Froguel P. A susceptibility locus for early-onset non-insulin dependent (type 2) diabetes mellitus maps to chromosome 20q, proximal to the phosphoenolpyruvate carboxykinase gene. *Hum Mol Genet.* 1997;6:1401–1408.
109. Ji L, Malecki M, Warram JH, Yang Y, Rich SS, Krolewski AS. New susceptibility locus for NIDDM is localized to human chromosome 20q. *Diabetes.* 1997;46:876–881.
110. Bowden DW, Sale M, Howard TD, Qadri A, Spray BJ, Rothschild CB, Akots G, Rich SS, Freedman BI. Linkage of genetic markers on human chromosomes 20 and 12 to NIDDM in Caucasian sib pairs with a history of diabetic nephropathy. *Diabetes.* 1997;46:882–886.
111. Ghosh S, Watanabe RM, Hauser ER, Valle T, Magnuson VL, Erdos MR, Langefeld CD, Balow Jr J, Ally DS, Kohtamaki K, Chines P, Birznieks G, Kaleta HS, Musick A, Te C, Tannenbaum J, Eldridge W, Shapiro S, Martin C, Witt A, So A, Chang J, Shurtleff B, Porter R, Kudelko K, Unni A, Segal L, Sharaf R, Blaschak-Harvan J, Eriksson J, Tenkula T, Vidgren G, Ehnholm C, Tuomilehto-Wolf E, Hagopian W, Buchanan TA, Tuomilehto J, Bergman RN, Collins FS, Boehnke M. Type 2 diabetes: evidence for linkage on chromosome 20 in 716 Finnish affected sib pairs. *Proc Natl Acad Sci USA.* 1999;96:2198–2203.
112. Li WD, Joo EJ, Furlong EB, Galvin M, Abel K, Bell CJ, Price RA. Melanocortin 3 receptor (MC3R) gene variants in extremely obese women. *Int J Obes Relat Metab Disord.* 2000;24:206–210.
113. Hani EH, Dupont S, Durand E, Dina C, Gallina S, Gantz I, Froguel P. Naturally occurring mutations in the melanocortin receptor 3 gene are not associated with type 2 diabetes mellitus in French Caucasians. *J Clin Endocrinol Metab.* 2001;86:2895–2898.
114. Lee YS, Poh LK, Loke KY. A novel melanocortin 3 receptor gene (MC3R) mutation associated with severe obesity. *J Clin Endocrinol Metab.* 2002;87:1423–1426.
115. Lee YS, Poh LK, Kek BL, Loke KY. The role of melanocortin 3 receptor gene in childhood obesity. *Diabetes.* 2007;56:2622–2630.
116. Valli-Jaakola K. *Molecular Genetic Studies of Melanocortin Receptors in Morbid Obesity.* Helsinki, Finland: University of Helsinki; 2007.

117. Mencarelli M, Walker GE, Maestrini S, Alberti L, Verti B, Brunani A, Petroni ML, Tagliaferri M, Liuzzi A, Di Blasio AM. Sporadic mutations in melanocortin receptor 3 in morbid obese individuals. *Eur J Hum Genet.* 2008;16:581–586.
118. Calton MA, Ersoy BA, Zhang S, Kane JP, Malloy MJ, Pullinger CR, Bromberg Y, Pennacchio LA, Dent R, McPherson R, Ahituv N, Vaisse C. Association of functionally significant Melanocortin-4 but not Melanocortin-3 receptor mutations with severe adult obesity in a large North American case-control study. *Hum Mol Genet.* 2009;18:1140–1147.
119. Mencarelli M, Dubern B, Alili R, Maestrini S, Benajiba L, Tagliaferri M, Galan P, Rinaldi M, Simon C, Tounian P, Hercberg S, Liuzzi A, Di Blasio AM, Clement K. Rare melanocortin-3 receptor mutations with in vitro functional consequences are associated with human obesity. *Hum Mol Genet.* 2011;20:392–399.
120. Zegers D, Beckers S, de Freitas F, Peeters AV, Mertens IL, Verhulst SL, Rooman RP, Timmermans JP, Desager KN, Massa G, Van Gaal LF, Van Hul W. Identification of three novel genetic variants in the melanocortin-3 receptor of obese children. *Obesity (Silver Spring).* 2011;19:152–159.
121. Zegers D, Beckers S, Hendrickx R, Van Camp JK, Van Hoorenbeeck K, Desager KN, Massa G, Van Gaal LF, Van Hul W. Prevalence of rare MC3R variants in obese cases and lean controls. *Endocrine.* 2013;44:386–390.
122. Zegers D, Beckers S, Mertens IL, Van Gaal LF, Van Hul W. Common melanocortin-3 receptor variants are not associated with obesity, although rs3746619 does influence weight in obese individuals. *Endocrine.* 2010;38:289–293.
123. Santos JL, De la Cruz R, Holst C, Grau K, Naranjo C, Maiz A, Astrup A, Saris WH, MacDonald I, Oppert JM, Hansen T, Pedersen O, Sorensen TI, Martinez JA. NUGENOB Consortium.Allelic variants of melanocortin 3 receptor gene (MC3R) and weight loss in obesity: a randomised trial of hypo-energetic high- versus low-fat diets. *PLoS One.* 2011;6:e19934.
124. Schalin-Jantti C, Valli-Jaakola K, Oksanen L, Martelin E, Laitinen K, Krusius T, Mustajoki P, Heikinheimo M, Kontula K. Melanocortin-3-receptor gene variants in morbid obesity. *Int J Obes Relat Metab Disord.* 2003;27:70–74.
125. Boucher N, Lanouette CM, Larose M, Perusse L, Bouchard C, Chagnon YC. A +2138InsCAGACC polymorphism of the melanocortin receptor 3 gene is associated in human with fat level and partitioning in interaction with body corpulence. *Mol Med.* 2002;8:158–165.
126. Obregon AM, Amador P, Valladares M, Weisstaub G, Burrows R, Santos JL. Melanocortin-3 receptor gene variants: association with childhood obesity and eating behavior in Chilean families. *Nutrition.* 2010;26:760–765.
127. Savastano DM, Tanofsky-Kraff M, Han JC, Ning C, Sorg RA, Roza CA, Wolkoff LE, Anandalingam K, Jefferson-George KS, Figueroa RE, Sanford EL, Brady S, Kozlosky M, Schoeller DA, Yanovski JA. Energy intake and energy expenditure among children with polymorphisms of the melanocortin-3 receptor. *Am J Clin Nutr.* 2009;90:912–920.
128. Wong J, Love DR, Kyle C, Daniels A, White M, Stewart AW, Schnell AH, Elston RC, Holdaway IM, Mountjoy KG. Melanocortin-3 receptor gene variants in a Maori kindred with obesity and early onset type 2 diabetes. *Diab Res Clin Pract.* 2002;58:61–71.
129. Farooqi IS, Keogh JM, Yeo GS, Lank EJ, Cheetham T, O'Rahilly S. Clinical spectrum of obesity and mutations in the melanocortin 4 receptor gene. *N Engl J Med.* 2003;348:1085–1095.
130. Tao YX. The melanocortin-4 receptor: physiology, pharmacology, and pathophysiology. *Endocr Rev.* 2010;31:506–543.
131. Yako YY, Fanampe BL, Hassan SM, Erasmus RT, van der Merwe L, Matsha TE. Negative association of MC3R variants with weight and blood pressure in Cape Town pupils aged 11–16 years. *S Afr Med J.* 2011;101:417–420.

132. Rutanen J, Pihlajamaki J, Vanttinen M, Salmenniemi U, Ruotsalainen E, Kuulasmaa T, Kainulainen S, Laakso M. Single nucleotide polymorphisms of the melanocortin-3 receptor gene are associated with substrate oxidation and first-phase insulin secretion in offspring of type 2 diabetic subjects. *J Clin Endocrinol Metab.* 2007;92:1112–1117.
133. Obregon AM, Diaz E, Santos JL. Effect of the melanocortin-3 receptor Thr6Lys and Val81Ile genetic variants on body composition and substrate oxidation in Chilean obese children. *J Physiol Biochem.* 2012;68:71–76.
134. Feng N, Young SF, Aguilera G, Puricelli E, Adler-Wailes DC, Sebring NG, Yanovski JA. Co-occurrence of two partially inactivating polymorphisms of MC3R is associated with pediatric-onset obesity. *Diabetes.* 2005;54:2663–2667.
135. Santoro N, Perrone L, Cirillo G, Raimondo P, Amato A, Brienza C, Del Giudice EM. Effect of the melanocortin-3 receptor C17A and G241A variants on weight loss in childhood obesity. *Am J Clin Nutr.* 2007;85:950–953.
136. Tao YX. Mutations in the melanocortin-3 receptor (MC3R) gene: impact on human obesity or adiposity. *Curr Opin Investig Drugs.* 2010;11:1092–1096.
137. Cieslak J, Majewska KA, Tomaszewska A, Skowronska B, Fichna P, Switonski M. Common polymorphism (81Val>Ile) and rare mutations (257Arg>Ser and 335Ile>Ser) of the MC3R gene in obese Polish children and adolescents. *Mol Biol Rep.* 2013;40:6893–6898.
138. Adams LA, Moller M, Nebel A, Schreiber S, van der Merwe L, van Helden PD, Hoal EG. Polymorphisms in MC3R promoter and CTSZ 3'UTR are associated with tuberculosis susceptibility. *Eur J Hum Genet.* 2011;19:676–681.
139. Hashemi M, Eskandari-Nasab E, Moazeni-Roodi A, Naderi M, Sharifi-Mood B, Taheri M. Association of CTSZ rs34069356 and MC3R rs6127698 gene polymorphisms with pulmonary tuberculosis. *Int J Tuberc Lung Dis.* 2013;17:1224–1228.
140. Park BL, Kim LH, Namgoong S, Kim JO, Kim JY, Chang HS, Park JS, Jang AS, Park SW, Kim do J, Kim KU, Kim YG, Uh ST, Seo KH, Kim YH, Park CS, Shin HD. Association analysis of melanocortin 3 receptor polymorphisms with the risk of pulmonary tuberculosis. *Lung.* 2014;192:857–862.
141. Tao YX. Molecular mechanisms of the neural melanocortin receptor dysfunction in severe early onset obesity. *Mol Cell Endocrinol.* 2005;239:1–14.
142. Tao YX. Mutations in melanocortin-4 receptor and human obesity. *Prog Mol Biol Transl Sci.* 2009;88:173–204.
143. Tao YX, Segaloff DL. Functional characterization of melanocortin-4 receptor mutations associated with childhood obesity. *Endocrinology.* 2003;144:4544–4551.
144. Tao YX. Inactivating mutations of G protein-coupled receptors and diseases: structure-function insights and therapeutic implications. *Pharmacol Ther.* 2006;111:949–973.
145. Tao YX, Conn PM. Chaperoning G protein-coupled receptors: from cell biology to therapeutics. *Endocr Rev.* 2014;35:602–647.
146. Tao YX, Segaloff DL. Functional characterization of melanocortin-3 receptor variants identify a loss-of-function mutation involving an amino acid critical for G protein-coupled receptor activation. *J Clin Endocrinol Metab.* 2004;89:3936–3942.
147. Rached M, Buronfosse A, Begeot M, Penhoat A. Inactivation and intracellular retention of the human I183N mutated melanocortin 3 receptor associated with obesity. *Biochim Biophys Acta.* 2004;1689:229–234.
148. Tao YX. Functional characterization of novel melanocortin-3 receptor mutations identified from obese subjects. *Biochim Biophys Acta.* 2007;1772:1167–1174.
149. Yang F, Huang H, Tao YX. Biased signaling in naturally occurring mutations in human melanocortin-3 receptor gene. *Int J Biol Sci.* 2015;11:423–433.
150. Wang SX, Fan ZC, Tao YX. Functions of acidic transmembrane residues in human melanocortin-3 receptor binding and activation. *Biochem Pharmacol.* 2008;76: 520–530.

151. Muller MJ, Bosy-Westphal A, Krawczak M. Genetic studies of common types of obesity: a critique of the current use of phenotypes. *Obes Rev.* 2010;11:612–618.
152. Konda Y, Gantz I, DelValle J, Shimoto Y, Miwa H, Yamada T. Interaction of dual intracellular signaling pathways activated by the melanocortin-3 receptor. *J Biol Chem.* 1994;269:13162–13166.
153. Nyan DC, Anbazhagan R, Hughes-Darden CA, Wachira SJ. Endosomal colocalization of melanocortin-3 receptor and beta-arrestins in CAD cells with altered modification of AKT/PKB. *Neuropeptides.* 2008;42:355–366.
154. Chai B, Li JY, Zhang W, Ammori JB, Mulholland MW. Melanocortin-3 receptor activates MAP kinase via PI3 kinase. *Regul Pept.* 2007;139:115–121.
155. He S, Tao YX. Defect in MAPK signaling as a cause for monogenic obesity caused by inactivating mutations in the melanocortin-4 receptor gene. *Int J Biol Sci.* 2014;10: 1128–1137.
156. Biebermann H, Krude H, Elsner A, Chubanov V, Gudermann T, Gruters A. Autosomal-dominant mode of inheritance of a melanocortin-4 receptor mutation in a patient with severe early-onset obesity is due to a dominant-negative effect caused by receptor dimerization. *Diabetes.* 2003;52:2984–2988.
157. Tao YX, Segaloff DL. Functional analyses of melanocortin-4 receptor mutations identified from patients with binge eating disorder and nonobese or obese subjects. *J Clin Endocrinol Metab.* 2005;90:5632–5638.
158. Fan ZC, Tao YX. Functional characterization and pharmacological rescue of melanocortin-4 receptor mutations identified from obese patients. *J Cell Mol Med.* 2009;13: 3268–3282.
159. Wang ZQ, Tao YX. Functional studies on twenty novel naturally occurring melanocortin-4 receptor mutations. *Biochim Biophys Acta.* 2011;1812:1190–1199.
160. Hinney A, Volckmar AL, Knoll N. Melanocortin-4 receptor in energy homeostasis and obesity pathogenesis. *Prog Mol Biol Transl Sci.* 2013;114:147–191.
161. Yang F, Tao YX. Functional characterization of nine novel naturally occurring human melanocortin-3 receptor mutations. *Biochim Biophys Acta.* 2012;1822: 1752–1761.
162. Tao YX. Constitutive activation of G protein-coupled receptors and diseases: insights into mechanism of activation and therapeutics. *Pharmacol Ther.* 2008;120:129–148.
163. Yang Z, Huang ZL, Tao YX. Functions of DPLIY motif and helix 8 of human melanocortin-3 receptor. *J Mol Endocrinol.* 2015;55:107–117.
164. Barak LS, Menard L, Ferguson SS, Colapietro AM, Caron MG. The conserved seven-transmembrane sequence $NP(X)_{2,3}Y$ of the G-protein-coupled receptor superfamily regulates multiple properties of the b_2-adrenergic receptor. *Biochemistry.* 1995;34: 15407–15414.
165. Hunyady L, Bor M, Baukal AJ, Balla T, Catt KJ. A conserved NPLFY sequence contributes to agonist binding and signal transduction but is not an internalization signal for the type 1 angiotensin II receptor. *J Biol Chem.* 1995;270:16602–16609.
166. Vaisse C, Clement K, Durand E, Hercberg S, Guy-Grand B, Froguel P. Melanocortin-4 receptor mutations are a frequent and heterogeneous cause of morbid obesity. *J Clin Invest.* 2000;106:253–262.
167. Schioth HB, Phillips SR, Rudzish R, Birch-Machin MA, Wikberg JE, Rees JL. Loss of function mutations of the human melanocortin 1 receptor are common and are associated with red hair. *Biochem Biophys Res Commun.* 1999;260:488–491.
168. Rettenbacher E, Tarnow P, Brumm H, Prayer D, Wermter AK, Hebebrand J, Biebermann H, Hinney A, Widhalm K. A novel non-synonymous mutation in the melanocortin-4 receptor gene (MC4R) in a 2-year-old Austrian girl with extreme obesity. *Exp Clin Endocrinol Diab.* 2007;115:7–12.

169. Sadeghi H, Robertson GL, Bichet DG, Innamorati G, Birnbaumer M. Biochemical basis of partial nephrogenic diabetes insipidus phenotypes. *Mol Endocrinol.* 1997;11:1806–1813.
170. Baldwin JM, Schertler GFX, Unger VM. An alpha-carbon template for the transmembrane helices in the rhodopsin family of G-protein-coupled receptors. *J Mol Biol.* 1997;272:144–164.
171. Mirzadegan T, Benko G, Filipek S, Palczewski K. Sequence analyses of G-protein-coupled receptors: similarities to rhodopsin. *Biochemistry.* 2003;42:2759–2767.
172. Tao YX, Huang H. Ipsen 5i is a novel potent pharmacoperone for intracellularly retained melanocortin-4 receptor mutants. *Front Endocrinol (Lausanne).* 2014;5:131.
173. Huang H, Tao YX. A small molecule agonist THIQ as a novel pharmacoperone for intracellularly retained melanocortin-4 receptor mutants. *Int J Biol Sci.* 2014;10: 817–824.

CHAPTER FIVE

Ghrelin Receptor Mutations and Human Obesity

W. Wang, Y.-X. Tao*

Department of Anatomy, Physiology, and Pharmacology, College of Veterinary Medicine, Auburn University, Auburn, Alabama, USA

* Corresponding author. E-mail address: taoyaxi@auburn.edu

Contents

Abstract

Growth hormone secretagogue receptor (GHSR) was originally identified as an orphan receptor in porcine and rat anterior pituitary membranes. In 1999, GHSR was deorphanized and shown to be a receptor for ghrelin, a peptide hormone secreted from the stomach. Therefore, GHSR is also called ghrelin receptor. In addition to regulating growth hormone secretion, ghrelin receptor regulates various physiological processes, including food intake and energy expenditure, glucose metabolism, cardiovascular functions, gastric acid secretion and motility, and immune function. Several human genetic studies conducted in populations originated from Europe, Africa, South America, and East Asia identified rare mutations and single nucleotide polymorphisms that might be associated with human obesity and short stature. Functional analyses of mutant GHSRs reveal multiple defects, including cell surface expression, ligand binding, and basal and stimulated signaling. With growing understanding in the functionality of naturally occurring *GHSR* mutations, potential therapeutic strategies including pharmacological chaperones and novel ligands could be used to correct the GHSR mutants.

Progress in Molecular Biology and Translational Science, Volume 140
ISSN 1877-1173
http://dx.doi.org/10.1016/bs.pmbts.2016.02.001

ABBREVIATIONS

CDGP	constitutional delay of growth and puberty
GH	growth hormone
GHD	growth hormone deficiency
GHRL	*ghrelin gene*
GHRP-6	growth hormone-releasing peptide-6
GHSR	growth hormone secretagogue receptor
GOAT	ghrelin *O*-acyltransferase
GPCR	G protein-coupled receptor
IP	inositol phosphate
MC4R	melanocortin-4 receptor
NPY	neuropeptide Y
SP-analog	[D-Arg^1, D-Phe^5, D-Trp^{7,9}, Leu^{11}]-substance P
SNP	single nucleotide polymorphism
TMD	transmembrane domain
WT	wild type

1. GHRELIN SYSTEM

The existence of the ghrelin system has been demonstrated in many species through a variety of biochemical, anatomical, and genetic studies. It is a complex and intricate regulatory system comprised of several peptides, including mature ghrelin and its derivative des-Gln^{14}-ghrelin, unacylated ghrelin, obestatin, and intron 1-ghrelin, receptors such as growth hormone secretagogue receptor (GHSR) and other unknown but necessary receptors,[1] as well as modifying enzyme ghrelin *O*-acyltransferase (GOAT).[2]

The agonists, ghrelin and des-Gln^{14}-ghrelin, were both originally identified from rat gastric extracts.[3,4] Mature ghrelin is a 28 amino acid peptide that shows high sequence homology across mammals, including human, rat, mouse, rhesus monkey, goat, and dog.[3,5–8] Ghrelin has also been identified in nonmammalian vertebrates, including fish, amphibians, and birds (reviewed in Ref.[9]). Ghrelin is encoded by the *ghrelin (GHRL)* gene, which, in addition to ghrelin, also encodes the 23 amino acid peptide obestatin.[10] To activate its only known functional receptor, GHSR type 1a (GHSR1a), ghrelin requires the attachment of 8-carbon medium-chain fatty acid, octanoic acid, to its serine 3 residue, a process called octanoylation, a rare posttranslational modification.[3] In 2008, two groups independently reported an enzyme that catalyzed *n*-octanoyl modification of ghrelin in vitro, and named the enzyme GOAT.[11,12] The optimal temperature and pH range of GOAT are 37–50°C and pH 7–8, respectively.[13] The GHSR has two isoforms, GHSR1a and GHSR1b. Both are rhodopsin-like G protein-coupled receptors (GPCRs) but only GHSR1a is functional in humans, therefore, GHSR1a, also called ghrelin receptor, will be the main focus of this review. Other receptors mediating unacylated ghrelin action, ghrelin receptor-like receptors, have also been proposed but the identities remain elusive so far.[1]

1.1 Ghrelin Peptide Hormones

The discovery of ghrelin leads to the deorphanization of GHSR1a. In 1999, Kojima and colleagues identified an endogenous ligand specific for GHSR1a from rat gastric extracts, a classical example of reverse pharmacology. The purified ligand is a peptide of 28 amino acids and specifically involved in the release of growth hormone (GH) both in vivo and in vitro. For this reason, they designated this endogenous GH-releasing peptide "ghrelin" (ghre- is the Proto-Indo-European root of the word "grow").[3]

The human *GHRL* gene is localized on chromosome 3p25-26.[14] Both human and mouse *GHRL* genes contain five exons, with exon 2 and exon 3 encoding the functional ghrelin peptide.[5] As the only known peptide that has a fatty acid attached, ghrelin is highly conserved among species, with human ghrelin differing from rat ghrelin in only two amino acids. The unique structure of ghrelin is the *n*-octanoyl ester at its serine-3 residue, which is essential for ghrelin binding to the GHSR1a and for eliciting the subsequent ghrelin activities including regulation of feeding, adiposity, and insulin secretion.[15] Another endogenous GHSR ligand, des-Gln^{14}-ghrelin, was also identified from rat stomach and its amino acid sequence is identical to that of ghrelin except that glutamine-14 in ghrelin is missing.[4] Analysis of the genomic structure revealed des-Gln^{14}-ghrelin is processed from the *GHRL* gene by alternative splicing.[16]

In situ hybridization, immunohistochemistry, and reverse transcription-polymerase chain reaction in mammals demonstrated that ghrelin is most abundantly expressed in stomach, followed by gastrointestinal organs (including duodenum, jejunum, ileum, and colon), pancreas, as well as hypothalamic arcuate nucleus, kidney, and placenta.[17–19] In addition, it is also detected in interstitial Leydig cells and in Sertoli cells at very low levels.[19] Ghrelin produced in these tissues might serve paracrine and autocrine functions, depending on whether GHSR1a is also expressed.

1.2 Ghrelin Receptor

Pioneering work by Bowers in the 1970s–80s led to the discovery of a series of short synthetic peptides, such as GH-releasing peptide-6 (GHRP-6) and hexarelin, that potently stimulate GH release when administered to children.[20] In 1995, scientists at Merck Research Laboratories developed a small molecule potent GH secretagogue, L-163,191 (MK-0677), that can stimulate GH release from cultured rat pituitary cells with EC_{50} of 1.3 nM.[21] Binding sites for MK-0677 were identified in porcine and rat pituitary membranes and this specific binding is Mg^{2+}-dependent, inhibited by guanosine triphosphate and leads to a pulsatile GH release.[22] These results suggest that the receptor mediating MK-0677 (as well as other small peptides such as GHRP-6 and hexarelin) action is a novel GPCR that is distinct from the receptor for growth hormone-releasing hormone that uses a G_s-cAMP signaling pathway. Shortly thereafter, the same group cloned the full-length sequence of this receptor and named it GHSR.[23] This orphan receptor was later renamed ghrelin receptor after the identification of ghrelin as the endogenous ligand (see Section 1.1).[3]

The human *GHSR* gene is located at chromosome 3q26.2 and contains two exons and a single intron. Exon 1 encodes transmembrane domains (TMDs) 1–5 and exon 2 encodes TMD6 and TMD7 of the receptor.[24] Two splice variants are known: the full-length isoform, GHSR1a, encodes a GPCR of 366 amino acids, whereas the truncated isoform, GHSR1b, is a protein of 289 amino acids that contains only five TMDs encoded by exon 1 (265 amino acids) and part of the intron (encoding 24 amino acids).[24] Several techniques including classical and real-time reverse transcription and polymerase chain reaction studies revealed that GHSR1a is primarily expressed in pituitary and several nuclei in the hypothalamus, particularly in the arcuate nucleus and mediobasal hypothalamus, and at much lower levels in the thyroid gland, pancreas, spleen, myocardium, and adrenal gland.[17,25] However, the unspliced GHSR1b is widely expressed.[17] In humans, only GHSR1a is a functional receptor whereas GHSR1b does not bind to ghrelin or other GHSs and does not activate downstream signaling.[16] Indeed, when coexpressed in vitro, GHSR1b heterodimerizes with GHSR1a and exerts dominant negative activity on GHSR1a cell surface expression.[26,27]

Consistent with the tissue distribution of GHSR1a, it is not surprising that ghrelin is a versatile regulator of several physiological processes, including food intake and energy expenditure,[28] GH secretion, glucose metabolism,[29] cardiovascular functions,[30] gastric acid secretion and motility,[31] and immune function.[32] However, it is the potent induction of GH secretion and the stimulation of appetite that are most intensively studied. As the only known circulating (peripheral) orexigenic factor, ghrelin is secreted from the stomach and circulates in the bloodstream under fasting conditions, indicating that it transmits a hunger signal from the periphery to the central nervous system.[33] Therefore, ghrelin has a potential benefit in antagonizing protein breakdown and weight loss in catabolic conditions such as cancer cachexia as well as renal and cardiac diseases.[15]

Studies in multiple cell lines demonstrated that depending on the agonists, the GHSR1a signals through diverse pathways mediated by $G\alpha_q/Ca^{2+}$,[34] $G\alpha_s$/adenylyl cyclase/cAMP,[35] $G\alpha_q$/ERK1/2,[36] $G\alpha_{12/13}$/RhoA-SRE (ras homolog gene family, member A/serum response element),[37] and β-arrestin/adaptor protein 2.[38]

Interestingly, human GHSR is characterized by a high degree of constitutive signaling activity in some signaling pathways.[34] In vitro studies showed GHSR possesses a basal activity of about half the maximal agonist-stimulated activity by analysis of inositol phosphate (IP) turnover ($G\alpha_q$ pathway).[39] In another study when human GHSR1a was expressed in GH-secreting cells mimicking more physiological condition, the constitutive activity is estimated at about 25% of maximal ghrelin-stimulated signaling.[40] A possible mechanism proposed by Holst et al. is that the TMD6 is pulled toward TMD3 by a salt-bridge formed between Arg^{281} and Glu^{124}, stabilizing the active conformation of the receptor through an inward movement of TMD6 and TMD7 at the extracellular end. Furthermore, both residues have previously been implicated to be important for ligand binding.[34] The extremely high basal activity of GHSR is of physiological importance in regulating both GH secretion and food intake as demonstrated with in vivo experiments.[40,41] Treatment with inverse agonist (that decrease the constitutive activity) led to decreased hypothalamic neuropeptide Y (NPY) expression, food intake, and body weight.[41] Therefore the application of inverse agonists (as well as neutral antagonists) of GHSR, especially small molecules that can cross the blood-brain barrier, might represent a promising strategy for obesity and diabetes treatment.[42,43]

Indeed, basal activity has been implicated as important in a number of GPCRs.[44–47] Some wild type (WT) GPCRs have significant constitutive activity. Constitutively active mutations in GPCRs can cause several diseases, such as congenital stationary night blindness, hyperthyroidism, familial male-limited pseudoprecocious puberty, and Jansen's metaphyseal chondrodysplasia.[44] Defective constitutive signaling in the melanocortin-4 receptor (MC4R) is suggested to be a potential cause of obesity in *MC4R* mutation carriers.[48–50]

2. GHRELIN SYSTEM AND ENERGY HOMEOSTASIS

Despite the versatile physiological roles of ghrelin described earlier, it is the regulation of food intake and energy expenditure that have been most extensively studied. In rodents, Nakazato et al. showed that peripheral administration of ghrelin leads to increased food consumption.[28] In addition, acute ghrelin administration increases respiratory quotient and fat deposition.[51] Further studies demonstrated that rodents with i.c.v. injection of ghrelin tend to augment significantly higher plasma GH level, food intake, and body weight gain compared to that of peripheral injection, suggesting that ghrelin/GHSR functions at the hypothalamic level.[28] Similarly, in humans, Wren and colleagues demonstrated that peripheral ghrelin administration induces the sensation of hunger and increases food intake in lean, obese, healthy, and malnourished individuals.[52] Ghrelin acts presynaptically by stimulating the NPY neurons and inhibiting anorexigenic proopiomelanocortin neurons, resulting in the release of NPY and other orexigenic neurotransmitters,[18] antagonizing leptin action. Kroemer et al. showed that in response to palatable food stimuli, the endogenous fasting levels of ghrelin correlate positively with hunger-modulated activity in limbic and paralimbic regions, areas known to be involved in visual processing, taste sensing, and reward.[53] Similar association was also observed in the hypothalamus, amygdala, and prefrontal cortex of women with active anorexia nervosa.[54]

Further evidences that the ghrelin system acts as a major factor in the regulation of energy homeostasis come from rodent studies with genetic manipulation of the *Ghsr* gene. In 2004, Smith's group showed that ghrelin treatment fails to increase food intake in *Ghsr* knockout mice, as compared with WT littermates.[55] Insulin and leptin levels are reduced with fasting, and this response remains intact in mice lacking the *Ghsr*. Furthermore, the body weights of congenic adult *Ghsr* knockout mice (C57BL6J) are modestly lower than WT controls when they are fed a normal chow diet.[55] In 2005, Zigman et al. demonstrated that mice of mixed genetic background (C57BL6J:129sv) are relatively resistant to diet-induced obesity.[56] In addition, Chen et al. demonstrated that NPY and Agouti-related peptide, well-known potent orexigenic peptides, are essential to mediate the appetite-stimulating effects of ghrelin since mice lacking both NPY and Agouti-related peptide fail to increase food intake following peripheral ghrelin injection.[57] Additionally, recent studies showed that the endocannabinoid system might also mediate the orexigenic effects of ghrelin.[58]

Due to the versatility of the ghrelin system and its multiple functions, dysregulation in some of the components of the ghrelin system, such as ghrelin, GOAT, and GHSR, might lead to pathogenesis of many endocrine-related diseases, including obesity, short stature, cancer, inflammatory, and cardiovascular diseases.[2] These components could also be used as diagnostic or therapeutic targets for these endocrine diseases. Next, we summarize the mutations in *GHSR* that have been identified so far.

3. NATURALLY OCCURRING *GHSR* MUTATIONS

3.1 Identification of *GHSR* Mutations

Previous genetic studies have shown that human chromosome 3q26-29, where the *GHSR* is located, is a quantitative trait locus strongly linked to multiple phenotypes related to obesity and the metabolic syndrome.[59] Due to its biological function and location, *GHSR* is an excellent candidate gene that might contribute to the pathogenesis of obesity. Following this assumption, several large-scale screening studies were conducted to identify mutations and single nucleotide polymorphisms (SNPs) in *GHSR* in different cohorts originated from Europe, Africa, and East Asia. To date, a total of one nonsense mutation (W2X), one inframe deletion (ΔQ36), and 12 missense mutations (S84I, P108L, I134T, V160M, A169T, C173R, V182A, A204E, R237W, D246A, F279L, and A358T), and 19 SNPs (rs2232165, rs495225, rs2232169, rs2232166, rs2232167, rs572169, rs4988509, rs509035, rs519384, rs512692, rs863441, rs10618418, rs2948694, rs2922126, rs474225, rs9819506, rs565105, rs6772676, and DQ864722), have been identified (Fig. 1) (the SNPs are not listed in Fig. 1).[59–68] As can be seen from Fig. 1, these mutations are scattered throughout the GHSR.

Since *GHSR* gene expression level is critical for regulating appetite and food intake, mutations or variants in the regulatory sequences, especially the promoter region, that enhance *GHSR* gene transcription might also result in obesity. Indeed, Mager and coworkers showed that two SNPs, rs863441 and rs9819506, in the promoter region of *GHSR* are involved in the change of body weight.[63] Subsequently, Gjesing and colleagues identified a novel -151C/T promoter mutation in the *GHSR*. This rare mutation increases the *GHSR* transcriptional activity and shows partial cosegregation with obesity or overweight in two unrelated families originated from different

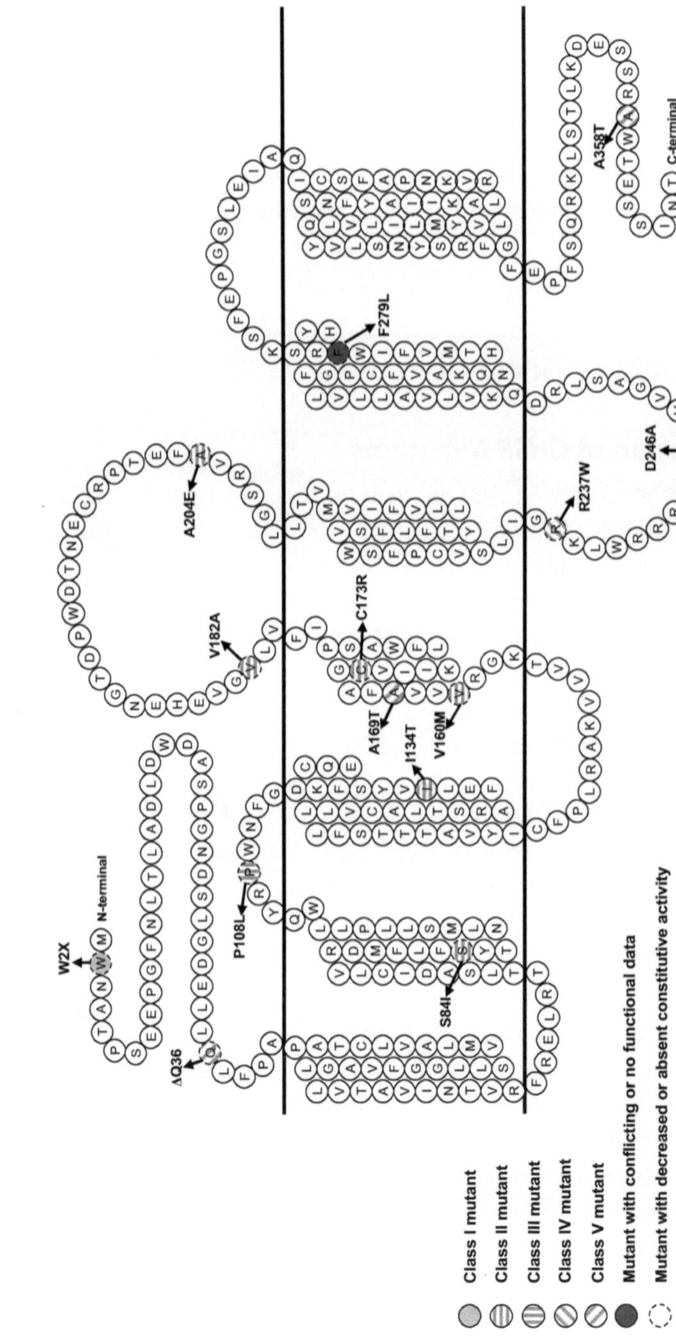

Figure 1 Schematic representation of naturally occurring *GHSR* mutations.

backgrounds, suggesting changes in *GHSR* expression might be important for appetite regulation and energy homeostasis in humans.[66]

Hinney and her colleagues first identified two missense *GHSR* mutations (A204E and F279L) through a systematic screening of the coding region of *GHSR* in a German cohort.[60] A204E mutation was found in one obese individual while F279L was identified from one child with short stature. Further cosegregation analysis showed that the mother of the short stature child, who also harbors the mutation, is short and obese (height, 1.57 m; BMI, 34.6 kg/m^2). A204E mutation was also identified by Pantel and colleagues from two unrelated Morocco families.[61] The patients present with short stature, with or without isolated GH deficiency (GHD), with some being overweight or obese.[61] Moreover, Hinney and coworkers also identified five silent SNPs (rs2232165, rs495225, rs2232169, rs572169, and rs4988509) that result in five synonymous variants (D20D, E57E, L149L, R159R, and P177P).[60] Interestingly, these five SNPs were also found in a population-based study samples of 15854 unrelated middle-aged Danes.[66]

Through comprehensive analysis of SNPs, linkage disequilibrium, and haplotype structure across the entire *GHSR* gene region in 178 pedigrees with at least two obese individuals as well as in an independent general population of 1418 whites, Baessler and colleagues identified a disequilibrium block consisting of five SNPs (rs509035, rs572169, rs519384, rs512692, and rs863441) and demonstrated a potential linkage between the five SNPs and haplotypes within the *GHSR* and human obesity.[59] However, a later study could not replicate this observation in a different population.[69] In 2009, Gueorguiev and coworkers reported three novel *GHSR* SNPs (rs2232166, rs10618418, and DQ864722) together with five previous reported SNPs through direct sequencing in 70 obese children.[70] Further genotype studies in 263 families consisting of 1275 French obese individuals and in 287 families consisting of 1059 general French subjects showed that rs572169 is associated with obesity.[70]

Several studies have also demonstrated that *GHSR* mutations are also associated with short stature. Pantel et al. reported a case of partial GHD in a short stature patient carrying two compound heterozygous genetic defects in the *GHSR*, a nucleotide substitution resulting in a premature termination codon (W2X) and a missense mutation (R237W), respectively.[65] Recently, Inoue and colleagues identified four novel heterozygous *GHSR* mutations (ΔQ36, P108L, C173R, and D246A) from 127 unrelated Japanese patients diagnosed with either GHD or short stature.[67] Each of the *GHSR* mutations was only found in one patient or one family, with the exception of ΔQ36 that was identified in three patients. Subsequently, Pugliese-Pires and colleagues reported two potential pathogenic *GHSR* mutations (S84I and V182A) in two Brazil female patients with idiopathic short stature, characterized by constitutional delay of growth and puberty (CDGP).[68] Two other *GHSR* mutations (A169T and A358T) were identified in two other male patients with CDGP.[68]

In addition to the candidate gene studies that are hypothesis driven, genome wide association studies that are not based on hypothesis but relying on advances in sequencing to investigate association of variants across the genome are also widely used to identify variants associated with particular trait, such as body weight. These studies do not provide convincing evidence of association of *GHSR* with obesity (reviewed by Liu et al.[71]). However, caveats such as not all variants in a gene are covered and the limit of the effect size that can be detected need to be bear in mind with genome wide association studies.[43]

3.2 Prevalence of *GHSR* Mutations

In the study of German cohorts consisting of 93 obese, 96 normal weight, and 94 underweight individuals, as well as 43 children with short stature, a prevalence of 1.1% pathogenic *GHSR* mutations in the severely obese German individuals was reported.[60] The patients in this cohort are unique in that they all have an age- and gender-specific BMI >90th percentile. In addition, they also found that 2.3% of short stature children, whose body height are <5th percentile, harbor *GHSR* mutations. Moreover, these mutations were not found in the control groups.[60]

In the study of Pantel and colleagues, one *GHSR* mutation, A204E, was found in heterozygous state in a patient from a group of 51 individuals with isolated GHD, and in homozygous state in a patient from a group of 41 individuals with idiopathic short stature, whereas it was not found in 100 controls. Therefore, the prevalence of this mutation in short stature patients is 2.2%. Furthermore, additional cosegregation studies in the families of these two short stature patients demonstrated that all other family members who have short stature also carry this mutation.[61]

In another small-scale study of 82 obese Danish individuals (mean BMI of ~35.5 kg/m^2), 1.2% of the obese patients were found to carry pathogenic *GHSR* mutations.[66] In a Czech group consisting of 289 unrelated obese children (aged 1–18 years, BMI above 97th percentile for age and sex), only one heterozygous carrier of the -151C/T mutation was identified, therefore, the prevalence in this cohort is 0.3%.[66] In a population-based study of 15,854 Danes, the -151C/T promoter mutation showed minor allele frequency of 0.01%.[66]

In a study of 127 unrelated Japanese short stature individuals diagnosed with either isolated GHD or idiopathic short stature, four pathogenic *GHSR* mutations, including ΔQ36, P108L, C173R, and D246A, were found in six patients. Therefore, the prevalence is 4.7%. ΔQ36 was presented in one normal individual while P108L, C173R, and D246A were not found in 188 controls; hence the controls have a mutation prevalence of 0.3%.[67] In 31 idiopathic short stature patients with CDGP from Brazil, five novel heterozygous *GHSR* mutations, including S84I, V182A, A169T, and A358T, and a mutation located 6 bp prior to the initiation codon, were identified from five CDGP patients, hence the prevalence is 16.1%.[68] None of these mutations were found in the controls.[68]

3.3 Functional Characterizations of *GHSR* Mutations

In order to prove the pathogenic role of a *GHSR* mutation identified from an obese or short stature patient, two additional lines of evidence are needed, including cosegregation of the mutation with the disease in the family and absence of the mutation in the corresponding controls, as well as functional defects of the mutant receptor. Several assays could be used to measure the GHSR functionality. Decreased cell surface expression, due to misfolding being detected by the cell's quality control system, is the most common defect in loss-of-function GPCR mutants. [72,73] Therefore it is important to measure the cell surface expression using flow cytometry or immunocytochemistry coupled with confocal microscopy. If decreased expression at the plasma membrane is detected, the same assays using permeabilized cells can be performed to confirm that receptors are synthesized but retained intracellularly. Nonsense mutations may result in decreased mRNA levels and corresponding decrease in protein translation. Because most of the antibodies against GPCRs (including GHSR) are not working, most of the assays measuring expression utilize antibodies against small epitope tags inserted at the N-terminus of the receptor. Ligand binding assay measures whether the receptor can bind radiolabeled GHSR specific ligand normally.

Since GHSR is primarily coupled to the $G\alpha_q$-mediated IP pathway, and it has significant constitutive activity, signaling assays that measure the IP level or reporter gene expression driven by increased IP level are used to measure GHSR basal and ligand-stimulated signaling activity.

Following the original identification of the two missense *GHSR* mutations, A204E and F279L, from two German obese patients, and subsequent identification of A204E from two unrelated pedigree with familial short stature,[60,61] several groups performed in vitro studies to elucidate the functional characteristics of these two mutations. Both mutants were demonstrated to be associated with loss of constitutive activity without affecting ghrelin affinity, potency, or efficacy.[61,62,67] Furthermore, [D-Arg^1, D-Phe^5, D-Trp^{7,9}, Leu^{11}]-substance P (SP-analog), a known GHSR inverse agonist, shows similar inhibitory activity at the F279L mutant compared to WT GHSR.[61,62,67] Another systematic structure-function analysis suggested that the aromatic residue F279, located on the inner face of TMD6 facing TMD7, is essential in holding GHSR in the active conformation.[74] Removal of the aromaticity at position 279 selectively decreases the constitutive activity.[74]

In addition to A204E and F279L, Kopin and colleagues also performed systematic pharmacological study of two other *GHSR* mutations, I134T and V160M. Using either HEK293 or COS-7 cells expressing WT or mutant GHSRs, with whole-cell enzyme-linked immunosorbent assay, they showed that the expression levels of both mutants are decreased compared to WT GHSR. They also measured basal and agonist-induced signaling as well as inverse agonist-mediated inhibition of constitutive activity by indirect luciferase reporter gene assays or direct measurement of IP production. V160M has decreased basal activity whereas I134T has normal basal activity compared to the WT GHSR. Both mutants show significantly decreased ghrelin-dependent signaling. V160M has markedly increased ghrelin potency whereas I134T is unresponsive to ghrelin. Interestingly, I134T responds to the SP-analog, although with a decreased efficacy. V160M responds to SP-analog in a similar manner compared to WT GHSR.[62]

Using similar techniques, functional characterization of R237W shows a similar level of cell surface expression, partial loss of constitutive activity, and preserved response to ghrelin stimulation compared to WT receptor.[65] Furthermore, Inoue and coworkers showed that ΔQ36, P108L, and D246A have normal cell surface expression, whereas C173R is totally trapped intracellularly.[67] ΔQ36 and D246A have decreased binding affinity to ghrelin, whereas P108L and C173R show virtually undetectable specific binding. ΔQ36, P108L, C173R, and D246A possess 67.5%, 32.0%, none, and 58.8% of constitutive signaling activity compared with the WT GHSR, respectively. ΔQ36 and P108L exert normal ghrelin-stimulated signaling while D246A shows impaired agonist-stimulated signaling. C173R, as expected, has no ghrelin-induced activation due to complete intracellular retention of the mutant receptor.[67] In addition, another functional study revealed that two pathogenic *GHSR* mutations (S84I and V182A) result in reduction in cell surface expression as well as decrease in basal activity. The S84I mutation is also associated with a defect in ghrelin potency. Two other missense mutations (A169T and A358T) turned out to be benign variants, with similar cell surface expression level, basal and ghrelin-stimulated signaling compared to WT GHSR.[68]

Table 1 summarizes the functional implications of all the naturally occurring *GHSR* mutations identified so far. These mutations are also classified into different classes as we proposed before.[45,72,73,75] This classification scheme is based on the life cycle of GPCRs, with Class I mutants depicting mutations that are not translated normally or prematurely degraded; Class II mutants are retained intracellularly, primarily in the endoplasmic reticulum; Class III mutants are those that reach cell surface but could not bind to the ligand; Class IV mutants are those that are expressed on the cell surface and can bind the ligand but could not generate normal signaling (defects in decreased basal signaling also belong to this Class); and Class V mutants have normal properties as the WT receptor.

It should be emphasized that the *GHSR* mutations identified so far are loss-of-function mutations or without obvious defect but not hyperactive. It is easy to reconcile loss of function in GHSR with short stature, since ghrelin/GHSR system is important for normal GH synthesis and secretion; defective GH release results in short stature. However, ghrelin is the only peripheral orexigenic hormone; animal experiments showed that hyper-functioning of the system results in obesity whereas hypo-functioning of the system results in protection from obesity (including high-fat induced).[56,76] Therefore the explanation for this paradox of loss-of-function *GHSR* mutations associated with obesity is not immediately apparent. However, similar paradox is also encountered with the MC4R. Whereas the vast majority of *MC4R* mutations identified from obese patients are associated with loss-of-function, including loss of basal signaling, consistent with MC4R physiology, several mutations that were also identified from obese patients result in constitutive activation, or hyper-functioning of the system, not consistent with MC4R physiology.[46,47,49,77] A generally accepted explanation is still lacking.

3.4 Potential Strategies to Correct *GHSR* Mutations

Based on studies in other GPCRs as well as other membrane proteins, we can identify potential strategies to correct the defects identified in the naturally occurring *GHSR* mutations.[72] Aminoglycosides and other drugs that can

Table 1 Summary of the Function Analysis of Naturally Occurring *GHSR* Mutations.

Mutation	Location	Cell Surface Expression	Ghrelin Binding	Signaling: Basal	Signaling: Ghrelin	Classification	References
W2X	N-term	ND	ND	ND	↓	I	[65]
ΔQ36	N-term	—	—	↓	—	V	[67]
S84I	TMD2	ND	NA	ND	↓	II	[68]
P108L	ECL1	—	ND	↓	—	III	[67]
I134T	TMD3	↓	NA	—	ND	II	[62]
V160M	TMD4	↓	NA	↓	—	II	[62]
A169T	TMD4	—	NA	—	—	V	[68]
C173R	TMD4	ND	ND	ND	ND	II	[67]
V182A	ECL2	↓	NA	↓	—	II	[68]
A204E	ECL2	↓	ND	ND	—	II	[61,62,67]
R237W	ICL3	—	—	↓	↓	IV	[65]
D246A	ICL3	↓	↓	↓	↓	II	[67]
F279L	TMD6	↓	—	↓	↓(—)[a]	II	[62,67]
A358T	C-term	—	NA	—	—	V	[68]

N-term, N-terminus; TMD, transmembrane domain; ECL, extracellular loop; ICL, intracellular loop; C-term, C-terminus; ND, not detectable; NA, not assessed.

"↓" Denotes the particular function is defective.

"—" Denotes the particular function is normal.

[a]Results from different studies are conflicting.

correct the premature termination might be able to correct defect in Class I mutants, such as W2X.[72] For Class II mutants, the most common mutation, small molecule ligands that can pass through the plasma and endoplasmic reticulum membranes and bind to misfolded mutants and correct the misfolding, the so-called pharmacological chaperones or pharmacoperones, might be able to rescue these mutants to the cell surface, potentially resulting in a functional receptor.[73] We have used this strategy to correct obesity-associated naturally occurring mutations in the *MC4R* that cause the mutant receptors to be retained intracellularly.[50,78,79,80] Both antagonists and agonists can serve as pharmacoperones for the MC4R. The same strategy might also be used to correct Class II GHSR mutants listed in Table 1.

For Class III and IV mutants that are trafficked to the cell surface but could not bind to the endogenous ligand ghrelin (Class III) or could bind to ghrelin but could not initiate signaling (Class IV), ligands that bind to different residues on GHSR might be able to correct these defects,[72] especially with the identification of many small molecule agonists of the GHSR.

4. SUMMARY

During the past two decades, extensive genetic and pharmacological studies have demonstrated that ghrelin/GHSR system plays important roles in multiple physiological processes, especially in GH secretion and appetite regulation. In addition, naturally occurring mutations in *GHSR* are associated with obesity and short stature from different backgrounds, suggesting GHSR might be another important pathogenic cause for obesity and short stature. Targeting GHSR and its ligand ghrelin might represent promising approaches toward treatment of obesity, diabetes, and short stature.

REFERENCES

1. Callaghan B, Furness JB. Novel and conventional receptors for ghrelin, desacyl-ghrelin, and pharmacologically related compounds. *Pharmacol Rev.* 2014;66:984–1001.
2. Gahete MD, Rincon-Fernandez D, Villa-Osaba A, Hormaechea-Agulla D, Ibanez-Costa A, Martinez-Fuentes AJ, Gracia-Navarro F, Castano JP, Luque RM. Ghrelin gene products, receptors, and GOAT enzyme: biological and pathophysiological insight. *J Endocrinol.* 2014;220:R1–R24.
3. Kojima M, Hosoda H, Date Y, Nakazato M, Matsuo H, Kangawa K. Ghrelin is a growth-hormone-releasing acylated peptide from stomach. *Nature.* 1999;402:656–660.
4. Hosoda H, Kojima M, Matsuo H, Kangawa K. Purification and characterization of rat des-Gln[14]-Ghrelin, a second endogenous ligand for the growth hormone secretagogue receptor. *J Biol Chem.* 2000;275:21995–22000.

5. Tanaka M, Hayashida Y, Iguchi T, Nakao N, Nakai N, Nakashima K. Organization of the mouse ghrelin gene and promoter: occurrence of a short noncoding first exon. *Endocrinology.* 2001;142:3697–3700.
6. Angeloni SV, Glynn N, Ambrosini G, Garant MJ, Higley JD, Suomi S, Hansen BC. Characterization of the rhesus monkey ghrelin gene and factors influencing ghrelin gene expression and fasting plasma levels. *Endocrinology.* 2004;145:2197–2205.
7. Ida T, Miyazato M, Lin XZ, Kaiya H, Sato T, Nakahara K, Murakami N, Kangawa K, Kojima M. Purification and characterization of caprine ghrelin and its effect on growth hormone release. *J Mol Neurosci.* 2010;42:99–105.
8. Tomasetto C, Wendling C, Rio MC, Poitras P. Identification of cDNA encoding motilin related peptide/ghrelin precursor from dog fundus. *Peptides.* 2001;22:2055–2059.
9. Kojima M, Ida T, Sato T. Structure of mammalian and nonmammalian ghrelins. *Vitam Horm.* 2008;77:31–46.
10. Zhang JV, Ren PG, Avsian-Kretchmer O, Luo CW, Rauch R, Klein C, Hsueh AJ. Obestatin, a peptide encoded by the ghrelin gene, opposes ghrelin's effects on food intake. *Science.* 2005;310:996–999.
11. Yang J, Brown MS, Liang G, Grishin NV, Goldstein JL. Identification of the acyltransferase that octanoylates ghrelin, an appetite-stimulating peptide hormone. *Cell.* 2008;132: 387–396.
12. Gutierrez JA, Solenberg PJ, Perkins DR, Willency JA, Knierman MD, Jin Z, Witcher DR, Luo S, Onyia JE, Hale JE, Ghrelin. octanoylation mediated by an orphan lipid transferase. *Proc Natl Acad Sci USA.* 2008;105:6320–6325.
13. Ohgusu H, Shirouzu K, Nakamura Y, Nakashima Y, Ida T, Sato T, Kojima M. Ghrelin *O*-acyltransferase (GOAT) has a preference for *n*-hexanoyl-CoA over *n*-octanoyl-CoA as an acyl donor. *Biochem Biophys Res Commun.* 2009;386:153–158.
14. Wajnrajch MP, Ten IS, Gertner JM, Leibel RL. Genomic organization of the human *GHRELIN* gene. *J Endocr Genet.* 2000;1:231–233.
15. Davenport AP, Bonner TI, Foord SM, Harmar AJ, Neubig RR, Pin JP, Spedding M, Kojima M, Kangawa K. International Union of Pharmacology. LVI. Ghrelin receptor nomenclature, distribution, and function. *Pharmacol Rev.* 2005;57:541–546.
16. Li Z, Li Y, Zhang W. Ghrelin receptor in energy homeostasis and obesity pathogenesis. *Prog Mol Biol Transl Sci.* 2013;114:45–87.
17. Gnanapavan S, Kola B, Bustin SA, Morris DG, McGee P, Fairclough P, Bhattacharya S, Carpenter R, Grossman AB, Korbonits M. The tissue distribution of the mRNA of ghrelin and subtypes of its receptor, GHS-R, in humans. *J Clin Endocrinol Metab.* 2002;87:2988–2991.
18. Cowley MA, Smith RG, Diano S, Tschop M, Pronchuk N, Grove KL, Strasburger CJ, Bidlingmaier M, Esterman M, Heiman ML, Garcia-Segura LM, Nillni EA, Mendez P, Low MJ, Sotonyi P, Friedman JM, Liu H, Pinto S, Colmers WF, Cone RD, Horvath TL. The distribution and mechanism of action of ghrelin in the CNS demonstrates a novel hypothalamic circuit regulating energy homeostasis. *Neuron.* 2003;37:649–661.
19. Kojima M, Kangawa K. Ghrelin: structure and function. *Physiol Rev.* 2005;85:495–522.
20. Bowers CY. Growth hormone-releasing peptide (GHRP). *Cell Mol Life Sci.* 1998;54: 1316–1329.
21. Patchett AA, Nargund RP, Tata JR, Chen MH, Barakat KJ, Johnston DB, Cheng K, Chan WW, Butler B, Hickey G, Jacks T, Schleim K, Pong SS, Chaung LYP, Chen HY, Frazier E, Leung KH, Chiu SHL, Smith RG. Design and biological activities of L-163,191 (MK-0677): a potent, orally active growth hormone secretagogue. *Proc Natl Acad Sci USA.* 1995;92:7001–7005.
22. Pong SS, Chaung LY, Dean DC, Nargund RP, Patchett AA, Smith RG. Identification of a new G-protein-linked receptor for growth hormone secretagogues. *Mol Endocrinol.* 1996;10:57–61.

23. Howard AD, Feighner SD, Cully DF, Arena JP, Liberator PA, Rosenblum CI, Hamelin M, Hreniuk DL, Palyha OC, Anderson J, Paress PS, Diaz C, Chou M, Liu KK, McKee KK, Pong SS, Chaung LY, Elbrecht A, Dashkevicz M, Heavens R, Rigby M, Sirinathsinghji DJ, Dean DC, Melillo DG, Patchett AA, Nargund R, Griffin PR, DeMartino JA, Gupta SK, Schaeffer JM, Smith RG, Van der Ploeg LH. A receptor in pituitary and hypothalamus that functions in growth hormone release. *Science.* 1996;273:974–977.
24. McKee KK, Palyha OC, Feighner SD, Hreniuk DL, Tan CP, Phillips MS, Smith RG, Van der Ploeg LH, Howard AD. Molecular analysis of rat pituitary and hypothalamic growth hormone secretagogue receptors. *Mol Endocrinol.* 1997;11:415–423.
25. Guan XM, Yu H, Palyha OC, McKee KK, Feighner SD, Sirinathsinghji DJ, Smith RG, Van der Ploeg LH, Howard AD. Distribution of mRNA encoding the growth hormone secretagogue receptor in brain and peripheral tissues. *Brain Res Mol Brain Res.* 1997;48:23–29.
26. Leung PK, Chow KB, Lau PN, Chu KM, Chan CB, Cheng CH, Wise H. The truncated ghrelin receptor polypeptide (GHS-R1b) acts as a dominant-negative mutant of the ghrelin receptor. *Cell Signal.* 2007;19:1011–1022.
27. Chow KB, Sun J, Chu KM, Tai Cheung W, Cheng CH, Wise H. The truncated ghrelin receptor polypeptide (GHS-R1b) is localized in the endoplasmic reticulum where it forms heterodimers with ghrelin receptors (GHS-R1a) to attenuate their cell surface expression. *Mol Cell Endocrinol.* 2012;348:247–254.
28. Nakazato M, Murakami N, Date Y, Kojima M, Matsuo H, Kangawa K, Matsukura S. A role for ghrelin in the central regulation of feeding. *Nature.* 2001;409:194–198.
29. Broglio F, Arvat E, Benso A, Gottero C, Muccioli G, Papotti M, van der Lely AJ, Deghenghi R, Ghigo E. Ghrelin, a natural GH secretagogue produced by the stomach, induces hyperglycemia and reduces insulin secretion in humans. *J Clin Endocrinol Metab.* 2001;86:5083–5086.
30. Shimizu Y, Nagaya N, Teranishi Y, Imazu M, Yamamoto H, Shokawa T, Kangawa K, Kohno N, Yoshizumi M. Ghrelin improves endothelial dysfunction through growth hormone-independent mechanisms in rats. *Biochem Biophys Res Commun.* 2003;310: 830–835.
31. Masuda Y, Tanaka T, Inomata N, Ohnuma N, Tanaka S, Itoh Z, Hosoda H, Kojima M, Kangawa K. Ghrelin stimulates gastric acid secretion and motility in rats. *Biochem Biophys Res Commun.* 2000;276:905–908.
32. Li WG, Gavrila D, Liu X, Wang L, Gunnlaugsson S, Stoll LL, McCormick ML, Sigmund CD, Tang C, Weintraub NL. Ghrelin inhibits proinflammatory responses and nuclear factor-κB activation in human endothelial cells. *Circulation.* 2004;109:2221–2226.
33. Cummings DE, Purnell JQ, Frayo RS, Schmidova K, Wisse BE, Weigle DS. A preprandial rise in plasma ghrelin levels suggests a role in meal initiation in humans. *Diabetes.* 2001;50:1714–1719.
34. Holst B, Cygankiewicz A, Jensen TH, Ankersen M, Schwartz TW. High constitutive signaling of the ghrelin receptor–identification of a potent inverse agonist. *Mol Endocrinol.* 2003;17:2201–2210.
35. Granata R, Settanni F, Biancone L, Trovato L, Nano R, Bertuzzi F, Destefanis S, Annunziata M, Martinetti M, Catapano F, Ghe C, Isgaard J, Papotti M, Ghigo E, Muccioli G. Acylated and unacylated ghrelin promote proliferation and inhibit apoptosis of pancreatic beta-cells and human islets: involvement of 3',5'-cyclic adenosine monophosphate/protein kinase A, extracellular signal-regulated kinase 1/2, and phosphatidyl inositol 3-Kinase/Akt signaling. *Endocrinology.* 2007;148:512–529.
36. Mousseaux D, Le Gallic L, Ryan J, Oiry C, Gagne D, Fehrentz JA, Galleyrand JC, Martinez J. Regulation of ERK1/2 activity by ghrelin-activated growth hormone secretagogue receptor 1A involves a PLC/PKC pathway. *Br J Pharmacol.* 2006;148:350–365.

37. Sivertsen B, Lang M, Frimurer TM, Holliday ND, Bach A, Els S, Engelstoft MS, Petersen PS, Madsen AN, Schwartz TW, Beck-Sickinger AG, Holst B. Unique interaction pattern for a functionally biased ghrelin receptor agonist. *J Biol Chem.* 2011;286:20845–20860.
38. Damian M, Marie J, Leyris JP, Fehrentz JA, Verdie P, Martinez J, Baneres JL, Mary S. High constitutive activity is an intrinsic feature of ghrelin receptor protein: a study with a functional monomeric GHS-R1a receptor reconstituted in lipid discs. *J Biol Chem.* 2012;287:3630–3641.
39. Holst B, Lang M, Brandt E, Bach A, Howard A, Frimurer TM, Beck-Sickinger A, Schwartz TW. Ghrelin receptor inverse agonists: identification of an active peptide core and its interaction epitopes on the receptor. *Mol Pharmacol.* 2006;70:936–946.
40. Lall S, Balthasar N, Carmignac D, Magoulas C, Sesay A, Houston P, Mathers K, Robinson I. Physiological studies of transgenic mice overexpressing growth hormone (GH) secretagogue receptor 1A in GH-releasing hormone neurons. *Endocrinology.* 2004;145:1602–1611.
41. Petersen PS, Woldbye DP, Madsen AN, Egerod KL, Jin C, Lang M, Rasmussen M, Beck-Sickinger AG, Holst B. In vivo characterization of high basal signaling from the ghrelin receptor. *Endocrinology.* 2009;150:4920–4930.
42. Holst B, Schwartz TW. Constitutive ghrelin receptor activity as a signaling set-point in appetite regulation. *Trends Pharmacol Sci.* 2004;25:113–117.
43. Cameron KO, Bhattacharya SK, Loomis AK. Small molecule ghrelin receptor inverse agonists and antagonists. *J Med Chem.* 2014;57:8671–8691.
44. Tao YX. Constitutive activation of G protein-coupled receptors and diseases: insights into mechanism of activation and therapeutics. *Pharmacol Ther.* 2008;120:129–148.
45. Tao YX. Molecular mechanisms of the neural melanocortin receptor dysfunction in severe early onset obesity. *Mol Cell Endocrinol.* 2005;239:1–14.
46. Tao YX, Huang H, Wang ZQ, Yang F, Williams JN, Nikiforovich GV. Constitutive activity of neural melanocortin receptors. *Methods Enzymol.* 2010;484:267–279.
47. Tao YX. Constitutive activity in melanocortin-4 receptor: biased signaling of inverse agonists. *Adv Pharmacol.* 2014;70:135–154.
48. Srinivasan S, Lubrano-Berthelier C, Govaerts C, Picard F, Santiago P, Conklin BR, Vaisse C. Constitutive activity of the melanocortin-4 receptor is maintained by its N-terminal domain and plays a role in energy homeostasis in humans. *J Clin Invest.* 2004;114:1158–1164.
49. Tao YX. Mutations in melanocortin-4 receptor and human obesity. *Prog Mol Biol Transl Sci.* 2009;88:173–204.
50. Tao YX. The melanocortin-4 receptor: physiology, pharmacology, and pathophysiology. *Endocr Rev.* 2010;31:506–543.
51. Tschop M, Smiley DL, Heiman ML. Ghrelin induces adiposity in rodents. *Nature.* 2000;407:908–913.
52. Wren AM, Seal LJ, Cohen MA, Brynes AE, Frost GS, Murphy KG, Dhillo WS, Ghatei MA, Bloom SR. Ghrelin enhances appetite and increases food intake in humans. *J Clin Endocrinol Metab.* 2001;86:5992–5995.
53. Kroemer NB, Krebs L, Kobiella A, Grimm O, Pilhatsch M, Bidlingmaier M, Zimmermann US, Smolka MN. Fasting levels of ghrelin covary with the brain response to food pictures. *Addict Biol.* 2013;18:855–862.
54. Holsen LM, Lawson EA, Christensen K, Klibanski A, Goldstein JM. Abnormal relationships between the neural response to high- and low-calorie foods and endogenous acylated ghrelin in women with active and weight-recovered anorexia nervosa. *Psychiatry Res.* 2014;223:94–103.
55. Sun Y, Wang P, Zheng H, Smith RG. Ghrelin stimulation of growth hormone release and appetite is mediated through the growth hormone secretagogue receptor. *Proc Natl Acad Sci USA.* 2004;101:4679–4684.

56. Zigman JM, Nakano Y, Coppari R, Balthasar N, Marcus JN, Lee CE, Jones JE, Deysher AE, Waxman AR, White RD, Williams TD, Lachey JL, Seeley RJ, Lowell BB, Elmquist JK. Mice lacking ghrelin receptors resist the development of diet-induced obesity. *J Clin Invest.* 2005;115:3564–3572.
57. Chen HY, Trumbauer ME, Chen AS, Weingarth DT, Adams JR, Frazier EG, Shen Z, Marsh DJ, Feighner SD, Guan XM, Ye Z, Nargund RP, Smith RG, Van der Ploeg LH, Howard AD, MacNeil DJ, Qian S. Orexigenic action of peripheral ghrelin is mediated by neuropeptide Y and agouti-related protein. *Endocrinology.* 2004;145:2607–2612.
58. Kola B, Farkas I, Christ-Crain M, Wittmann G, Lolli F, Amin F, Harvey-White J, Liposits Z, Kunos G, Grossman AB, Fekete C, Korbonits M. The orexigenic effect of ghrelin is mediated through central activation of the endogenous cannabinoid system. *PLoS One.* 2008;3:e1797.
59. Baessler A, Hasinoff MJ, Fischer M, Reinhard W, Sonnenberg GE, Olivier M, Erdmann J, Schunkert H, Doering A, Jacob HJ, Comuzzie AG, Kissebah AH, Kwitek AE. Genetic linkage and association of the growth hormone secretagogue receptor (ghrelin receptor) gene in human obesity. *Diabetes.* 2005;54:259–267.
60. Wang HJ, Geller F, Dempfle A, Schauble N, Friedel S, Lichtner P, Fontenla-Horro F, Wudy S, Hagemann S, Gortner L, Huse K, Remschmidt H, Bettecken T, Meitinger T, Schafer H, Hebebrand J, Hinney A. Ghrelin receptor gene: identification of several sequence variants in extremely obese children and adolescents, healthy normal-weight and underweight students, and children with short normal stature. *J Clin Endocrinol Metab.* 2004;89:157–162.
61. Pantel J, Legendre M, Cabrol S, Hilal L, Hajaji Y, Morisset S, Nivot S, Vie-Luton MP, Grouselle D, de Kerdanet M, Kadiri A, Epelbaum J, Le Bouc Y, Amselem S. Loss of constitutive activity of the growth hormone secretagogue receptor in familial short stature. *J Clin Invest.* 2006;116:760–768.
62. Liu G, Fortin JP, Beinborn M, Kopin AS. Four missense mutations in the ghrelin receptor result in distinct pharmacological abnormalities. *J Pharmacol Exp Ther.* 2007;322: 1036–1043.
63. Mager U, Degenhardt T, Pulkkinen L, Kolehmainen M, Tolppanen AM, Lindstrom J, Eriksson JG, Carlberg C, Tuomilehto J, Uusitupa M. Finnish Diabetes Prevention Study Group.Variations in the ghrelin receptor gene associate with obesity and glucose metabolism in individuals with impaired glucose tolerance. *PLoS One.* 2008;3:e2941.
64. Dossus L, McKay JD, Canzian F, Wilkening S, Rinaldi S, Biessy C, Olsen A, Tjonneland A, Jakobsen MU, Overvad K, Clavel-Chapelon F, Boutron-Ruault MC, Fournier A, Linseisen J, Lukanova A, Boeing H, Fisher E, Trichopoulou A, Georgila C, Trichopoulos D, Palli D, Krogh V, Tumino R, Vineis P, Quiros JR, Sala N, Martinez-Garcia C, Dorronsoro M, Chirlaque MD, Barricarte A, van Duijnhoven FJ, Bueno-de-Mesquita HB, van Gils CH, Peeters PH, Hallmans G, Lenner P, Bingham S, Khaw KT, Key TJ, Travis RC, Ferrari P, Jenab M, Riboli E, Kaaks R. Polymorphisms of genes coding for ghrelin and its receptor in relation to anthropometry, circulating levels of IGF-I and IGFBP-3, and breast cancer risk: a case-control study nested within the European Prospective Investigation into Cancer and Nutrition (EPIC). *Carcinogenesis.* 2008;29: 1360–1366.
65. Pantel J, Legendre M, Nivot S, Morisset S, Vie-Luton MP, le Bouc Y, Epelbaum J, Amselem S. Recessive isolated growth hormone deficiency and mutations in the ghrelin receptor. *J Clin Endocrinol Metab.* 2009;94:4334–4341.
66. Gjesing AP, Larsen LH, Torekov SS, Hainerova IA, Kapur R, Johansen A, Albrechtsen A, Boj S, Holst B, Harper A, Urhammer SA, Borch-Johnsen K, Pisinger C, Echwald SM, Eiberg H, Astrup A, Lebl J, Ferrer J, Schwartz TW, Hansen T, Pedersen O. Family and population-based studies of variation within the ghrelin receptor locus in relation to measures of obesity. *PLoS One.* 2010;5:e10084.

67. Inoue H, Kangawa N, Kinouchi A, Sakamoto Y, Kimura C, Horikawa R, Shigematsu Y, Itakura M, Ogata T, Fujieda K. Japan Growth Genome Consortium.Identification and functional analysis of novel human growth hormone secretagogue receptor (GHSR) gene mutations in Japanese subjects with short stature. *J Clin Endocrinol Metab.* 2011;96: E373–E378.
68. Pugliese-Pires PN, Fortin JP, Arthur T, Latronico AC, Mendonca BB, Villares SM, Arnhold IJ, Kopin AS, Jorge AA. Novel inactivating mutations in the GH secretagogue receptor gene in patients with constitutional delay of growth and puberty. *Eur J Endocrinol.* 2011;165:233–241.
69. Garcia EA, King P, Sidhu K, Ohgusu H, Walley A, Lecoeur C, Gueorguiev M, Khalaf S, Davies D, Grossman AB, Kojima M, Petersenn S, Froguel P, Korbonits M. The role of ghrelin and ghrelin-receptor gene variants and promoter activity in type 2 diabetes. *Eur J Endocrinol.* 2009;161:307–315.
70. Gueorguiev M, Lecoeur C, Meyre D, Benzinou M, Mein CA, Hinney A, Vatin V, Weill J, Heude B, Hebebrand J, Grossman AB, Korbonits M, Froguel P. Association studies on ghrelin and ghrelin receptor gene polymorphisms with obesity. *Obesity (Silver Spring).* 2009;17:745–754.
71. Liu B, Garcia EA, Korbonits M. Genetic studies on the ghrelin, growth hormone secretagogue receptor (GHSR) and ghrelin O-acyl transferase (GOAT) genes. *Peptides.* 2011;32:2191–2207.
72. Tao YX. Inactivating mutations of G protein-coupled receptors and diseases: structure-function insights and therapeutic implications. *Pharmacol Ther.* 2006;111:949–973.
73. Tao YX, Conn PM. Chaperoning G protein-coupled receptors: from cell biology to therapeutics. *Endocr Rev.* 2014;35:602–647.
74. Holst B, Holliday ND, Bach A, Elling CE, Cox HM, Schwartz TW. Common structural basis for constitutive activity of the ghrelin receptor family. *J Biol Chem.* 2004;279: 53806–53817.
75. Tao YX, Segaloff DL. Functional characterization of melanocortin-4 receptor mutations associated with childhood obesity. *Endocrinology.* 2003;144:4544–4551.
76. Wortley KE, del Rincon JP, Murray JD, Garcia K, Iida K, Thorner MO, Sleeman MW. Absence of ghrelin protects against early-onset obesity. *J Clin Invest.* 2005;115: 3573–3578.
77. Mo XL, Tao YX. Activation of MAPK by inverse agonists in six naturally occurring constitutively active mutant human melanocortin-4 receptors. *Biochim Biophys Acta.* 2013;1832:1939–1948.
78. Fan ZC, Tao YX. Functional characterization and pharmacological rescue of melanocortin-4 receptor mutations identified from obese patients. *J Cell Mol Med.* 2009;13: 3268–3282.
79. Tao YX, Huang H. Ipsen 5i is a novel potent pharmacoperone for intracellularly retained melanocortin-4 receptor mutants. *Front Endocrinol (Lausanne).* 2014;5:131.
80. Huang H, Tao YX. A small molecule agonist THIQ as a novel pharmacoperone for intracellularly retained melanocortin-4 receptor mutants. *Int J Biol Sci.* 2014;10:817–824.

CHAPTER SIX

Epigenetics of Obesity

A. Lopomo*[,†], E. Burgio[‡,§], L. Migliore*[,1]

[*] Department of Translational Research and New Technologies in Medicine and Surgery, Medical Genetics Laboratories, University of Pisa, Pisa, Italy
[†] Doctoral School in Genetics, Oncology, and Clinical Medicine; University of Siena; Siena, Italy
[‡] European Cancer and Environment Research Institute (ECERI), Bruxelles, Belgium
[§] ISDE International Society of Doctors for Environment (Scientific Office), Arezzo, Italy
[1] Corresponding author. E-mail address: lucia.migliore@med.unipi.it

Contents

Abstract

Obesity is a metabolic disease, which is becoming an epidemic health problem: it has been recently defined in terms of Global Pandemic. Over the years, the approaches through family, twins and adoption studies led to the identification of some causal genes in monogenic forms of obesity but the origins of the pandemic of obesity cannot be considered essentially due to genetic factors, because human genome is not likely to change in just a few years. Epigenetic studies have offered in recent years valuable tools for the understanding of the worldwide spread of the pandemic of obesity. The involvement of epigenetic modifications—DNA methylation, histone tails, and miRNAs modifications—in the development of obesity is more and more evident. In the epigenetic literature, there are evidences that the entire embryo-fetal and perinatal period of development plays a key role in the programming of all human organs and tissues. Therefore, the molecular mechanisms involved in the epigenetic programming require a new and general pathogenic paradigm, the Developmental Origins of Health and Disease theory, to explain the current epidemiological transition,

Progress in Molecular Biology and Translational Science, Volume 140
ISSN 1877-1173
http://dx.doi.org/10.1016/bs.pmbts.2016.02.002

that is, the worldwide increase of chronic, degenerative, and inflammatory diseases such as obesity, diabetes, cardiovascular diseases, neurodegenerative diseases, and cancer. Obesity and its related complications are more and more associated with environmental pollutants (obesogens), gut microbiota modifications and unbalanced food intake, which can induce, through epigenetic mechanisms, weight gain, and altered metabolic consequences.

1. INTRODUCTION

Obesity is a metabolic disease, which is rapidly increasing worldwide. The most recent data of World Health Organization state that more than one third of adults over 18 years are overweight (38% of men and 40% of women).[1] The worldwide prevalence of obesity nearly doubled between 1980 and 2014. In 2014, 11% of men and 15% of women worldwide were obese. Consequently, more than half a billion adults are now classified as obese. In all WHO regions, women are more likely to be obese than men. The prevalence of overweight and obesity is higher in the United States of America and lowest in South-East Asia. In Europe, Eastern Mediterranean regions, and United States of America, over 50% of women are overweight, and in all the three regions approximately half of overweight women are obese. The global prevalence of overweight and obesity in children under 5 years has increased from about 5% in 2000 to 6% in 2010, and 6.3% in 2013,[2] especially in Africa and Asia where the prevalence increased from 11 to 19% in some countries in southern Africa and from 3 to 7% in South-East Asia. It is estimated that the prevalence of overweight in children under 5 years is going to rise to 11% worldwide by 2025, if current trends continue.[2] There has been an increasing global acknowledgment of the need for effective strategies to prevent and control childhood overweight and obesity. In 2012, the World Health Assembly fixed the goal of no increase in childhood overweight by 2025.[3] Since obesity is an increasing problem in the world, we should try to better understand what are the causes of such a huge epidemiological change, recently defined in terms of Global Pandemic.[4] Some authors suggest that the current epidemic of obesity should be more properly considered as having a communicable rather than noncommunicable course, establishing a sort of "socially contagious feature of globalization."[5] Other authors observed that, in terms of evolutionary change, this is the first time that an entire species faces a dramatic change of its phenotype.[6] The main problem is that obesity is not just limited to one group of people or to one range of age, but it is prevalent in all ages and ethnic groups inducing an increase in other pathologies, such as type 2 diabetes (T2D), cardiovascular diseases, stroke, osteoarthritis, insulin resistance, high blood pressure, asthma, liver and respiratory diseases, reproductive problems, cancer.[7] Variety of cancers are associated to obesity (breast, kidney, colorectum, endometrium, esophagus, and pancreas cancers). In the United States, overweight and obesity could account for 14% of all deaths from cancer in men and 20% in women.[8]

In the past decades obesity has been considered the result of lack of balance between energy intake and expenditure, driven by an easy access to high-calorie food and reduced energy expenditure. From an evolutionary perspective the current pandemic should be essentially ascribed to the current adoption of a sedentary lifestyle, coupled with the high availability of foods with high caloric content.[9]

Over the years, the approaches through family, twins and adoption studies led to the identification of some causal genes in monogenic forms of obesity. To date, there are eight well-established monogenic obesity genes: leptin (*LEP*), leptin receptor (*LEPR*), proopiomelanocortin (*POMC*), prohormone convertase 1 (*PCSK1*), melanocortin 4 receptor (*MC4R*), single-minded homologue 1 (*SIM1*), brain-derived neurotrophic factor (*BDNF*), and neurotrophic tyrosine kinase receptor type 2 (*NTRK2*). Mutations in these eight genes are known to cause early onset of obesity and hyperphagia and may account for up to 10% of severely obese children.[10] Nevertheless, susceptibility variants identified in the last years, mainly through the GWAS approach, only explain a small part of the individual variation in risk, in common forms of obesity.[11] Anyway the rate of growth of the "epidemic" makes implausible the explanations essentially based on genetic changes, as genomes cannot change through mutations in just a few years.

In an evolutionary perspective the main questions are: how has natural selection favored the spread of genes that increase the risk for an obese phenotype and how has such a susceptibility (predisposition) to obesity evolved?[12]

The first hypothesis was advanced in the second half of the last century by the American geneticist J. Neel who proposed the "thrifty genotype hypothesis."[13] He theorized that during human evolution, genes who enabled individuals to efficiently collect and process food and deposit fat during periods of food abundance, would have acquired an important role for survival in adverse nutritional conditions. So the genes that promoted efficient fat accumulation would be extremely advantageous for primitive humans because they allow their owners to survive famine periods. Since in modern societies food is easily available, these genes have become disadvantageous resulting in the spread of obesity and diabetes.[14] Some studies identified genes under positive selection that have a role on obesity; for examples, a study suggested that the high frequency of the risk allele of the Gly482Ser variant in the *PPARGC1A* gene in Polynesians populations remains a thrifty allele in the Pacific populations.[15] Another study provided evidence for a positive selection of *TRIB2* gene, which influences visceral fat accumulation in East Asians.[16]

Neel's theory was quite simplistic and was widely criticized. The assumption that famines were common and severe enough to select thrifty genes during the 2.5 million years of human Paleolithic era was refuted by some anthropological evidence; the hypothesis had a critical point because if accumulating extra adipose tissue was advantageous in the past populations, many people with these thrifty genotypes in modern society do not develop the obese phenotype, despite the environmental change favoring fat storage. Moreover many of the populations that later developed high levels of obesity seemed to have no clear history of famine or starvation.[17–19] Even if the specific mechanisms proposed by Neel have been refuted, the basic assumption of his hypothesis remained relevant, and in subsequent years the "*thrifty phenotype hypothesis,*" derived from challenges posed to the thrifty gene hypothesis, was proposed.[20] In the early 1990s, Lucas had enunciated the hypothesis of the *fetal programming*, namely a genomic, adaptive, and predictive adjustment during the embryo-fetal period.[21]

At the same time, Hales and coworkers came to the hypothesis of the "*thriftyphenotype*" and of the fetal origin of adult diseases, according to which the embryo-fetal period would be critical to determine many chronic and degenerative diseases.[22,23] The two hypothesis of Neel (thrifty genotype) and Hales (thrifty phenotype) were complementary and claimed the same hypothesis of fetal programming; there is no doubt that the genome cannot radically change in a few decades and the genomic structures more adaptable to ever changing nutritional intake have been positively selected, moreover, conditions of poor nutrition, maternal fetal stress, and exposure to toxic substances are capable to interfere with the programming of organs and tissues, playing a complex pathogenetic role.

In the following years, in the light of epigenetic studies, it became evident that the entire embryo-fetal and perinatal period of development plays a key role in the programming of all human organs and tissues. The molecular mechanisms involved in this process as effects of the environmental signals on epigenetic programming can be at the basis of a new and general pathogenic paradigm, the Developmental Origins of Health and Disease (DOHaD) theory, able to explain the current epidemiological transition, consisting of the worldwide increase of chronic, degenerative, and inflammatory diseases such as obesity, diabetes, cardiovascular, neurodegenerative disease, and cancer.

It is now emerging that obesity and its related complications are also associated with other factors such as environmental pollutants (obesogens), gut microbiota, and specific nutrients, which can increase susceptibility to weight gain and to other metabolic consequences through epigenetic changes.

Thanks to epigenetics many pathogenic pathways described in the second half of the last century are starting to be elucidated throughout molecular mechanisms. In recent years, epigenetic studies have also offered tools for the understanding of the worldwide spread of the pandemic obesity.[24]

2. HOW EPIGENETIC CHANGES ARE RELATED TO OBESITY

Epigenetics is a term that was first introduced by developmental biologist Conrad Hal Waddington in 1942 and usually refers to heritable changes in genetic expression without changing the DNA sequence itself. The most important epigenetic mechanisms include DNA methylation, histone modification, and noncoding RNAs (ncRNAs) and can be passed on mitotically (through cell division) or meiotically (transgenerational inheritance). Environmental factors influence the epigenetic programming of parental gametes, fetus, and early postnatal development and may influence epigenetic programming so that also epigenetic regulation of tissue gene expression may be altered in response to environmental exposures throughout the life course into adulthood.[25] An example of obesity that can be modulated through an epigenetic mechanism is the animal model of the agouti mouse with the agouti viable yellow (A_{vy}) mutation. The agouti locus controls coat color in mouse, and a mutation leads to color change in yellow and to ectopic expression of the agouti protein, which binds to the melanocortin 4 receptor (MC4R) in the hypothalamus, and antagonistically disrupts its function, and induces hyperphagic obesity. Methylation controls the extent to which the gene is expressed; therefore, the resulting obesity phenotype depends on methylation.[26]

Dietary factors may influence epigenetic regulation of body weight, in fact, in agouti mice, methyl supplementation of the maternal diet influences methylation of the agouti viable yellow allele and offspring obesity.[27] Folate, required in DNA synthesis and repair, provide the pool of S-adenosylmethionine (SAM), the methyl group donor in DNA methylation and histone methylation reaction; it was observed that folate deficiency is associated to DNA hypomethylation in the agouti variable yellow mouse model, which shows a yellow coat color as result of folate deficient diet.[28] When pregnant mice were fed with methyl donor rich diets, the offspring had a higher rate of wild-type coat color compared with the control group treated with normal diet.[29] Interestingly, the effect of methyl donor on DNA methylation A_{vy} model is associated with an obese phenotype that improved during generations.[30,31] Moreover heterozygous agouti viable yellow mice (A_{vy}) exposed to the flavonoid genistein in utero via the maternal diet resulted in an altered coat color toward pseudo-agouti and conferred protection against obesity in adulthood, by altering the epigenome.[32] The protective effects were also observed in monkeys, showing improved body weight, insulin sensitivity, and lipid profiles by DNA methylation changes in liver and muscle.[33]

In humans, Prader Willi Syndrome (PWS) is the most well characterized example of obesity caused by an epigenetic mechanism.[34] PWS is due to a paternal deletion or uniparental disomy (both chromosomes from only one parent) of chromosome 15. In 25% of cases, this lack of expression is due to an epigenetic silencing on the maternal and paternal copy of the genes (instead of just the maternal copy). PWS is characterized by severe hyperphagia, infertility, mild learning and behavioral problems, and obesity, but the obese phenotype is not passed on due to infertility.[34] Many genes have been implicated in human obesity, and their effects may be modified through epigenetic mechanisms.[35] For example, mutations in both leptin (*LEP*) and proopiomelanocortin (*POMC*) cause human obesity, and both genes have CpG islands, which methylation can control their expression.[36]

3. EPIGENETICS STUDIES IN HUMAN AND ANIMALS

3.1 DNA Methylation

There are many important genes involved in the development of obesity and metabolic disorders, whose expression is under the influence of epigenetic regulation. Studies are beginning to appear on the assessment of changes in the pattern of methylation of specific genes related with obesity. DNA methylation represents one of the most important epigenetic mechanisms for the regulation of gene expression, and so far the most widely studied; it involves the covalent attachment of a methyl group to the 5' position of cytosine present in DNA sequence of genome by DNA methyltransferases (DNMTs). This methylation impedes the binding of transcription factors, which results in repression of a particular locus; hypermethylation of CpG islands is usually associated to transcriptional repression, whereas hypomethylation is associated to transcriptional activation. Change in global DNA methylation is considered a hallmark of many diseases, particularly in cancer, but the association between global DNA methylation and obesity is in general inconsistent. However, candidate gene methylation studies in animal models and humans have demonstrated methylation changes in promoters of various genes that are implicated in obesity, appetite control and metabolism, insulin signaling, immunity, inflammation, growth, and circadian clock regulation.[37–40] Specific DNA methylation patterns have been found to be associated with obesity in a genome-wide DNA methylation analysis in leucocytes and adipose tissue and in analyses of specific genes, such as the circadian clock genes (*CLOCK*: clock circadian regulator; *BMAL1*: aryl hydrocarbon receptor nuclear translocator-like; *PER2*: period circadian 2), whose methylation status in human leucocytes is associated with obesity.[41–43] The circadian clock system gives instructions about 24 h rhythmicity on gene expression in quite all cells, including adipocytes.[44] Peripheral blood leukocytes of obese people have hypermethylation in *UBASH3A* (ubiquitin-associated and SH3 domain-containing protein A) and *TRIM3* (tripartite motif containing 3) genes.[45]

A study reported an association of increased BMI in adults of European origin with increased methylation at the *HIF3A* (hypoxia-inducible factor 3a) locus in blood cells and in adipose tissue; this factor regulates a wide variety of cellular and physiological responses to reduced oxygen concentrations by controlling expression of many target genes. During obesity the hypoxia response has been reported, so this finding provides direct evidence in people that perturbation of HIF signaling that plays a key part in metabolism, energy expenditure, and obesity, could have an important role in mediation of the downstream adverse responses to increased BMI.[43] The association of *HIF3A* DNA methylation with weight and adiposity is detectable early in life, in fact higher methylation levels associated with greater infant weight and adiposity were observed, even if this was slightly smaller than reported for adult BMI.[46]

Leptin, a hormone produced by adipocytes in response to increased fat reserves, has a role in the regulation of the adipose-tissue; in diet-induced obese mice, the hypermethylation of CpGs at the leptin promoter, working together with histone modification, could play a feedback role in maintaining leptin concentrations within a normal range.[47] *LEP* promoter methylation was found associated in a tissue-specific manner with maternal (prepregnancy obesity, pregnancy smoking, and gestational weight gain) and infant factors (small for gestational age, LEP genotype and gender) in nonpathological pregnancies. In particular methylation of *LEP* was lower in infants born to prepregnancy obese mothers.[48] Maternal metabolic status before and during pregnancy and gestational diabetes mellitus can alter *LEP* placental DNA methylation profile at birth, potentially contributing to metabolic programming of obesity.[49] Not only *LEP* but also *ADIPOQ* epigenetic status is associated with obesity and there are evidences linking LDL-cholesterol levels with both *LEP* and *ADIPOQ* DNA methylation levels. LDL-cholesterol levels were found to be positively correlated with *LEP* DNA methylation levels in blood and tissue and with *ADIPOQ* DNA methylation levels in tissue.[50] *ADIPOQ* encodes adiponectin, the hormone that plays an important role in glucose regulation, inflammation, and fatty acid oxidation. It was observed that sleep fragmentation during late gestation, a common feature of women pregnancy, induces epigenetic modifications in *AdipoQ* in male offspring mouse adipocytes, indicating that altered gestational environments, provoked by sleep fragmentation, could cause long-lasting metabolic consequences in the next generation.[51]

An important candidate gene in obesity is *PGC1α* (peroxisome proliferator-activated receptor γ coactivator 1 α), which belong to a family of transcriptional coactivators involved in regulation of cellular energy metabolism including both carbohydrate and lipid metabolism; *PGC1α* promoter methylation in human umbilical cord and human muscle have been correlated to maternal pregestational BMI and to high-fat overfeeding, respectively.[52,53] IGF-2 (insulin-like growth factor 2) promotes the growth and division of various kinds of cells, its expression is specific for fetal tissue during development and only the paternal *IGF-2* allele is transcribed in normal human tissue, so a study observed that paternal obesity is associated with a decreased methylation levels of differentially methylated regions of *IGF-2*.[54] The methylation percentage of *IGF-2* promoter was higher in overweight infants than in lean infants.[55] On the contrary caloric restriction causes hypomethylation of *Igf2* locus in the adrenal glands of sheeps.[56]

IRS-1 (insulin receptor substrate 1), involved in insulin signaling, has altered methylation status in adipose tissue of subjects with type 2 diabetes compared with healthy controls, suggesting that its expression could be regulated by DNA promoter methylation.[57] Also the methylation status of genes encoding for insulin can be altered; is well known that insulin is an hormone secreted by pancreatic β cells in response to blood glucose levels, which expression starts in the early embryonic development of pancreas, so the demethylation of the promoter can have a role in pancreatic β cell maturation and tissue-specific expression.[58]

LY86 encodes for a protein, noted as lymphocyte antigen 86, also known as protein MD-1. One CpG site in *LY86* gene shows higher methylation in the obese and in the younger population; moreover, the methylation levels of this region were significantly correlated with adiposity indices, insulin resistance, and inflammatory markers, suggesting that *LY86* methylation may be involved in the (patho)physiological regulation of the innate immune system and inflammation.[59,60] In a mice model, it was noticed that MD-1 may contribute to high-fat, diet-induced obesity, adipose tissue inflammation, and insulin resistance.[61]

Recently a study on obesity in parents before conception in relation to DNA methylation patterns at multiple human imprinted genes critical for normal growth and development was performed. Paternal obesity was significantly associated with lower methylation levels at the mesoderm-specific transcript gene (*MEST*), paternally expressed gene 3 (*PEG3*) and neuronatin gene (*NNAT*). Changes in methylation levels related to maternal obesity were instead detected at pleiomorphic adenoma gene-like 1 (*PLAGL1*) and at the maternally expressed gene 3 (*MEG3*).[62]

Two appetite-regulatory genes, studied in weight regain process, showed altered methylation, particularly weight regainers showed higher methylation patterns than nonregainers in *POMC* and lower levels on neuropeptide Y (*NPY*). Total baseline *NPY* methylation was associated with weight-loss regain, baseline plasma ghrelin levels and leptin/ghrelin ratio; lower methylation levels in *NPY* promoter were associated with higher risk of weight regain while lower methylation levels of *POMC* were associated with weight loss maintenance.[63]

In order to analyze the effects of dams fed a high-fat, high-sucrose diet during pregnancy and lactation and weaned to high-sucrose diet, a study observed offspring of high-sucrose diet had heavier body weight, impaired glucose tolerance, decreased insulin sensitivity, higher serum LEP level, and hypomethylation of *POMC* promoter in the hypothalamus. Not only the expression of *POMC* but also of *MC4R* were significantly increased in offspring exposed to high-sucrose diet during gestation, lactation, even if *MC4R* was not affected by altered methylation levels.[64] In nonmated females rats, high fat diet decreased Pomc/leptin ratio, associated with *Pomc* promoter hypermethylation, while during lactation high-fat diet dams showed body weight decrease and loss of *Pomc* promoter hypermethylation. Their offspring displayed increase of body weight, and *Pomc* promoter hypermethylation; this demonstrates a long-term effect of maternal high fat diet on CpG methylation of the *Pomc* promoter in the offspring.[65] A study investigated the DNA methylation of CpG dinucleotides in the proximal promoter region of *Npy* and *Pomc* genes in order to investigate the onset mechanism of obesity in newborn rats reared with high carbohydrate milk. Altered methylation of specific CpG dinucleotides proximal to the transcription start site was observed for the *Npy* gene in the hypothalami of 16 and 100 day old high carbohydrate rats compared with their methylation status in mother fed rats.[66]

To assess the effects of excessive weight and obesity on gene-specific methylation levels of promoter regions, a study analyzed methylation status of genes involved in inflammation and oxidative stress, such as *IL6* (interleukin 6), *TNF* (tumor necrosis factor), *TFAM* (mitochondrial transcription factor A), and *GLUT4* (glucose transport 4) in blood cell-derived DNA from healthy women volunteers with a range of body mass indices (BMIs). They observed that samples from obese individuals, with $BMI > 30\ kg\ m^{-2}$, showed significantly increased hypermethylation only for *IL6* gene compared to normal weight and overweight samples, suggesting that analysis of aberrant DNA methylation of *IL6* gene promoter could be useful as molecular biomarker for obesity risk assessment.[67]

The importance of epigenetic regulation in obesity is also shown by the study of Wang and coworkers in an in vitro model (a murine cell line with MeCP2 specifically deleted in POMC neurons, which regulate energy homeostasis, in response to leptin signaling). The researchers demonstrated that MeCP2 positively regulates *POMC* expression in the hypothalamus. Absence of MeCP2 in POMC neurons leads to increased DNA methylation of the *POMC* promoter, which induces *POMC* expression downregulation and lead to obese mice showing a high degree of leptin resistance.[68]

All these studies are important in the search of epigenetic biomarkers to indicate obesity status or risk to develop obesity since the early months of life. For example, a high *POMC* methylation status in cord blood was associated with lower birth weight, and children with higher *POMC* methylation in cord blood showed higher triglycerides and insulin levels in blood. Thus, *POMC* methylation status in cord blood may be an early predictive marker of metabolic syndrome.[69]

4. HISTONE MODIFICATIONS

Histone proteins, at which DNA wrapped around, are made of a globular domain and an N-terminal tail domain. Modification, such as acetylation, methylation, phosphorilation, sumoylation, ubiquitination, at N-terminal amino acids of histones, facilitate the entry of various transcriptional factors, which activate or repress gene expression, for example, histone acetylation activates gene expression, while histone methylation can activate or repress the state of chromatin depending on the specific lysine involved. These histone modifications are involved in the epigenetic regulation of adipogenesis and can play an important role in obesity development. Modulation of five key regulatory genes of adipogenesis, preadipocyte factor-1 (*Pref-1*), CCAAT-enhancer-binding protein β (*C/EBPβ*), *C/EBPα*, *PPARγ*, and adipocyte protein 2 (*aP2*), is regulated by histone modifications during adipocyte differentiation.[70] Investigation of histone tail modifications on hypothalamic chromatin extracts from 16 day old rats indicated decreased acetylation of lysine 9 in histone 3 (H3K9) for the *Pomc* gene and increased acetylation for the same residue for the *Npy* gene, without changes in histone methylation (H3K9) in both genes in high-carbohydrate rats. These epigenetic modifications could contribute to the altered gene expression of the *Npy* and *Pomc* genes in the hypothalamus of high-carbohydrate rats and could be a mechanism leading to the development of obesity.[71] Increase of histone H3 lysine 9 and 18 acetylation at *Tnfa* (tumor necrosis factor α) and *Ccl2* (monocyte chemotactic protein 1) genes in the liver of obese mouse treated with high fat diet observed[71], while caloric restriction increased histone H4 acetylation of *Glut4* gene in adipose tissue of obese mice, causing an increased expression of gene.[72]

Enzymes involved in histone modification can have a role in obesity. In response to high-fat diets and fasting, the medial hypothalamus changes the expression of neuropeptides regulating feeding, metabolism, and reproductive behaviors, regulating also the expression of histone deacetylases (HDACs), involved in the epigenetic control of gene expression in response to a variety of environmental factors. A study observed that mice treated for four weeks with a high-fat diet resulted in the increased expression of HDAC5 and

HDAC8, such as fasting induced increase of HDAC3 and HDAC4 expression levels and decrease HDAC10 and HDAC11 levels.[73] The H3K9-specific demethylase, Jhdm2a (also known as Jmjd1a and Kdm3a), has an important role in nuclear hormone receptor-mediated gene activation and male germ cell development, its disruption is associated to obesity.[74]

5. miRNA

Recent findings indicate that microRNAs (miRNAs) are involved in the regulatory network of adipocyte differentiation and obesity development. Adipogenesis is partially regulated by several adipocyte-selective miRNAs and transcription factors involved in the regulation of proliferation and differentiation of human adipose-derived mesenchymal stem cells (hMSCs-Ad). The levels of miR-148a, miR-26b, miR-30, and miR-199a increased during the differentiation of hMSCs-Ad, in particular expression levels of miR-148a increase in adipose tissues from obese people and mice fed with high fat diet. Ectopic expression of miR-148a accelerated differentiation and partially rescued inhibition of adipogenesis. These observations indicate miR-148a as a biomarker of obesity in human subjects and mouse model, thereby promoting adipocyte differentiation.[75]

Another miRNA, miR-26b, is involved in the adipogenesis process, in fact, a study found that expression of the miR-26b was increased in mature adipocytes and was gradually upregulated during adipocyte differentiation. By loss-of-function approach to silence miR-26b stably in human preadipocytes, it was found that miR-26b inhibition suppressed adipocyte differentiation, as not only a decrease of lipid droplets was found, but also a reduced ability of this miR to decline mRNA levels of adipocyte-specific molecular markers and triglyceride accumulation.[76] Regarding the role of miR-26b in the proliferation of human preadipocytes, a study observed human preadipocytes that overexpress miR-26b exhibited increased triglyceride content in the adipocytes and upregulation of the protein expression levels of adipogenesis-associated marker genes, such as *PPAR-γ* (peroxisome proliferator activated receptor γ), during differentiation. Human preadipocytes, overexpressing miR-26b, show also a slow rate of growth and remain in the G1 phase, in fact miR-26b downregulate the protein expression of cyclin D2.[77] These data indicate that miR-26b promotes differentiation, attenuates cell proliferation, may be involved in adipogenesis and could be targeted for therapeutic intervention in obesity.[76,77] MiR-200a, miR-200b, and miR-429 are upregulated in the hypothalamus of genetically obese and leptin deficient ob/ob mice. The treatment with leptin downregulates these miRNAs in ob/ob hypothalamus and the hypothalamic silencing of miR-200a increases the expression level of *LEPR* (leptin receptor) and *ISR-2* (insulin receptor substrate 2). These observations, linking the alteration of leptin and insulin signaling to the upregulation of hypothalamic miR-200a, suggest that miR-200a could be another target for the treatment of obesity.[78] Indeed, miRNA-130b inhibits adipogenesis and lipogenesis and reduces fat deposition in recipient adipocytes by targeting *PPAR-γ*.[79]

Diet could influence miRNA expression, for example, miRNA-168a, a miRNA abundantly found in rice, binds to human and mouse low-density lipoprotein receptor adaptor protein 1 (LDLRAP1) mRNA and inhibits its expression in the liver, which in turn results in elevated plasma LDL-cholesterol levels. This observation provides a link between diet and epigenetic regulation of metabolism.[80]

6. EPIGENETIC EFFECTS OF MATERNAL AND PATERNAL DIET ON FETUS DEVELOPMENT

Metabolic disease results from a complex interaction of many factors, including genetic, physiologic, behavioral, and environmental influences. The worldwide shift toward an obese phenotype, occurred in a period of one or two generations, suggests that environment, behavioral effects or epigenetic factors play a role in obesity; some evidences support the idea that obesity may have its origins in utero and may be evident in multiple generations, in fact, is clearly evident that the intrauterine environment influences development of the fetus.[81] Plasticity during gestational development is crucial for adapting the fetus to its anticipated environment. A nutrient-deprived intrauterine environment influences offspring body fatness in animals, and the effect may be exaggerated if the offspring are exposed to a high-fat diet postweaning.[82] The confirmation of this hypothesis derives from two large epidemiological studies.[83,84] The studies were performed on two large cohorts of subjects exposed in utero to serious nutritional deficits during the Second World War, who later lived in totally opposite conditions, returning to normal nutrition in the case of the Dutch cohort exposed to the so-called Dutch Famine in 1944, and, conversely, persisting conditions of poor nutrition in the case of children who survived the dramatic siege of Leningrad. In the case of Dutch cohort, offspring of mothers exposed to famine during the first trimester had, six decades later, less DNA methylation of the imprinted *IGF2* gene compared with their unexposed same sex siblings. In the case of Russian cohort, survivors of the starvation during the sieges of Leningrad and Stalingrad showed an incidence of chronic diseases much lower than in Dutch survivors.[84,85] This behavioral difference could be explained in that Russian children, programmed for a life characterized by stress and nutritional deficiencies had been able to better face their difficult lives; whereas the Dutch children, programmed in the same way, had enjoyed a better postnatal life and a diet richer than expected and, although at the beginning recovered a good weight, then they were sick because of the mismatch between their programming and the relatively rich diet they had in their adult life. So the idea of *fetal programming*, proposed by Hales and Baker in the 1990s, it has now been accepted worldwide. The results of these studies contributed to the understanding of the molecular mechanisms underlying the epigenetic programmatic mismatch, concerning the organs involved in metabolic organization in subjects exposed in utero and postnatal life to quite different environmental and nutritional conditions. This data reinforcing the idea that very early mammalian development is a crucial period for establishing and maintaining epigenetic marks.[86] It would also seem that the type of childbirth may influence obesity; Cesarean section delivery appears to be a risk factor for turning obese[87,88] and evidence from observational studies suggests such risk is stronger at early ages.[88–90] Cesarean section and antibiotic use during pregnancy may alter normal maternal-offspring microbiota exchange, thereby contributing to aberrant microbial colonization of the infant gut and increased susceptibility to obesity later in life; a study observed cesarean section and exposure to antibiotics in the second or third trimester were associated with higher offspring risk of childhood obesity.[91]

In utero, nutrition, environmental exposures and other factors may permanently alter offspring gene expression via epigenetic mechanisms and hence alter the structure and function of cells and organs leading to metabolic abnormalities.[92]

As anticipated above, suboptimal nutritional conditions of parents can increase risk of metabolic disorder when offspring become adulthood. Maternal status is fundamental for baby health, for his life-long conditions, and also for the maintenance of methylation patterns, in fact, maternal nutrition before and during pregnancy may affect the establishment of CpG methylation and the life-long expression of epigenetically modified alleles.[93,94] Maternal prepregnancy obesity and increased weight gain during pregnancy are associated with higher birth weight newborns and an increased risk of obesity and diabetes in later life;[95,96] this observation suggests that during the developmental period, fetuses and neonates are vulnerable to alteration of maternal nutrition.[92]

The most sensitive time windows for the developmental programming of adiposity seem the gestation and lactation periods. During these stages, plasma levels of circulating factors, as well as adipose tissue hormone sensitivity, show perturbations in the offspring of females suffering from malnutrition, resulting in enduring adipose tissue programming.[97] In effect, many studies have reported the role of caloric restriction, total energy intake, and diet composition during development and lactation to have a major impact on epigenetic modulation of obesity development.[98] Godfrey and coworkers performed a study on two prospective cohorts by using DNA extracted from umbilical cord tissue obtained at birth in children who were assessed for adiposity nine years later to measure methylation status level in the promoters of candidate genes.[99] Methylation of *RXR-A* (retinoid X receptor-α), and *eNOS* (endothelial nitric oxide synthase) was found to correlate with higher adiposity in later childhood.[99] These observations confirm that modification of epigenetic marks may be crucial in fetal programming of later obesity.

Also the duration of breastfeeding is associated to lower *LEP* methylation in very young children. Breast milk contains unique growth factors and signaling molecules, as well as leptin. Lower methylation of *LEP* leads to increased expression and higher concentrations of leptin that, interacting with almost all neuropeptides that are involved in the regulation of energy balance and food intake, play an important role in the programming of metabolic pathways.[100] Furthermore, some evidence suggests that intake of a sufficient amount of leptin from breast milk prevents obesity in adult life.[101]

Caloric restriction and low-protein diets have been reported to induce epigenetic changes and metabolic modification that persist later in development. Animal models of maternal malnutrition have highlighted that maternal low-protein diet results in global DNA hypomethylation in the liver of rat fetus,[102] and in hypomethylation of hepatic *GR* (glucocorticoid receptor) and *PPAR-α* (peroxisome proliferator-activated receptor α), influencing carbohydrate and lipid metabolism in juvenile and adult offspring.[103] A low protein diet is associated with impaired fetal growth, the development of obesity, insulin resistance and diabetes in the offspring.[104] Maternal protein undernutrition during pregnancy and lactation could lead to progressive epigenetic silencing at the enhancer region of *Hnf4a*, which weakens the promoter-enhancer interaction and results in a permanent reduction in *Hnf4a* expression,[105] and to DNA hypomethylation of promoter *Lep* in adipose tissue of Balc/c mice.[106] Studies have been carried up in which intrauterine growth restricted (IUGR) offspring was induced by maternal undernutrition or uterine artery ligation. In experimental models of IUGR, it was observed a decrease in postnatal *IGF1* mRNA variants, H3 acetylation and the gene elongation mark histone 3 trimethylation of lysine 36 of the *IGF1* gene (H3Me3K36) in the rodents,[107,108] and changes in hepatic *IGF1* mRNA expression and histone H3K4 methylation in the rat.[109] It was also noticed a decreased expression of the Pdx1 (pancreatic duodenal homeobox factor-1), a transcription factor that regulates pancreatic development, whose reduced activity was associated with alterations in histone modifications in rats.[110] Offspring of mice fathers fed with low protein diet exhibited in elevated hepatic expression of many genes involved in glucose metabolism and lipid biosynthesis, including a substantial increase in methylation at an intergenic CpG island of the *Pparα* gene.[111]

All these epigenetic effects induced by caloric restriction and low protein diet could be transmitted to next generations with probability to cause metabolic disease, in fact, a study observed that in utero undernourishment environment of F1 embryos alters the germline DNA methylome of F1 adult males in a locus specific manner, particularly differentially methylated regions are hypomethylated and enriched in nucleosome-retaining regions.[112]

Also energy excess in maternal diet is associated with an altered metabolic phenotype and can elicit epigenetic changes in offspring. In animal models high fat diets affect DNA methylation at metabolic-related genes, such as melanocortin receptor 4 in mouse brain and leptin in rat adipose tissue.[113,114] Maternal high fat diet induces hypermethylation of *Pomc* promoter and obesity in postweaning rats linking maternal fat intake to altered epigenetic regulation of genes involved in polyunsaturated fatty acid synthesis;[115] changes methylation and gene expression of dopamine and opioid-related genes, altering appetite regulation and inducing preference for energy dense foods in postnatal life;[116] alters hepatic metabolism in the neonate in a sex-specific manner;[117] modulates fetal sirtuin 1 (SIRT1) histone and protein deacetylase activity in nonhuman primates.[118]

A study on pigs demonstrated that maternal diets with high and low protein contents effectively interferes with global DNA methylation through changes in DNMT1, DNMT2, and DNMT3 levels in the liver and skeletal muscle of newborn offspring.[119]

Overnutrition during the suckling period results in increased body weight gains, hyperphagia and adult-onset obesity as well as increased levels of serum insulin, glucose, and leptin. A study observed that overnutrition during suckling period leads to hypermethylation of specific CpG sites in the proximal promoter region of the *Irs1* and *Glut4* genes, which correlates with the reduction in Irs1 and Glut4 mRNA levels in skeletal muscle of adult rats, suggesting that epigenetic modifications in genes involved in the insulin signaling pathway could result in the development of insulin resistance in skeletal muscle of rats.[120] In an ovine model of maternal obesity fetal muscle miRNA expression resulted altered and let-7 g downregulation may enhance intramuscular adipogenesis during fetal muscle development.[121]

Similarly, alterations in paternal diet are also associated with altered DNA methylation in the offspring, particularly paternal diet induces impaired glucose metabolism and insulin homeostasis in offspring through epigenetic mechanisms. *Il13ra2* (IL13 receptor-α2) promoter was hypomethylated in female offspring after high fat feeding of their fathers. Il13ra2 is part of the Jak-Stat signaling pathway, modulates growth and invasion of various pancreatic cancer cell lines and is upregulated by TNF-α.[122]

Changes in maternal assumption of nutrients involved in folate metabolism, such as folic acid, choline, vitamin B12 are associated to epigenetic changes. Increase of maternal folic acid content in the rat can induce differential changes in mRNA expression and promoter methylation of *Pepck* (phosphoenolpyruvate carboxykinase).[123] Choline, required for successful completion of fetal development and involved in one-carbon transfer could have a role during pregnancy, in fact, maternal choline supply in the rat modifies fetal histone and DNA methylation.[124]

Also in human studies maternal micronutrient levels were proved to be essential for the one-carbon metabolism involved in DNA methylation and an imbalance in these nutrients can influence DNA methylation patterns in offspring. Regarding vitamin B12, its deficiency can result in global hypomethylation, as B12 is required for the synthesis of methionine, from homocysteine, and SAM, which is the methyl donor required for the maintenance of methylation patterns in DNA. Increased maternal vitamin B12 levels during pregnancy are associated with decreased global DNA methylation in newborns while increased serum B12 levels in newborns are associated with reduced methylation of the *IGFBP3* gene, involved in intrauterine growth.[125]

Correlations between maternal gestational diabetes and a phenotype of diabetes and obesity in offspring have been investigated. In a rat model, an induced phenotype of maternal hyperglycemia in the third trimester of pregnancy induces a programmed phenotype of glucose intolerance and impaired insulin secretion in F1 offspring and similar consequences transmitted to the F2 generation.[126] A study investigated genome-wide epigenetic variation in fetal tissues exposed to maternal gestational diabetes. By analysis of cord blood and placenta, they observed there are methylation variable positions achieving genome-wide significance with methylation differences. Pathway analysis suggests that gestational diabetes exposure may have functional impact on placental endocytic processes, increasingly understood to play a role in fetal growth and metabolism, and multiple other extra- and intracellular signaling pathways involved in growth and metabolism.[127]

7. TRANSGENERATIONAL EFFECTS

The studies mentioned above support the DOHaD theory, which focuses on the association between perinatal nutrition and the onset of diseases in the adult, such as obesity, insulin resistance, impaired glucose tolerance, and type 2 diabetes.[104] DOHaD suggests that the fetus adapts in response to cues from the intrauterine environment, adjusting the homeostatic systems in order to survive but these adaptations may be disadvantageous in postnatal life, leading to an increased risk of chronic diseases in adulthood and the inheritance of risk factors passing across generations.[81] Epigenetics is likely to be an important molecular basis of malnutrition during early life and glucose metabolism disorders in later life.[104] Studies in twins evidenced that differences in the environmental exposure lead to different epigenetic patterns in the somatic tissues of individuals; twins in which DNA methylation and histone acetylation patterns diverged more strongly in adulthood were those with more marked life history differences.[128] Developmental programming should be seen as a transgenerational phenomenon and is therefore often viewed as a form of epigenetic inheritance, either via the maternal or paternal line. Transgenerational epigenetic transmission of traits allows future generations to be maximally competitive in their environment.[129] So adaptive gene programs acquired during the parental lifespan persist in the subsequent generation, enabling future generations to better survive in a potentially adverse environment. Evidence suggests that environmental exposures, such as poor early life nutrition, result in maladaptive parental responses that can be passed to offspring. These epigenetic traits have the potential to result in a population-wide manifestation of a phenotype over several generations and might explain the rapid diffusion of obesity. Recent studies have shown that newborns of obese parents have altered DNA methylation patterns at imprinted genes and paternal obesity has been shown to be associated with *IGF2* hypomethylation in newborns.[130,131] Mouse models of diet-induced obesity have also been used to investigate the effect of male obesity on embryo quality showing that paternal obesity has significant negative effects on the embryo at early developmental stages, resulting in delayed development, reduced placental size, smaller offspring, and reduced the cleavage rate of zygotes.[132–134] The association between paternal obesity and the offspring's methylation status suggests the susceptibility of the developing sperm for environmental insults.[130] Diet-induced paternal obesity modulated sperm microRNA content and germ methylation status which are potential signals that program offspring health and initiate the transmission of obesity to future generations.[134] Paternal obesity was shown to initiate metabolic disturbances in two generations of mice albeit with incomplete penetrance to the F2 generation. Studies in F1 sperm have suggested a role for altered *IGF2* and *H19* expression in transmission of a phenotype to the F2 offspring.[135] However, not all studies reporting a paternal line transmission have reported epigenetic alterations in the F1 sperm.[136] In a Drosophila model Ost and co-workers showed that as little as 2 days of dietary intervention in fathers (a high-sugar diet causing an increase in their triglyceride content) elicits obesity in offspring. From a transcriptome analysis of embryos generated from fathers fed the high-sugar diet, a dysregulation of transcripts encoding two proteins—one of them is called Su(var)—that are involved in the dynamic shaping of chromatin conformation, was detected. This dysregulation involved down-regulation of enzymes known to change chromatin structure and gene regulation.[137]

Maternal obesity adversely affects oocyte quality, embryo development, and the health of the offspring. Maternal obesity could have negative effects on oocyte quality and the embryo development of the offspring altering the DNA methylation status of imprinted genes and metabolism-related genes. It was showed that DNA methylation patterns of several metabolism related genes are changed in oocytes of obese mice and in oocytes and liver of their offspring, particularly the DNA methylation level of the *leptin* promoter was increased and the *Ppar-α* promoter was reduced; this alterations was also observed in the liver of female offspring from dams fed the high-fat diet.[138]

8. OBESOGENS

In the ambit of epigenetics of obesity, it is important to analyze the role that endocrine disrupting chemicals may play interfering with the body's adipose tissue biology, endocrine hormone system, central hypothalamic-pituitary-adrenal axis, and major homeostatic mechanisms involved in weight control.

The concept of their involvement in obesity was introduced in 2006 by Grun and Blumberg that formulated the obesogen hypothesis, according to which a part of the cause of obesity epidemic would be the spread in the environment, especially in food chain, of xenobiotics able to act as endocrine disruptors, mainly during fetal programming;[139] they could act in different ways, such as promoting hyperplasia of the adipocyte pool, facilitating adipogenic pathways that activate the hyperplasia during periods of increased physiological development, perturbing the lipid homeostasis.[140] Already in 2002, Paula Baillie Hamilton noticed the coincidence in time between the beginning of the obesity epidemic and the worldwide spread of a large number of new industrial chemicals suggesting that endocrine disruptors chemicals, such as bisphenol A, organophosphate pesticides, bisphenyls and polybrominated bisphenyls, phthalates, heavy metals, and solvents, could have damaged many of the body's natural weight-control mechanisms.[141]

Obesogens, such as heavy metals, solvents, pesticides, PCBs, organic phosphate, phthalates, organotins, diethylstiberol, can be defined from a functional point of view as chemicals that promote adipogenesis by perturbing various endocrine axis, generally targeting nuclear receptors and affecting directly or indirectly adipocytes physiology and more generally the regulation of energy homeostasis.[142–144] Some factors, such as sites of actions, dose-response curves, timing of exposure, gender, genetic susceptibility, prevent the full understanding of the role that chemical obesogens play. They include either mimetic substance of lipophilic hormones or inhibitors of endogenous hormones so the action sites are diverse and the interaction complex, especially for compounds that have multiple molecular targets; the dose-response curves are not monotonic, since it is becoming clear that when it occurs, the effect of low doses cannot be predicted on the basis of the effects exerted at high doses; exposure during fetal development represents a windows of heightened sensitivity where long term effects can be established only in a small fraction of the population.[144,145]In utero exposure to endocrine-disrupting compounds, including dichlorodiphenyltrichloroethane (DDT) and its metabolite dichlorodiphenylethylene (DDE), increases the risk of obesity at 9 years of age in boys but not in girls, providing support for the chemical obesogen hypothesis.[146] The concentration of persistent organic pollutants (POPs), other endocrine disruptors, accumulates in adipose tissue, correlates with gene expression of obesity marker genes, such as leptin and adiponectin, suggesting that POPs are able to influence the association between obesity and the development of associated pathologies.[147] Although diethyl-hexyl-phthalate (DEHP) did not show any effect on adipocyte differentiation in the murine 3T3-L1 cell model, it induced the expression of transcriptional factor, for example, PPARγ, and downstream target genes required for adipogenesis in vivo.[148] Epigenetic changes EDC-mediated could play a role in the developmental origins of obesity. Some studies focused on changes in DNA methylation after developmental EDC exposure. Exposure of the murine 3T3-L1 preadipocyte cell line to a variety of EDCs results in enhanced differentiation of adipocytes in vitro accompanied by global DNA hypomethylation.[149] Cultured myotubes, when exposed to palmitate or oleate, show hypomethylation of the *PGC1α* promoter.[150] Even if we are still far from a complete knowledge of epigenetic changes induced by obesogens, in the last years their potential long lasting and transgenerational effects are becoming clear.[151] Bisphenol A (BPA), a xenoestrogen lipophilic compound, found in food, beverage containers, baby bottles, and dental materials, accumulates into adipose tissue; it was recently demonstrated that people with high BPA plasma values presented markers of low grade inflammation, higher visceral adiposity and higher prevalence of metabolic syndrome and insulin resistance.[152] BPA was associated to hypomethylation in Agouti mouse, particularly, it induces hypomethylation and consequently increased expression of the Agouti gene in prenatally exposed mice, which at birth had yellow coating rather than brown and the tendency to develop obesity, diabetes, and tumors.[32] Moreover, agouti mice females were more likely to have offspring with the same phenotype in the following generation. This experiment evidences that prenatal exposure to synthetic estrogen agonists can interfere with epigenetic marks, thereby leading to endocrinological consequences.[153]

Higher levels of prenatal exposure to polycyclic aromatic hydrocarbon (PAH) have been associated with childhood obesity in epidemiological studies. A study observed offspring of dams exposed to greater PAH during gestation had increased weight, fat mass, as well as higher gene expression of

PPARγ, *Cox2*, and adiponectin and lower DNA methylation of *PPARγ*, this was extended through the grand-offspring mice.[154]

Moreover, in order to test the hypothesis that environmental toxicants, acting as PPARγ agonists, influence adipogenesis and osteogenesis, Watt and coworkers observed that all the toxicants tested, such as organotins tributyltin and triphenyltin, a ubiquitous phthalate metabolite and two brominated flame retardants, activated PPARγ1 and 2, increased adipogenesis, and suppressed osteogenesis.[155]

Widely diffuse environmental compounds such a mixture of plastic derived compounds, BPA and phthalates, and a hydrocarbon mixture involving jet fuel (JP-8) can promote epigenetic transgenerational inheritance of adult onset diseases, including obesity. F0 generation female rats was exposed during pregnancy to a plastic mixture (BPA, DEHP, and DBP, dibutyl phthalate) in a period ranging from 8 to 14 days of gonadal sex determination in the embryos. From the analysis in F1 and F3 generation rats of the incidence of adult onset disease, a significant increase in the incidence of total disease/abnormalities in both male and female rats of F1 and F3 generations was observed. In particular, in the F3 generation, pubertal abnormalities, obesity, testis disease, and ovarian disease were increased and in F1, kidney and prostate diseases were increased frequently. Analysis of the plastics lineage F3 generation sperm epigenome identified 197 differential DNA methylation regions in gene promoters, termed epimutations, part of which correlate with the pathologies identified.[156] In another experiment, female rats were exposed to a hydrocarbon mixture involving jet fuel (JP-8) during the fetal gonadal development period. The F1 generation showed an increased incidence of kidney abnormalities in both females and males, prostate and pubertal abnormalities in males, and primordial follicle loss and polycystic ovarian disease in females. The jet fuel lineage had an increased incidence of primordial follicle loss and polycystic ovarian disease in females as well as obesity in both males and females also in the first transgenerational generation (the F3 generation). Moreover, analysis of the F3 generation sperm epigenome identified 33 differentially methylated DNA regions.[157] Similarly, exposition of F0 generation to DDT induces obesity and sperm epimutations in F3.[158]

Also elements contained in food can alter epigenetic mechanisms, for example, N-3 polyunsaturated fatty acids (n-3 PUFAs) are negatively associated with body leptin levels and reduce the expression of leptin. n-3 PUFAs may affect epigenetic processes because methyl groups are required for the metabolism of docosahexaenoic acid (DHA). In the adipose tissue of diet induced obese (DIO) mice, methylation of the CpG island and the binding of methyl-CpG-binding domain protein 2 (MBD2) and DNA methyltransferases (DNMTs) at the leptin promoter are increased and RNA Pol II is decreased. Additionally, histones H3 and H4 are hypoacetylated, lysine 4 of histone H3 (H3K4) is hypomethylated and the binding of histone deacetylases (HDACs) 1, 2 and 6 increased at the leptin promoter in the DIO mice. These modifications may serve a feedback role to maintain leptin concentrations within a normal range.[47]

Even if much remains to be discovered about molecular mechanisms activated by environmental obesogens, taken into account the already existing data on the effects of obesogens, and the multiple potential targets with which they might interfere daily, it seems likely that the exposure to obesogens can have an important role in the obesity pandemic.

9. EPIGENETIC CHANGES IN RESPONSE TO DIETARY INTERVENTION

Therapeutic strategies for counteracting excess body weight, which are able to remodel DNA methylation profiles concomitant with the reduction of body weight and the DNA methylation patterns, could be useful as biomarkers to predict responsiveness to caloric restriction in obese people. A study involving healthy sedentary men and women exposed to an acute bout of exercise showed a decrease in whole genome methylation and hypomethylation on the promoter of the *PGC1α*, *PDK4*, *PPARδ*, *TFAM*, and *MEF2A* genes associated with exercise-induced dose-dependent expression.[159] Genome-wide DNA methylation patterns of human adipose tissue that potentially affect adipocyte metabolism are changed after a 6-month intervention of exercise.[160]

A multidisciplinary intervention in obese or overweight adolescents revealed differential methylation levels in *AQP9* (aquaporin 9), *DUSP22* (dual specific phosphatase 22), *HIPK3* (homeodomain-interacting protein kinase 3), *TNNT1* (troponin T type 1), and *TNNI3* (troponin I type 1) between high and low responders.[161] The DNA methylation and expression levels of several genes, which are related to metabolic processes and mitochondrial functions, such as *PGC1α* and *PDK4* (pyruvate dehydrogenase kinase, isozyme 4), are altered in the skeletal muscle of obese people and after Roux-en-Y gastric bypass (RYGB), a type of weight-loss surgery, are normalized to levels observed in normal-weight, healthy controls.[162] The baseline methylation levels of *CLOCK* and *PER2* genes correlated with the degree of weight loss after treatment, suggesting that methylation of *CLOCK* and *PER2* could be used as biomarkers of weight-loss success.[163] Other potential epigenetic biomarkers of weight loss after an energy restriction intervention are the *ATP10A* and *WT1* genes.[164] Promoter methylation of *TNF* (tumor necrosis factor-α) gene could be involved in the predisposition to lose body weight after following a balanced hypocaloric diet.[165]

About weight regain responsiveness after weight loss intervention, it was observed that weight gainers had higher methylation levels in *POMC* gene in leucocytes and lower methylation levels at the promoter of the *NPY* gene than did nongainers; so epigenetic regulation of *NPY* and *POMC* may be used as biomarkers for predicting weight regain after dieting.[63]

These studies suggest that exercise and multidisciplinary intervention in humans could alter the DNA methylation status of specific genes and these changes may be used as epigenetic markers to predict the weight loss response in obese humans.

A dietary supplementation with apple extracts rich in the polyphenols chlorogenic acid, prevented body weight gain and ameliorated hyperglycaemia, hyperleptinaemia, and insulin resistance in rats fed a high-fat sucrose diet for 8 weeks. These results were associated with decreased methylation of two CpG sites in the leptin promoter of rat epididymal adipocytes.[166] Polyphenols and other plant compounds are considered as potential therapeutic agents to treat obesity-mediated inflammation and oxidative stress. EGCG, genistein, curcumin, and resveratrol, act through epigenetic mechanisms and have been demonstrated to trigger the antiinflammatory machinery and ameliorate some of the symptoms accompanying metabolic syndrome.[167]

10. CONCLUDING REMARKS

Our genome cannot be changed in short times to predispose entire world to obesity, so environment may have interacted with the genome to influence human health and disease inducing epigenetic modifications, in fact, increasing evidences on animal and human studies support the involvement of epigenetic status of our genes in obesity epidemic. The obesity spread can be explained by the evidence that environmental factors, such as lifestyle and nutrition, affect the epigenetic programming of parental gametes, the fetus, and the early postnatal development, so that the epigenetic marks induced in utero and in early life could determine a significant increase of obesity (and of other complex disease, such as T2D and cardiovascular disease) and could be transmitted transgenerationally. This epigenetic memory might fill the gap of missing genetic heritability for obesity but also for other complex diseases. So a good knowledge about the exact mechanism of epigenetic inheritance and the identification of molecular patterns that are either transmitted or deleted between the generations is becoming very important.

Nowadays, the attention on obesity is even higher, not only for its increasing prevalence worldwide, but also because it can be the cause of many other pathologies and epigenetic changes. For example, there are many recent data about an alteration of reproductive capacity in obese individuals. In a study on a rodent model of diet-induced obesity, the sperm of obese mice was found with decreased motility and reduced fertilization capacity, as well as with increased DNA damage and oxidative stress.[168] Moreover obesity was associated to altered spermatozoa physiology, aberrant mitochondrial function and changes in both spermatozoa RNA content and seminal vesicle fluid constitution,[169,170] in fact, an increased mRNA levels of cytochrome c oxidase subunit IV isoform 1 of the terminal enzyme in the mitochondrial respiratory chain in sperm of obese mice, increased levels of insulin and leptin, and decreased levels of estradiol in seminal plasma of obese males were observed.[170,171] Studies on humans do not reveal important associations, in fact, a recent meta-analysis did not find any significant difference in sperm concentration between normal and obese men, even if by a dichotomized analysis of concentration and sperm count obesity resulted associated with an increased incidence of oligospermia and azoospermia.[172]

The identification of epigenetic alterations at the very beginning of obesity development is important in order to predict disease trajectories and to choose eventually the most effective therapy. The reversible nature of epigenetic modifications makes them attractive targets for a possible epigenetic therapy of obesity, in fact growing evidences about the use of "epigenetic drugs" (compound able to interfere with epigenetic mechanisms) in the treatment of obesity are emerging.

REFERENCES

1. Global status report on non communicable diseases 2014. Global target 7: Halt the rise in diabetes and obesity. ISBN 978 92 4 156485 4. http://who.int/iris/bitstream/10665/148114/1/9789241564854_eng.pdf?ua=1
2. UNICEF—WHO—The World Bank. Joint child malnutrition estimates

3. Resolution WHA65.6. Maternal, infant and young child nutrition. In: Sixty-fifth World Health Assembly, Geneva: World Health Organization; 2012
4. Swinburn BA, Sacks G, Hall KD, et al. The global obesity pandemic: shaped by global drivers and local environments. *Lancet.* 2011;378:804–814.
5. Bornstein SR, Ehrhart-Bornstein M, Wong ML, Licinio J. Is the worldwide epidemic of obesity a communicable feature of globalization? *Exp Clin Endocrinol Diabetes.* 2008;116:S30–S32.
6. Bartolomucci A, Parmigiani S, Rodgers RJ, Vidal-Puig A, Allan SE, Siegel V. The obese species: a special issue on obesity and metabolic disorders foreword. *Dis Model Mech.* 2012;5:563–564.
7. Mokdad AH, Ford ES, Bowman BA. Prevalence of obesity, diabetes, and obesity-related health risk factors, 2001. *JAMA.* 2003;289:76–79.
8. Calle EE, Rodriguez C, Walker-Thurmond K, Thun MJ. Overweight, obesity, and mortality from cancer in a prospectively studied cohort of U.S. adults. *N Engl J Med.* 2003;348:1625–1638.
9. Lopez KN, Knudson JD. Obesity: from the agricultural revolution to the contemporary pediatric epidemic. *Congenit Heart Dis.* 2012;7:189–199.
10. D'Angelo CS, Koiffmann CP. Copy number variants in obesity-related syndromes: review and perspectives on novel molecular approaches. *J Obes.* 2012;2012:.
11. Fall T, Ingelsson E. Genome-wide association studies of obesity and metabolic syndrome. *Mol Cell Endocrinol.* 2014;382:740–757.
12. Wells JC. The evolution of human fatness and susceptibility to obesity: an ethological approach. *Biol Rev Camb Philos Soc.* 2006;81:183–205.
13. Neel JV. Diabetes mellitus: a "thrifty" genotype rendered detrimental by "progress"? *Am J Hum Genet.* 1962;14:353–362.
14. Neel JV. Diabetes mellitus: a "thrifty" genotype rendered detrimental by "progress"? 1962. *Bull World Health Organ.* 1993;77:694–703.
15. Myles S, Lea RA, Ohashi J, et al. Testing the thrifty gene hypothesis: the Gly482Ser variant in PPARGC1A is associated with BMI in tongans. *BMC Med Genet.* 2011;12:10.
16. Nakayama K, Ogawa A, Miyashita H, et al. Positive natural selection of TRIB2, a novel gene that influences visceral fat accumulation East Asia. *Hum Genet.* 2013;132: 201–217.
17. Baschetti R. Diabetes epidemic in newly westernized populations: is it due to thrifty genes or to genetically unknown foods? *J R Soc Med.* 1998;91:622–625.
18. Speakman JR. A nonadaptive scenario explaining the genetic predisposition to obesity: the "predation release" hypothesis. *Cell Metab.* 2007;6:5–12.
19. Speakman JR. Obesity: the integrated roles of environment and genetics. *J Nutr.* 2004;134:2090S–2105S.
20. Hales CN, Barker DJ. The thrifty phenotype hypothesis. *Br Med Bull.* 2001;60:5–20.
21. Lucas A. Programming by early nutrition in man. *Ciba Found Symp.* 1991;156:38–50.
22. Hales CN, Barker DJ, Clark PM, et al. Fetal and infant growth and impaired glucose tolerance at age 64. *BMJ.* 1991;303:1019–1022.
23. Hales CN, Barker DJ. Type 2 (non-insulin-dependent) diabetes mellitus: the thrifty phenotype hypothesis. *Diabetologia.* 1992;35:595–601.
24. Gluckman P, Hanson M. Developmental origins of disease paradigm: a mechanistic and evolutionary perspective. *Pediatr Res.* 2004;56:311–317.
25. Dhurandhar EJ, Keith SW. The aetiology of obesity beyond eating more and exercising less. *Best Pract Res Clin Gastroenterol.* 2014;28:533–544.
26. Wolff GL, Roberts DW, Mountjoy KG. Physiological consequences of ectopic agouti gene expression: the yellow obese mouse syndrome. *Physiol Genomics.* 1999;1:151–163.
27. Waterland RA, Jirtle RL. Transposable elements: targets for early nutritional effects on epigenetic gene regulation. *Mol Cell Biol.* 2003;23:5293–5300.

28. Michaud EJ, van Vugt MJ, Bultman SJ, Sweet HO, Davisson MT, Woychik RP. Differential expression of a new dominant agouti allele (Aiapy) is correlated with methylation state and is influenced by parental lineage. *Genes Dev.* 1994;8: 1463–1472.
29. Wolff GL, Kodell RL, Moore SR, Cooney CA. Maternal epigenetics and methyl supplements affect agouti gene expression in Avy/a mice. *FASEB J.* 1998;12: 949–957.
30. Waterland RA, Travisano M, Tahiliani KG, Rached MT, Mirza S. Methyl donor supplementation prevents transgenerational amplification of obesity. *Int J Obes.* 2008;32:1373–1379.
31. Cropley JE, Suter CM, Beckman KB, Martin DI. Germ-line epigenetic modification of the murine A vy allele by nutritional supplementation. *Proc Natl Acad Sci USA.* 2006;103:17308–17312.
32. Dolinoy DC. The agouti mouse model: an epigenetic biosensor for nutritional and environmental alterations on the fetal epigenome. *Nutr Rev.* 2008;66:S7–S11.
33. Howard TD, Ho SM, Zhang L, et al. Epigenetic changes with dietary soy in cynomolgus monkeys. *PloS One.* 2011;6:e26791.
34. Goldstone AP. Prader-Willi syndrome: advances in genetics, pathophysiology and treatment. *Trends Endocrinol Metab.* 2004;15:12–20.
35. Rankinen T, Zuberi A, Chagnon YC, et al. The human obesity gene map: the 2005 update. *Obesity (Silver Spring).* 2006;14:529–644.
36. Melzner I, Scott V, Dorsch K, et al. Leptin gene expression in human preadipocytes is switched on by maturation-induced demethylation of distinct CpGs in its proximal promoter. *J Biol Chem.* 2002;277:45420–45427.
37. Okamura M, Inagaki T, Tanaka T, Sakai J. Role of histone methylation and demethylation in adipogenesis and obesity. *Organogenesis.* 2010;6:24–32.
38. Pinnick KE, Karpe F. DNA methylation of genes in adipose tissue. *Proc Nutr Soc.* 2011;70:57–63.
39. Drummond EM, Gibney ER. Epigenetic regulation in obesity. *Curr Opin Clin Nutr Metab Care.* 2013;16:392–397.
40. van Dijk SJ, Molloy PL, Varinli H, Morrison JL, Muhlhausler BS. Members of EpiSCOPE.Epigenetics and human obesity. *Int J Obes (Lond).* 2015;39:85–97.
41. Carless MA, Kulkarni H, Kos MZ, et al. Genetic effects on DNA methylation and its potential relevance for obesity in Mexican Americans. *PLoS One.* 2013;8:e73950.
42. Xu X, Su S, Barnes VA, et al. A genome-wide methylation study on obesity: differential variability and differential methylation. *Epigenetics.* 2013;8:522–533.
43. Dick KJ, Nelson CP, Tsaprouni L, et al. DNA methylation and body-mass index: a genome-wide analysis. *Lancet.* 2014;383:1990–1998.
44. Milagro FI, Mansego ML, De Miguel C, Martinez JA. Dietary factors, epigenetic modifications and obesity outcomes: progresses and perspectives. *Mol Asp Med.* 2013;34:782–812.
45. Bouchard L, Rabasa-Lhoret R, Faraj M, et al. Differential epigenomic and transcriptomic responses in subcutaneous adipose tissue between low and high responders to caloric restriction. *Am J Clin Nutr.* 2010;91:309–320.
46. Pan H, Lin X, Wu Y, et al. HIF3A association with adiposity: the story begins before birth. *Epigenomics.* 2015;26:1–13.
47. Shen W, Wang C, Xia L, Fan C, Dong H, Deckelbaum RJ, Qi K. Epigenetic modification of the leptin promoter in diet-induced obese mice and the effects of N-3 polyunsaturated fatty acids. *Sci Rep.* 2014;4:5282.
48. Lesseur C, Armstrong DA, Paquette AG, Koestler DC, Padbury JF, Marsit CJ. Tissue-specific Leptin promoter DNA methylation is associated with maternal and infant perinatal factors. *Mol Cell Endocrinol.* 2013;381:160–167.

49. Lesseur C, Armstrong DA, Paquette AG, Li Z, Padbury JF, Marsit CJ. Maternal obesity and gestational diabetes are associated with placental leptin DNA methylation. *Am J Obstet Gynecol.* 2014;211(654):e1–e9.
50. Houde AA, Légaré C, Biron S, et al. Leptin and adiponectin DNA methylation levels in adipose tissues and blood cells are associated with BMI, waist girth and LDL-cholesterol levels in severely obese men and women. *BMC Med Genet.* 2015;16:29.
51. Khalyfa A, Mutskov V, Carreras A, Khalyfa AA, Hakim F, Gozal D. Sleep fragmentation during late gestation induces metabolic perturbations and epigenetic changes in adiponectin gene expression in male adult offspring mice. *Diabetes.* 2014;63: 3230–3241.
52. Gemma C, Sookoian S, Alvarinas J, et al. Maternal pregestational BMI is associated with methylation of the PPARGC1A promoter in newborns. *Obesity (Silver Spring).* 2009;17:1032–1039.
53. Brons C, Jacobsen S, Nilsson E, et al. Deoxyribonucleic acid methylation and gene expression of PPARGC1A in human muscle is influenced by high-fat overfeeding in a birth-weight-dependent manner. *J Clin Endocrinol Metab.* 2010;95:3048–3056.
54. Soubry A, Schildkraut JM, Murtha A, et al. Paternal obesity is associated with IGF2 hypomethylation in newborns: results from a Newborn Epigenetics Study (NEST) cohort. *BMC Med.* 2013;11:29–38.
55. Perkins E, Murphy SK, Murtha AP, et al. Insulin-like growth factor 2/H19 methylation at birth and risk of overweight and obesity in children. *J Pediatr.* 2012;161:31–39.
56. Zhang S, Rattanatray L, MacLaughlin SM, et al. Periconceptional undernutrition in normal and overweight ewes leads to increased adrenal growth and epigenetic changes in adrenal IGF2/H19 gene in offspring. *FASEB J.* 2010;24:2772–2782.
57. Nilsson E, Jansson PA, Perfilyev A, et al. Altered DNA methylation and differential expression of genes influencing metabolism and inflammation in adipose tissue from subjects with type 2 diabetes. *Diabetes.* 2014;63:2962–2976.
58. Kuroda A, Rauch TA, Todorov I, et al. Insulin gene expression is regulated by DNA methylation. *PLoS One.* 2009;4:e6953.
59. Su S, Zhu H, Xu X, et al. DNA methylation of the LY86 gene is associated with obesity, insulin resistance, and inflammation. *Twin Res Hum Genet.* 2014;17:183–191.
60. Sasaki S, Nagai Y, Yanagibashi T, et al. Serum soluble MD-1 levels increase with disease progression in autoimmune prone MRL(lpr/lpr) mice. *Mol Immunol.* 2012;49: 611–620.
61. Watanabe Y, Nakamura T, Ishikawa S, et al. The radioprotective 105/MD-1 complex contributes to diet-induced obesity and adipose tissue inflammation. *Diabetes.* 2012;61:1199–1209.
62. Soubry A, Murphy SK, Wang F, et al. Newborns of obese parents have altered DNA methylation patterns at imprinted genes. *Int J Obes (Lond).* 2015;39:650–657.
63. Crujeiras AB, Campion J, Díaz-Lagares A, et al. Association of weight regain with specific methylation levels in the NPY and POMC promoters in leukocytes of obese men: a translational study. *Regul Pept.* 2013;186:1–6.
64. Zheng J, Xiao X, Zhang Q, et al. Maternal and post-weaning high-fat, high-sucrose diet modulates glucose homeostasis and hypothalamic POMC promoter methylation in mouse offspring. *Metab Brain Dis.* 2015.
65. Marco A, Kisliouk T, Tabachnik T, Meiri N, Weller A. Overweight and CpG methylation of the Pomc promoter in offspring of high-fat-diet-fed dams are not "reprogrammed" by regular chow diet in rats. *FASEB J.* 2014;28:4148–4157.
66. Mahmood S, Smiraglia DJ, Srinivasan M, Patel MS. Epigenetic changes in hypothalamic appetite regulatory genes may underlie the developmental programming for obesity in rat neonates subjected to a high-carbohydrate dietary modification. *J Dev Orig Health Dis.* 2014;4:479–490.

67. Na YK, Hong HS, Lee WK, Kim YH, Kim DS. Increased methylation of interleukin 6 gene is associated with obesity in Korean women. *Mol Cells.* 2015;38:452–456.
68. Wang X, Lacza Z, Sun YE, Han W. Leptin resistance and obesity in mice with deletion of methyl-CpG-binding protein 2 (MeCP2) in hypothalamic pro-opiomelanocortin (POMC) neurons. *Diabetologia.* 2014;57:236–245.
69. Yoo JY, Lee S, Lee HA, et al. Can proopiomelanocortin methylation be used as an early predictor of metabolic syndrome? *Diabetes Care.* 2014;37:734–739.
70. Zhang Q, Ramlee MK, Brunmeir R, Villanueva CJ, Halperin D, Xu F. Dynamic and distinct histone modifications modulate the expression of key adipogenesis regulatory genes. *Cell Cycle.* 2012;11:4310–4322.
71. Mikula M, Majewska A, Ledwon JK, Dzwonek A, Ostrowski J. Obesity increases histone H3 lysine 9 and 18 acetylation at Tnfa and Ccl2 genes in mouse liver. *Int J Mol Med.* 2014;34:1647–2547.
72. Wheatley KE, Nogueira LM, Perkins SN, Hursting SD. Differential effects of calorie restriction and exercise on the adipose transcriptome in diet-induced obese mice. *J Obes.* 2011;2011:265417.
73. Funato H, Oda S, Yokofujita J, Igarashi H, Kuroda M. Fasting and high-fat diet alter histone deacetylase expression in the medial hypothalamus. *PLoS One.* 2011;6:e18950.
74. Tateishi K, Okada Y, Kallin EM, Zhang Y. Role of Jhdm2a in regulating metabolic gene expression and obesity resistance. *Nature.* 2009;458:757–761.
75. Graham C, Mullen A, Whelan K. Obesity and the gastrointestinal microbiota: a review of associations and mechanisms. *Nutr Rev.* 2015;73:376–385.
76. Song G, Xu G, Ji C, et al. The role of microRNA-26b in human adipocyte differentiation and proliferation. *Gene.* 2014;533:481–487.
77. Xu G, Ji C, Song G, et al. Obesity-associated microRNA-26b regulates the proliferation of human preadipocytes via arrest of the G1/S transition. *Mol Med Rep.* 2015;12: 3648–3654.
78. Crèpin D, Benomar Y, Riffault L, Amine H, Gertler A, Taouis M. The over-expression of miR-200a in the hypothalamus of ob/ob mice is linked to leptin and insulin signaling impairment. *Mol Cell Endocrinol.* 2014;384:1–11.
79. Pan S, Yang X, Jia Y, Li R, Zhao R. Microvesicle-shuttled miR-130b reduces fat deposition in recipient primary cultured porcine adipocytes by inhibiting PPAR-g expression. *J Cell Physiol.* 2014;229:631–639.
80. Zhang L, Hou D, Chen X, et al. Exogenous plant MIR168a specifically targets mammalian LDLRAP1: evidence of cross-kingdom regulation by microRNA. *Cell Res.* 2012;22:107–126.
81. Vickers MH. Early life nutrition, epigenetics and programming of later life disease. *Nutrients.* 2014;6:2165–2178.
82. Vickers MH, Breier BH, Cutfield WS, Hofman PL, Gluckman PD. Fetal origins of hyperphagia, obesity, and hypertension and postnatal amplification by hypercaloric nutrition. *Am J Physiol Endocrinol Metab.* 2000;279:E83–E87.
83. Painter RC, Osmond C, Gluckman P, Hanson M, Phillips DI, Roseboom TJ. Transgenerational effects of prenatal exposure to the Dutch famine on neonatal adiposity and health in later life. *BJOG.* 2008;115:1243–1249.
84. Stanner SA, Bulmer K, Andre's C, Lantseva OE, Borodina V, Poteen VV, Yudkin JS. Does malnutrition in utero determine diabetes and coronary heart disease in adulthood? Results from the Leningrad siege study, a cross sectional study. *BMJ.* 1997;315: 1342–1348.
85. Bell C. Long term mortality after starvation during the Leningrad siege: no evidence that starvation around puberty causes later cardiovascular disease. *BMJ.* 2004;328:346.
86. Barker DJ. The developmental origins of adult disease. *Eur J Epidemiol.* 2003;18: 733–736.

87. Darmasseelane K, Hyde MJ, Santhakumaran S, Gale C, Modi N. Mode of delivery and offspring body mass index, overweight and obesity in adult life: a systematic review and meta-analysis. *PLoS One.* 2014;9:e1046.
88. Kuhle S, Tong OS, Woolcott CG. Association between caesarean section and childhood obesity: a systematic review and meta-analysis. *Obes Rev.* 2015;16:295–303.
89. Pei Z, Heinrich J, Fuertes E, et al. Cesarean delivery and risk of childhood obesity. *J Pediatr.* 2014;164(5):1068–1073.
90. Carrillo-Larco RM, Miranda JJ, Bernabé-Ortiz A. Delivery by caesarean section and risk of childhood obesity: analysis of a Peruvian prospective cohort. *PeerJ.* 2015;3:e1046.
91. Mueller NT, Whyatt R, Hoepner L, et al. Prenatal exposure to antibiotics, cesarean section and risk of childhood obesity. *Int J Obes (Lond).* 2015;39:665–670.
92. Desai M, Jellyman JK, Ross MG. Epigenomics, gestational programming and risk of metabolic syndrome. *Int J Obes (Lond).* 2015;39:633–641.
93. Waterland RA, Michels KB. Epigenetic epidemiology of the developmental origins hypothesis. *Annu Rev Nutr.* 2007;27:363–388.
94. Waterland RA, Dolinoy DC, Lin JR, Smith CA, Shi X, Tahiliani KG. Maternal methyl supplements increase offspring DNA methylation at axin fused. *Genesis.* 2006;44: 401–406.
95. Boney CM, Verma A, Tucker R, Vohr BR. Metabolic syndrome in childhood: association with birth weight, maternal obesity, and gestational diabetes mellitus. *Pediatrics.* 2005;115:e290–e296.
96. Armitage JA, Poston L, Taylor PD. Developmental origins of obesity and the metabolic syndrome: the role of maternal obesity. *Front Horm Res.* 2008;36:73–84.
97. Lukaszewski MA, Eberlè D, Vieau D, Breton C. Nutritional manipulations in the perinatal period program adipose tissue in offspring. *Am J Physiol Endocrinol Metab.* 2013;305:E1195–E1207.
98. Pico C, Palou M, Priego T, Sánchez J, Palou A. Metabolic programming of obesity by energy restriction during the perinatal period: different outcomes depending on gender and period, type and severity of restriction. *Front Physiol.* 2012;3:436.
99. Godfrey KM, Sheppard A, Gluckman PD, et al. Epigenetic gene promoter methylation at birth is associated with child's later adiposity. *Diabetes.* 2011;60:1528–1534.
100. Obermann-Borst SA, Eilers PH, Tobi EW, et al. Duration of breastfeeding and gender are associated with methylation of the LEPTIN gene in very young children. *Pediatr Res.* 2013;74:344–349.
101. Pico C, Oliver P, Sánchez J, et al. The intake of physiological doses of leptin during lactation in rats prevents obesity in later life. *Int J Obes (Lond).* 2007;31:1199–1209.
102. Rees WD, Hay SM, Brown DS, Antipatis C, Palmer RM. Maternal protein deficiency causes hypermethylation of DNA in the livers of rat fetuses. *J Nutr.* 2000;130: 1821–1826.
103. Burdge GC, Slater-Jefferies J, Torrens C, Phillips ES, Hanson MA, Lillycrop KA. Dietary protein restriction of pregnant rats in the F0 generation induces altered methylation of hepatic gene promoters in the adult male offspring in the F1 and F2 generations. *Br J Nutr.* 2007;97:435–439.
104. Zheng J, Xiao X, Zhang Q, Yu M. DNA methylation: the pivotal interaction between early-life nutrition and glucose metabolism in later life. *Br J Nutr.* 2014;112: 1850–1857.
105. Sandovici I, Smith NH, Nitert MD, et al. Maternal diet and aging alter the epigenetic control of a promoter-enhancer interaction at the Hnf4a gene in rat pancreatic islets. *Proc Natl Acad Sci USA.* 2011;108:5449–5454.
106. Jousse C, Parry L, Lambert-Langlais S, et al. Perinatal undernutrition affects the methylation and expression of the leptin gene in adults: implication for the understanding of metabolic syndrome. *FASEB J.* 2011;25:3271–3278.

107. Zinkhan EK, Fu Q, Wang Y, et al. Maternal hyperglycemia disrupts histone 3 lysine 36 trimethylation of the IGF-1 gene. *J Nutr Metab.* 2012;2012:930364.
108. Fu Q, McKnight RA, Yu X, Wang L, Callaway CW, Lane RH. Uteroplacental insufficiency induces site-specific changes in histone H3 covalent modifications and affects DNA-histone H3 positioning in day 0 IUGR rat liver. *Physiol Genomics.* 2004;20:108–116.
109. Tosh DN, Fu Q, Callaway CW, et al. Epigenetics of programmed obesity: alteration in IUGR rat hepatic IGF1 mRNA expression and histone structure in rapid vs. delayed postnatal catch-up growth. *Am J Physiol Gastrointest Liver Physiol.* 2010;299: G1023–G1029.
110. Park JH, Stoffers DA, Nicholls RD, Simmons RA. Development of type 2 diabetes following intrauterine growth retardation in rats is associated with progressive epigenetic silencing of Pdx1. *J Clin Invest.* 2008;118:2316–2324.
111. Carone BR, Fauquier L, Habib N, et al. Paternally induced transgenerational environmental reprogramming of metabolic gene expression in mammals. *Cell.* 2010;143: 1084–1096.
112. Radford EJ, Ito M, Shi H, et al. In utero effects. In utero undernourishment perturbs the adult sperm methylome and intergenerational metabolism. *Science.* 2014;345:1255903.
113. Widiker S, Karst S, Wagener A, Brockmann GA. High-fat diet leads to a decreased methylation of the Mc4r gene in the obese BFMI and the lean B6 mouse lines. *J Appl Genet.* 2010;51:193–197.
114. Milagro FI, Campión J, García-Díaz DF, Goyenechea E, Paternain L, Martínez JA. High fat diet-induced obesity modifies the methylation pattern of leptin promoter in rats. *J Physiol Biochem.* 2009;65:1–9.
115. Marco A, Kisliouk T, Weller A, Meiri N. High fat diet induces hypermethylation of the hypothalamic Pomc promoter and obesity in post-weaning rats. *Psychoneuroendocrinology.* 2013;38:2844–2853.
116. Vucetic Z, Kimmel J, Totoki K, Hollenbeck E, Reyes TM. Maternal high-fat diet alters methylation and gene expression of dopamine and opioid-related genes. *Endocrinology.* 2010;151:4756–4764.
117. Strakovsky RS, Zhang X, Zhou D, Pan YX. The regulation of hepatic Pon1 by a maternal high-fat diet is gender specific and may occur through promoter histone modifications in neonatal rats. *J Nutr Biochem.* 2014;25:170–176.
118. Suter MA, Chen A, Burdine MS, et al. A maternal high-fat diet modulates fetal SIRT1 histone and protein deacetylase activity in nonhuman primates. *FASEB J.* 2012;26: 5106–5114.
119. Altmann S, Murani E, Schwerin M, Metges CC, Wimmers K, Ponsuksili S. Maternal dietary protein restriction and excess affects offspring gene expression and methylation of non-SMC subunits of condensin I in liver and skeletal muscle. *Epigenetics.* 2012;7:239–252.
120. Liu HW, Mahmood S, Srinivasan M, Smiraglia DJ, Patel MS. Developmental programming in skeletal muscle in response to overnourishment in the immediate postnatal life in rats. *J Nutr Biochem.* 2013;24:1859–1869.
121. Yan X, Huang Y, Zhao JX, et al. Maternal obesity downregulates microRNA let-7 g expression, a possible mechanism for enhanced adipogenesis during ovine fetal skeletal muscle development. *Int J Obes (Lond).* 2013;37:568–575.
122. Ng SF, Lin RC, Laybutt DR, Barres R, Owens JA, Morris MJ. Chronic high-fat diet in fathers programs β-cell dysfunction in female rat offspring. *Nature.* 2010;467:963–966.
123. Hoile SP, Lillycrop KA, Grenfell LR, Hanson MA, Burdge GC. Increasing the folic acid content of maternal or post-weaning diets induces differential changes in phosphoenolpyruvate carboxykinase mRNA expression and promoter methylation in rats. *Br J Nutr.* 2012;108:852–857.

124. Davison JM, Mellott TJ, Kovacheva VP, Blusztajn JK. Gestational choline supply regulates methylation of histone H3, expression of histone methyltransferases G9a (Kmt1c) and Suv39h1 (Kmt1a), and DNA methylation of their genes in rat fetal liver and brain. *J Biol Chem.* 2009;284:1982–1989.
125. McKay JA, Groom A, Potter C, et al. Genetic and non-genetic influences during pregnancy on infant global and site specific DNA methylation: role for folate gene variants and vitamin B12. *PLoS One.* 2012;7:e33290.
126. Gauguier D, Bihoreau MT, Ktorza A, Berthault MF, Picon L. Inheritance of diabetes mellitus as consequence of gestational hyperglycemia in rats. *Diabetes.* 1990;39: 734–739.
127. Finer S, Mathews C, Lowe R, et al. Maternal gestational diabetes is associated with genome-wide DNA methylation variation in placenta and cord blood of exposed offspring. *Hum Mol Genet.* 2015;24:3021–3029.
128. Fraga MF, Ballestar E, Paz MF, et al. Epigenetic differences arise during the lifetime of monozygotic twins. *Proc Natl Acad Sci USA.* 2005;102:10604–10609.
129. Dunn GA, Bale TL. Maternal high-fat diet effects on third-generation female body size via the paternal lineage. *Endocrinology.* 2011;152:2228–2236.
130. Soubry A, Murphy SK, Wang F, et al. Newborns of obese parents have altered DNA methylation patterns at imprinted genes. *Int J Obes (Lond).* 2015;39:650–657.
131. Soubry A, Schildkraut JM, Murtha A, et al. Paternal obesity is associated with IGF2 hypomethylation in newborns: results from a Newborn Epigenetics Study (NEST) cohort. *BMC Med.* 2013;11:29.
132. Mitchell M, Bakos HW, Lane M. Paternal diet-induced obesity impairs embryo development and implantation in the mouse. *Fertil Steril.* 2011;95:1349–1353.
133. Binder NK, Hannan NJ, Gardner DK. Paternal diet-induced obesity retards early mouse embryo development, mitochondrial activity and pregnancy health. *PLoS One.* 2012;7:e52304.
134. Fullston T, Ohlsson Teague EM, et al. Paternal obesity initiates metabolic disturbances in two generations of mice with incomplete penetrance to the F2 generation and alters the transcriptional profile of testis and sperm microRNA content. *FASEB J.* 2013;27:4226–4243.
135. Ding GL, Wang FF, Shu J, et al. Transgenerational glucose intolerance with Igf2/H19 epigenetic alterations in mouse islet induced by intrauterine hyperglycemia. *Diabetes.* 2012;61:1133–1142.
136. Drake AJ, Liu L, Kerrigan D, Meehan RR, Seckl JR. Multigenerational programming in the glucocorticoid programmed rat is associated with generation-specific and parent of origin effects. *Epigenetics.* 2011;6:1334–1343.
137. Öst A, Lempradl A, Casas E, et al. Paternal diet defines offspring chromatin state and intergenerational obesity. *Cell.* 2014;159:1352–1364.
138. Ge ZJ, Luo SM, Lin F, et al. DNA methylation in oocytes and liver of female mice and their offspring: effects of high-fat-diet-induced obesity. *Environ Health Perspect.* 2014;122:154–164.
139. Grün F, Blumberg B. Environmental obesogens: organotins and endocrine disruption via nuclear receptor signaling. *Endocrinology.* 2006;147:S50–S55.
140. Grün F. Obesogens. *Curr Opin Endocrinol Diabetes Obes.* 2010;17:453–459.
141. Baillie-Hamilton PF. Chemical toxins: a hypothesis to explain the global obesity epidemic. *J Altern Complement Med.* 2002;8:185–192.
142. Grün F, Blumberg B. Perturbed nuclear receptor signalling by environmental obesogens as emerging factors in the obesity crisis. *Rev Endocr Metab Disord.* 2007;8: 161–171.
143. Somm E, Schwitzgebel VM, Toulotte A, et al. Perinatal exposure to bisphenol a alters early adipogenesis in the rat. *Environ Health Perspect.* 2009;117:1549–1555.

144. Kirchner S, Kieu T, Chow C, Casey S, Blumberg B. Prenatal exposure to the environmental obesogen tributyltin predisposes multipotent stem cells to become adipocytes. *Mol Endocrinol.* 2010;24:526–539.
145. Vandenberg LN, Colborn T, Hayes TB, et al. Hormones and endocrine-disrupting chemicals: low-dose effects and nonmonotonic dose responses. *Endocr Rev.* 2012;33: 378–455.
146. Warner M, Wesselink A, Harley KG, Bradman A, Kogut K, Eskenazi B. Prenatal exposure to dichlorodiphenyltrichloroethane and obesity at 9 years of age in the CHAMACOS study cohort. *Am J Epidemiol.* 2014;179:1312–1322.
147. Pereira-Fernandes A, Dirinck E, Dirtu AC, et al. Expression of obesity markers and Persistent Organic Pollutants levels in adipose tissue of obese patients: reinforcing the obesogen hypothesis? *PLoS One.* 2014;9:e84816.
148. Hao C, Cheng X, Guo J, Xia H, Ma X. Perinatal exposure to diethyl-hexyl-phthalate induces obesity in mice. *Front Biosci (Elite Ed).* 2013;5:725–733.
149. Bastos Sales L, Kamstra JH, Cenijn PH, van Rijt LS, Hamers T, Legler J. Effects of endocrine disrupting chemicals on in vitro global DNA methylation and adipocyte differentiation. *Toxicol In Vitro.* 2013;27:1634–1643.
150. Barrès R, Osler ME, Yan J, et al. Non-CpG methylation of the PGC-1alpha promoter through DNMT3B controls mitochondrial density. *Cell Metab.* 2009;10:189–198.
151. Grün F, Blumberg B. Minireview: the case for obesogens. *Mol Endocrinol.* 2009;23: 1127–1134.
152. Savastano S, Tarantino G, D'Esposito V, et al. Bisphenol-A plasma levels are related to inflammatory markers, visceral obesity and insulin-resistance: a cross-sectional study on adult male population. *J Transl Med.* 2015;13:169.
153. Fleisch AF, Wright RO, Baccarelli AA. Environmental epigenetics: a role in endocrine disease? *J Mol Endocrinol.* 2012;49:R61–R67.
154. Yan Z, Zhang H, Maher C, et al. Prenatal polycyclic aromatic hydrocarbon, adiposity, peroxisome proliferator-activated receptor (PPAR) c methylation in offspring, grand-offspring mice. *PLoS One.* 2014;9:e110706.
155. Watt J, Schlezinger JJ. Structurally-diverse. PPARγ-activating environmental toxicants induce adipogenesis and suppress osteogenesis in bone marrow mesenchymal stromal cells. *Toxicology.* 2015;331:66–77.
156. Manikkam M, Tracey R, Guerrero-Bosagna C, Skinner MK. Plastics derived endocrine disruptors (BPA, DEHP and DBP) induce epigenetic transgenerational inheritance of obesity. Reproductive disease and sperm epimutations. *PLoS One.* 2013;8: e55387.
157. Tracey R, Manikkam M, Guerrero-Bosagna C, Skinner MK. Hydrocarbons (jet fuel JP-8) induce epigenetic transgenerational inheritance of obesity, reproductive disease and sperm epimutations. *Reprod Toxicol.* 2013;36:104–116.
158. Skinner MK, Manikkam M, Tracey R, Guerrero-Bosagna C, Haque M, Nilsson EE. Ancestral dichlorodiphenyltrichloroethane (DDT) exposure promotes epigenetic transgenerational inheritance of obesity. *BMC Med.* 2013;11:228.
159. Barres R, Yan J, Egan B, et al. Acute exercise remodels promoter methylation in human skeletal muscle. *Cell Metab.* 2012;15:405–411.
160. Ronn T, Volkov P, Davegardh C, et al. A six months exercise intervention influences the genome-wide DNA methylation pattern in human adipose tissue. *PLoS Genet.* 2013;9: e1003572.
161. Moleres A, Campión J, Milagro FI, et al. Differential DNA methylation patterns between high and low responders to a weight loss intervention in overweight or obese adolescents: the EVASYON study. *FASEB J.* 2013;27:2504–2512.
162. Barres R, Kirchner H, Rasmussen M, et al. Weight loss after gastric bypass surgery in human obesity remodels promoter methylation. *Cell Rep.* 2013;3:1020–1027.

163. Milagro FI, Gómez-Abellán P, Campión J, Martínez JA, Ordovás JM, Garaulet M. CLOCK, PER2 and BMAL1 DNA methylation: association with obesity and metabolic syndrome characteristics and monounsaturated fat intake. *Chronobiol Int.* 2012;29:1180–1194.
164. Milagro FI, Campión J, Cordero P, et al. A dual epigenomic approach for the search of obesity biomarkers: DNA methylation in relation to diet-induced weight loss. *FASEB J.* 2011;25:1378–1389.
165. Campión J, Milagro FI, Goyenechea E, Martínez JA. TNF-alpha promoter methylation as a predictive biomarker for weight-loss response. *Obesity (Silver Spring).* 2009;17: 1293–1297.
166. Boque N, de la Iglesia R, de la Garza AL, et al. Prevention of diet-induced obesity by apple polyphenols in Wistar rats through regulation of adipocyte gene expression and DNA methylation patterns. *Mol Nutr Food Res.* 2013;8:1473–1478.
167. Remely M, Lovrecic L, de la Garza AL, et al. Therapeutic perspectives of epigenetically active nutrients. *Br J Pharmacol.* 2015;172:2756–2768.
168. Bakos HW, Mitchell M, Setchell BP, Lane M. The effect of paternal diet-induced obesity on sperm function and fertilization in a mouse model. *Int J Androl.* 2011;34: 402–410.
169. Fariello RM, Pariz JR, Spaine DM, Cedenho AP, Bertolla RP, Fraietta R. Association between obesity and alteration of sperm DNA integrity and mitochondrial activity. *BJU Int.* 2012;110:863–867.
170. Binder NK, Sheedy JR, Hannan NJ, Gardner DK. Male obesity is associated with changed spermatozoa Cox4i1 mRNA level and altered seminal vesicle fluid composition in a mouse model. *Mol Hum Reprod.* 2015;21:424–434.
171. Leisegang K, Bouic PJ, Menkveld R, Henkel RR. Obesity is associated with increased seminal insulin and leptin alongside reduced fertility parameters in a controlled male cohort. *Reprod Biol Endocrinol.* 2014;12:34.
172. Sermondade N, Faure C, Fezeu L, et al. BMI in relation to sperm count: an updated systematic review and collaborative meta-analysis. *Hum Reprod Update.* 2013;19: 221–231.

CHAPTER SEVEN

From Genome-Wide Association Study to Phenome-Wide Association Study: New Paradigms in Obesity Research

Y.-P. Zhang*, Y.-Y. Zhang†, D.D. Duan‡,1

* Pediatric Heart Center, Beijing Anzhen Hospital, Capital Medical University, Beijing, China
† Department of Cardiology, Changzhou Second People's Hospital, Changzhou, Jiangsu, China
‡ Laboratory of Cardiovascular Phenomics, Center for Cardiovascular Research, Department of Pharmacology, and Center for Molecular Medicine, University of Nevada School of Medicine, Reno, NV, United States
1 Corresponding author. E-mail address: dduan@medicine.nevada.edu

Contents

Progress in Molecular Biology and Translational Science, Volume 140
ISSN 1877-1173
http://dx.doi.org/10.1016/bs.pmbts.2016.02.003

Abstract

Obesity is a condition in which excess body fat has accumulated over an extent that increases the risk of many chronic diseases. The current clinical classification of obesity is based on measurement of body mass index (BMI), waist–hip ratio, and body fat percentage. However, these measurements do not account for the wide individual variations in fat distribution, degree of fatness or health risks, and genetic variants identified in the genome-wide association studies (GWAS). In this review, we will address this important issue with the introduction of phenome, phenomics, and phenome-wide association study (PheWAS). We will discuss the new paradigm shift from GWAS to PheWAS in obesity research. In the era of precision medicine, phenomics and PheWAS provide the required approaches to better definition and classification of obesity according to the association of obese phenome with their unique molecular makeup, lifestyle, and environmental impact.

ABBREVIATIONS

BBS	Bardet–Biedl syndrome
BDNF	Brain-derived neurotrophic factor
BF%	Body fat percentage
BIA	Bioelectrical impedance analysis
BMI	Body mass index
BVI	Body volume index
CDW	Clinical data warehouse
CVD	Cardiovascular disease
DZ	Dizygotic
EHR	Electronic health record
eMERGE	Electronic Medical Records and Genomics
EMR	Electronic medical record
EWAS	Epigenome wide association studies
FTO	Fat mass and obesity-associated gene
GIANT	Genetic investigation of anthropometric traits
GWAS	Genome-wide association study
miRNA	MicroRNA
MPD	Mouse phenome database
MRI	Magnetic resonance imaging
MZ	Monozygotic
NCDs	Noncommunicable diseases
NGS	Next-generation sequencing
NWO	Normal weight obesity
PheWAS	Phenome-wide association study
PWS	Prader–Willi syndrome
QTL	Quantitative trait loci
SNPs	Single-nucleotide polymorphisms
T2D	Type-2 diabetes
VAT	Visceral adipose tissue
WC	Waist circumference
WHR	Waist–hip ratio

1. INTRODUCTION

Obesity is an abnormal accumulation of body fat (>20% over an individual's body weight), which causes many health and medical problems worldwide.[1–5] Available data suggest that the increase in the prevalence of obesity began to emerge during the 1980s and ever since more countries have joined the global obesity pandemic.[6–8] In 2013, the global estimated prevalence of overweight [body mass index (BMI) > 25 kg/m^2] and obesity (BMI > 30 kg/m^2) in men and women was 36.9 and 38.0%, respectively. In 1989, the worldwide estimate for the prevalence of overweight and obesity among adults (>20 years) was around 857 million individuals, compared to the 2.1 billion in 2013.[9] These values represent an increase of ~41% in 33 years. The global rise of the prevalence of overweight and obesity is fueled by a shift in dietary habits owing to the widespread availability of low-cost, hypercaloric foods and contributes to a significant increase in the morbidity and mortality of many chronic diseases such as cancer, cardiovascular disease (CVD), type-2 diabetes (T2D), and metabolic syndrome (Table 1).[2,5,10,11] To date the strategies for obesity prevention and management have proven quite inefficient.[12–17] The prevalence of obesity is escalating worldwide due to rapid changes in lifestyles (such as diet and physical activity), increase in aging populations, and, most importantly, our insufficient understanding of the etiological determinants for obesity.[18–20]

In the past two decades, extensive studies have been focused on the association of obesity with genetic variations and lifestyle.[21,22] It has been demonstrated that human adiposity is highly heritable with the estimated 20–90% genetic contribution to individual differences in relative body weight as estimated by BMI.[23–25] In recent years different robust techniques and approaches for detecting genetic sequence variations have been used to understand the molecular makeups of obesity in a big-data fashion.[21,26–37] Genome-wide association study (GWAS) has been applied to investigate the impact of common genetic variants on obesity.[9,30,38–45] The identification of genetic variants associated with human adiposity and BMI has quickly evolved from hypothesis-driven candidate gene study of a single gene to an "omic" scale identification of multigenetic variants.[21,30,33–37] These studies with major focus on the genotypes of obesity, however, have further widened the gap between the narrowly defined obese phenotypes and the broader genetic variants at the omic level. The genetic variants identified in the GWAS do not have direct correlations with the current clinical classification of obesity, which is not based on its etiology.[46,47] In fact, the etiology of obesity can be very diverse although clinically we do not classify human adiposity according to their causes. Multiple sociocultural, socioeconomic, lifestyle, behavioral, and biological factors may contribute to the establishment and perpetuation of obese phenotypes or traits.[46–51] Recognition of the relationship between the diverse clinical forms of obesity and their different etiologies, including genetic and environmental mechanisms, would permit more specific treatment regimens and increase the likelihood of success.[52]

In the era of precision medicine, obesity needs to be redefined in the "omic" scale by extending obese phenotypes into much greater details (or deep phenotyping) using the underlying molecular causes and other factors in addition to traditional signs and symptoms such as increased body weight and BMI.[53,54] In this review, we will address this important issue with the introduction of phenome, phenomics, and phenome-wide association study (PheWAS), which provides novel approaches to better definition and classification of obesity according to the association of phenotypic characterizations of obesity with their unique molecular makeups, lifestyles, and environmental impacts. We will discuss the new paradigm shift from GWAS to PheWAS in obesity research.

2. CURRENT CLINICAL CLASSIFICATION OF OBESITY

Clinically, obesity is defined as a condition of excessive accumulation of adipose tissue or body fat more than 20% over an individual's body weight, which is associated with adverse health outcomes.[1–5] As shown in Table 1, the current classification of obesity is based on measurement of BMI, waist circumferences (WCs) or the waist–hip ratio (WHR), and total cardiovascular risk factors.[48,55,56] Alternative methods in an effort to better take into account different body shapes [such as body volume index (BVI)[57]] and the lean and fat body compartments [such as bioelectrical impedance analysis (BIA)[58]] have also been developed. The classification based on these measurements, however, does not reflect the causes of obesity. The failure of the current classification and diagnosis to address specific causes of obesity in individual patients, which may vary from a low resting metabolic rate to poor lifestyle and environmental factors, is a major reason why the current treatment (mainly symptomatically lowering energy intake) is unsuccessful. Therefore, etiology-based classification of obesity was proposed although

Table 1 Classification of Overweight and Obesity by BMI, Waist Circumference (WC), and Associated Disease Risks.

	BMI (kg/m^2)	Obesity Class	Disease Risk[a] Relative to Normal Weight and WC[b]: WC: Men 102 cm (40 in.) or Less; Women 88 cm (35 in.) or Less	WC: Men > 102 cm (40 in.); Women > 88 cm (35 in.)
Underweight	<18.5		—	—
Normal	18.5–24.9		—	—
Overweight	25.0–29.9		Increased	High
Obesity	30.0–34.9	I	High	Very high
	35.0–39.9	II	Very high	Very high
Extreme obesity	40.0+	III	Extremely high	Extremely high

[a]Disease risk for T2D, hypertension, and CVD.
[b]Increased WC also can be a marker for increased risk, even in persons of normal weight.
Adapted from National Heart, Lung, and Blood Institute, https://www.nhlbi.nih.gov/health/educational/lose_wt/BMI/bmi_dis.htm

not accepted for common use in clinics.[52,59] For example, as early as 1989, Alemany proposed to classify obesity based on etiology into hypothalamic, bulimic, digestive, hyperinsulinemic, hypothermogenic, and hypothyroid obesities. It was also pointed out that these conditions should not be treated therapeutically in the same way, as the causes of development of the illness are not equal.[59] In the era of precision medicine, these issues may need to be revisited and the individualized causes of obesity should be more clearly determined prior to any treatment given to the patient.

2.1 BMI

BMI is the most widely used anthropometric index to diagnose obesity.[60] It is defined as the weight divided by the square of height of the subject, that is, BMI = body weight/height^2 (kg/m^2). BMI is closely related to both percentage body fat and total body fat. But it does not differentiate between body fat and muscle mass. Therefore, people who have big muscle bulk will have a high BMI but are not overweight or obese. In children, a healthy weight varies with age and sex,[61] therefore obesity in children and adolescents is defined not as an absolute number but in relation to a historical normal group, such that obesity is a BMI greater than the 95th percentile. The sensitivity of BMI for diagnosing obesity and overweight varied considerably; specificity was less variable.[62] Another inability of BMI is that it does not provide any metabolic or etiologic information of the individuals with obesity. These limitations put BMI in the center of the controversies regarding the diagnosis and treatment of obesity.[63] Other indices such as WC and WHR have also been used as predictors of T2D and cardiovascular events associated with obesity in many metabolic and epidemiological studies.[64–66]

2.2 Waist Circumference and Waist–Hip Ratio

Visceral adipose tissue (VAT) has been found to promote dyslipidemia, insulin resistance, and hypertension.[67–70] Therefore, abdominal obesity is closely associated with cardiometabolic risks. Abdominal VAT stores can be measured by computerized axial tomography, magnetic resonance imaging (MRI), and dual energy X-ray absorptiometry, but these techniques are too expensive and not feasible for everyday clinical use. Anthropometric measurement of abdominal obesity using WC and WHR are the most common proxy measures of VAT. Both measurements are well correlated with VAT. It seems that WC is more strongly associated with VAT, while WHR may be a better predictor of CVD risk as hip circumference is inversely associated with the development of cardiometabolic risk factors and CVD.[64–66]

2.3 Body Fat Percentage

Based on various theoretical approaches to the relationships between body fat percentage (BF%) and health, different recommendations for ideal BF% have been developed. Epidemiologically, BF% in an individual varies according to sex and age.[71] For example, a National Health and Nutrition Examination Survey of Americans from 1999 to 2004 found that females had higher mean percentage body fat than males at all ages. Male/female differences were smallest at age 8–11 years (3.9%) but increased to 12.0% at age 16–19 years. In males, mean percentage body fat ranged from 22.9% at age 16–19 years to 30.9% at age 60–79 years. In females, mean percentage body fat ranged from 32.0% at age 8–11 years to 42.4% at age 60–79 years.[71] Romero-Corral et al. reported that some subjects with a normal BMI but high BF% content, so called normal weight obesity (NWO), have a higher prevalence of cardiometabolic dysregulation and cardiovascular mortality.[72] However, in clinical practice, NWO is not commonly recognized as a unique obese phenotype with high risk for cardiovascular mortality, metabolic dysregulation, and poor functional outcomes.[73] The incorporation of BF% and fat distribution with BMI measurement in the clinical settings and the genetic mechanisms may allow more accurate identification of adiposity variants and adiposity-related metabolic status and long-term risk. This may also help identify more cases of NWO in clinics and implicate early lifestyle changes and behavioral modifications in these patients on an individual basis that are essential to the treatment of obesity in terms of precision medicine.

2.4 Obesity in Childhood, Adolescence, and Adulthood[47,74–78]

Recent systematic review and metaanalysis found that while persistence of obesity from children and adolescents to adulthood was high, most obesity-related adult morbidity occurs in adults who had a healthy childhood weight.[62] Around 55% of obese children go on to be obese in adolescence, around 80% of obese adolescents will still be obese in adulthood, and around 70% will be obese over age 30. Therefore, action to reduce and prevent obesity in these adolescents is needed. However, the majority (70%) of obese adults were not obese in childhood or adolescence, so targeting obesity reduction solely at obese or overweight children may not substantially reduce the overall burden of adult obesity and related morbidities.[79] The gene-by-age effects were attributed to the inconsistent replication of GWAS finding of 32 loci influencing BMI variants in European-American adults with those findings in other studies.[80]

2.5 Obesity-Related Conditions

Obesity is characterized by an imbalanced excess accumulation of body fat resulting from a mismatch between energy intake and expenditure that represents a conglomerate of traits, each one influenced by numerous variables such as behavior, diet, environment, social structures, metabolic factors involving various genetic and nongenetic factors.[81] It is not a surprise why obesity has such a far-ranging negative effect on health with multiple obesity-related conditions, which cost over 150 billion dollars and cause an estimated 300,000 premature deaths each year in the United States. The obesity-associated health problems and clinical phenotypes include, but are not limited to hypertension, heart disease (atherosclerosis and coronary artery disease), diabetes (obesity is the major cause of T2D and resistance to insulin), joint problems, sleep apnea and respiratory problems, asthma, cancer (breast cancer, colon, gallbladder, prostate, and uterus cancers), and metabolic syndrome. In addition, overweight and obesity are also highly related to psychosocial problems and bias, discrimination, and even torment.

The correlation of these conditions with obesity and the underlying etiological and molecular mechanisms have not been well studied and remain poorly understood.

3. GENETICS, HERITABILITY, AND GENOME-WIDE ASSOCIATION STUDY OF OBESITY[47,82–84]

While the exposure to obesogenic and other environmental factors is the main cause of the increase in the prevalence of high BMI over the last 30 years,[19] the clear differences in obesity susceptibility among individuals exposed to the same obesogenic environment implicate important genetic risk factors. Since the heritability for BMI is high the differences in obesity and others traits could arise primarily as a consequence of genetic factors. Screening for a number of monogenic obesity variants will provide more informed prognosis and help in the identification of at-risk individuals who could benefit from early intervention, in evaluation of the outcomes of current obesity treatments, and in personalization of the clinical management of obesity.

The field of the genetics of obesity was dominated by candidate gene studies examining the association of concrete polymorphisms in one or a few candidate genes with obesity and/or obesity-related phenotypes.[85–87] Candidate genes are those with higher prior probability for phenotypic involvement on the basis of different criteria including biology, pharmacology, transgenic and knockout murine models relevant to obesity.[88] The approaches used in the detection and analysis of a candidate gene in body weight regulation include linkage studies, candidate gene association studies, and GWAS.[89,90] These studies resulted in the suggestion of numerous genes involved in the development of monofactorial forms of obesity.[83] A large number of common variants have been associated with adiposity levels in recent GWAS and each of the variants account for only a small proportion of the predicted heritability.[30,91] Based on the genetic etiology three main categories of obesity are considered: monogenic nonsyndromic, monogenic syndromic, and polygenic obesity. For the monogenic forms of obesity, the gene causing the phenotypic obesity is clearly identified, whereas for the polygenic (or common) obesity the *loci* architecture underlying the phenotype remains to be characterized, which is usually thought to be complex to depend on genetic variations at several susceptibility loci with variable contributions from environmental factors such as age,[80,92] diet,[93–95] and physical activity.[95–97]

3.1 Monogenic Obesity

Monogenic obesity results from the mutation of a single gene and is rare, affecting ~5% of the population.[98] In 2002, Rankinen et al. reviewed the evidence from the rodent and human obesity cases caused by single-gene mutations, Mendelian disorders exhibiting obesity as a clinical feature; quantitative trait loci (QTLs) uncovered in human genome-wide scans and in crossbreeding experiments in various animal models, association and linkage studies with candidate genes and other markers.[99] More than 200 types of human obesity are associated with homozygous forms of a single-gene mutation. There are two forms of Mendelian inheritance of obesity: syndromic and nonsyndromic obesities. The majority of monogenic forms of obesity are characterized by an early onset of the disease and an extreme phenotype.[100] Twin studies have been used to model the genetic component of a given trait, due to the fact that monozygotic (MZ) twins are genetically identical, while nonidentical dizygotic (DZ) twins share only 50% of their genetic material.[101] Family and twin studies proved to be very successful in the detection of obesity-related mutations.[102] Studies in twins and adopted children found that genetic factors could have a much stronger effect than environmental factors on the BMI trends in children up to the age of 18 years. The identification of inborn deficiency of the mostly adipocyte-derived satiety hormone leptin in extremely obese children from consanguineous families paved the way to the first pharmacological therapy for obesity based on a molecular genetic finding.[102] Over the past two decades, several gene mutations have been reported to cause autosomal recessive and dominant forms of obesity. More than 200 single-gene mutations have been found to cause human obesity.[98,103] These mutations are rare and lead to extreme obesity with an early onset obesity and other endocrine disorders. There are eight well-known gene mutations in monogenic nonsyndromic, involving *LEP, LEPR*, *POMC*, *PCSK1*, *MC4R*, *BDNF, NTRK2*, and *SIM1*.[98] All these genes code for proteins with a central role in the leptin–melanocortin signaling pathway. Later, six more leptin–-melanocortin pathway genes including *POMC*, *PCSK1*, *MC4R*, *BDNF, NTRK2*, and *SIMI* were identified and validated.[98]

Syndromic obesity is defined as those obesity cases that occur in a distinct set of associated clinical phenotypes, such as mental retardation or organ-specific developmental abnormalities.[104] The genetic basis of these disorders is extremely heterogeneous.

WAGR syndrome is a rare genetic disorder characterized by a deletion at chromosome 11p13 in a region containing the Wilm's tumor 1 (*WT1*) and paired box 6 (*PAX6*) genes.[100] A specific type of WAGR has been associated with a deletion in the brain-derived neurotrophic factor (*BDNF*) gene, which results in an obese phenotype. Prader–Willi syndrome (PWS) can have several etiologies, characterized by central obesity, neonatal hypotonia, hyperphagia, hypothalamic hypogonadism, and mild mental retardation, with such abnormalities as short stature and peculiar facial features.[100] Most of the cases were associated with loss of expression from paternal deletions of the 15q11.2-q12 chromosomal region. Bardet–Biedl syndrome (BBS) is characterized by early onset obesity, which is associated with progressive conerod dystrophy, morphological finger abnormalities, dyslexia, learning disabilities, and progressive renal disease. BBS has extensive genetic heterogeneity with at least 14 *loci* and several mutations identified within these *loci*.[42] Alström (ALMS) and Cohen syndromes are associated with childhood mild truncal obesity and small stature. Both of them are autosomal recessive and genetically homogenous. ALMS is caused by a balanced translocation of chromosome 2p13 that disrupts *ALMS1* gene or by a small number of mutations in this gene. Cohen syndrome results from mutations in the *COH1* gene, located at chromosome 8q22, which encodes a transmembrane protein of unknown function.[98,100]

Despite the rarity of these monogenic forms of human obesity, their underlying genetic bases substantially helped our understanding of the pathogenesis of obesity and shed molecular light on several pathways and mechanisms involved in the development of obesity. Moreover, the study of monogenic forms of obesity has altered our perception of obesity as an endogenous disorder with variant molecular bases.

3.2 Polygenic Obesity

Polygenic obesity, also known as "common obesity," is defined as a result of the combined effect of variants in multiple genes acting in concert with environmental risk factors. However, the genetic and molecular mechanisms involved in body weight regulation are complex.[99,105] The polygenic multifactorial condition reflects the additive contribution of many genes conferring different degrees of susceptibility to obesity, with heritability levels ranging from 25 to 70% for BMI. When total adiposity is taken into account, heritability of abdominal obesity is on the order of 50%.[106]

In contrast to monogenic obesity, each polymorphism in polygenic obesity leads to a variant that confers susceptibility, requiring additionally the presence of other variants and an obesogenic environment to determine the obese phenotype.[107]

Recently, the next-generation sequencing (NGS) technology has been used to analyze 26 susceptible genes for obesity in Pakistani children with early onset obesity.[108] They found two new *LEPR* mutations at the homozygous state: a splice site mutation in exon 15 (c.2396-1 G > T), and a nonsense mutation in exon 10 (c.1675 G > A). From 524 severely obese and 527 lean Swedish children Sällman et al. amplified the entire region of the fat mass and obesity-associated gene (*FTO*) gene (412 kilobase pairs) and detected 705 single-nucleotide polymorphisms (SNPs), from which 19 were novel BMI- and obesity-associated polymorphisms within the first intron of the *FTO* gene.[109] An interesting finding was the fact that 10 of them have a stronger association with obesity ($p < 0.007$) when compared with the commonly studied rs9939609 polymorphism ($p < 0.012$). This study concluded that within the entire region of the *FTO* gene the first intron was the only one associated with obesity.[109] Using NGS, Bonnefond et al. blindly reanalyzed mutations in 40 patients carrying a known causal mutation for those subtypes according to diagnostic laboratories and reidentified all causal mutations in each patient associated with an almost-perfect sequencing of the targets except for one variant (mean of 98.6%).[110]

3.3 MicroRNA and Obesity

MicroRNAs (miRNAs) are involved in the posttranscriptional regulation of gene expression by binding to complementary sequences located in target mRNAs and leading to their translation repression or degradation. Numerous studies have described miRNA-induced shifts in metabolic pathways under various obesity-related disease settings[111] and in the development of obesity.[112–115] Microarray studies have highlighted an altered profile of miRNA expression in human and animal models of diabetes and obesity.[112,114,116] Emerging evidences suggest that miRNAs play significant roles in lipid metabolism.[113,117] They are also involved in many functional aspects of adipocyte differentiation and potentially contri-bute to the pathogenesis of obesity.[115,117–119] These studies revealed that miRNAs may represent biomarkers for obesity, and could also be implicated in the molecular mechanisms leading to this disease.[112,114–116,119–126]

3.4 Genome-Wide Association Study of Obesity

Using powerful statistical methods to identify loci associated with a particular phenotype, GWAS allows the scanning of numerous polymorphisms across the entire genome for common disease-associating SNPs in a hypothesis-free manner in large cohorts of familial-unrelated people. Since the start of the GWAS era in 2005, there have been five waves of GWAS' discoveries for BMI, which had identified more than 52 loci associated with obesity-related traits.[81,91] The first *loci* identified through GWAS was the *FTO* gene.[127] Later, Frayling et al. conducted a GWAS to test the correlation between polymorphisms across the entire human genome and T2D.[128] They found that a common variant, the rs9939609 polymorphism, located in the first intron of the *FTO* gene predisposes to diabetes through an increased BMI. The finding has been independently replicated and has consistently confirmed the association of rs9939609 polymorphism with the etiology of common obesity in several populations.[129–134]

Following the discovery of the *FTO* locus, investigators enhanced GWAS by increasing the sample size, improving statistical power to uncover additional obesity susceptibility *loci*. A large-scale international consortium, called the Genetic Investigation of Anthropometric Traits (GIANT) emerged. The association data of 16,876 Caucasians from seven GWAS for BMI were combined in a metaanalysis.[135] This study confirmed the strong association of obesity with polymorphisms in the *FTO* gene, and identified one new locus near the *MC4R* gene whose mutations are known to be the common cause of extreme childhood obesity.[129,135] The *MC4R* was the second gene significantly associated with common obesity.[129,135–138]

In the third wave of discoveries, a metaanalysis was performed using 15 GWAS for BMI in Caucasians ($n > 32{,}000$) and replicated in another 14 studies for a second-stage sample of 59,082 individuals.[137] They not only confirmed the association of the *FTO* and *MC4R* genes, but also found six new genes positively associated with obesity: *MTCH2*, *GNPDA2*, *KCTD15*, *SH2B1*, *NEGR1*, and *TMEM18*.[137]

In 2010, as the fourth wave of GWAS, the GIANT consortium expanded to include 249,796 individuals of European origin, and revealed 18 new *loci* associated with BMI near or in: *PRKD1*, *SLC39A8*, *GPRC5B*, *MAP2K5*, *QPCTL*, *RBJ*, *LRRN6C*, FLJ35779, *CADM2*, *TMEM160*, *FANCL*, *LRP1B*, *TNNI3 K*, *MTIF3*, TFAP2B, *ZNF608*, NRXN3, *RPL27A*, *PTBP2*, and *NUDT3*.[139] By 2011, therefore, GWAS had identified 32 genetic *loci* unequivocally associated with phenotypes of BMI.

The most recent and fifth wave expanded the GIANT metaanalysis, to comprise 263,407 individuals of European ancestry.[120] Besides confirming all the 32 BMI-associated *loci* previously identified by the fourth wave, they found seven new *loci*, *ZZZ3*, *RPTOR*, *ADCY9*, *GNAT2*, *MRPS33P4*, *HS6ST3*, and *HNF4G*, explaining an additional 0.09% of the variability in BMI.[120]

Although genetics of obesity and GWAS in obesity research have been studied extensively, none of these genes or combinations could be firmly validated in clinics. The effect sizes of obesity variants identified in GWAS are currently too small to be used for diagnosis and treatment in clinical settings.

4. LIFESTYLE AND ENVIRONMENTAL IMPACTS AND EPIGENETICS OF OBESITY

Obesity only emerges if food consumption exceeds the energy expenditure on a lasting basis, resulting in a prolonged positive energy balance. As described earlier, numerous studies support that the personal genetic profile could be a cause for individual differences in the predisposition to weight gain and obesity. It is interesting that most of the genes involved in the susceptibility of obesity are also related to food intake and regulation of energy balance. Over the last 30 years, the prevalence of obesity in many countries has increased threefold. It seems difficult to conjugate with the notion that genetics are the primary cause of obesity as revealed by twin and adoption studies. A number of studies indicate that the childhood obesity epidemic which has emerged in the last 30 years is a disease resulting from complex multifactorial interaction of susceptibility genes with an obesogenic environment.[93] Gene–diet interactions, in particular the specific genes that were identified through GWAS to be associated with childhood obesity (FTO, MC4R, and NPC1), may have a prominent role in promoting childhood obesity. The increase in the prevalence of obesity could be attributed primarily to high-calorie food intake together with the sedentary lifestyle of modern societies, or to environmental changes. Moreover, epigenetic mechanisms, in which environmental factors cause changes in the expression of genes thus the genetic background, could also help in explaining the observed increase in obesity prevalence.

4.1 Lifestyle and Environment on Obesity

Heritability represents the proportion of phenotypic variation among individuals due to genetic contribution. Hence, it is not surprising that one important risk factor for childhood and adolescent obesity is parental obesity. Whitaker et al. found that when both parents are obese there is an increase of more than double of the risk for childhood obesity.[140] However, most of the studies found a small to medium effect of parental obesity as risk factor for childhood obesity.[140–143] Other studies have found a stronger effect for maternal obesity compared to paternal obesity, which may reflect prenatal and postnatal environmental factors.[144–147] Moreover, maternal weight gain in pregnancy has been positively associated with BMI of the children into adulthood.[145,146,148]

Accumulated evidence from epidemiological studies and clinical trials have demonstrated the roles of lifestyle/dietary and genetic factors in the development of obesity.[149] The recognition that nutrients have the ability to interact and modulate molecular mechanisms underlying an organism's physiological functions has prompted the emerging fields of nutrigenetics, which studies the effect of genetic variation on nutrient requirements,[150,151] and nutrigenomics, which studies the interrelationships among diet, genetic makeup, and physiological responses at genome-wide level and in a systematic manner.[152,153] The nutrigenetics and nutrigenomics unravel the complex relationships between bioactive molecules, genetic polymorphisms and biological system and "dietary signatures" in specific cells, tissues, and organisms. Both are powerful approaches to understand how these signals influence homeostasis and health and thus regulate the progress of diet-related chronic diseases such as obesity,[154,155] metabolic diseases,[156] diabetes,[150,157] cancers,[158–162] inflammation,[163,164] and CVD.[165–169] The core concepts of nutrigenomics also imply that the individual genetic background can influence nutrient status, metabolic response to diets, and predisposition to diet-related diseases and thus may give rise to personalized nutrition and dietary recommendations.[156,170,171] The ingestion of nutrients introduces some bioactive molecules that carry information from the external environment.[172] Many dietary components can modulate epigenetic phenomena by inhibiting enzymes such as DNA methyltransferases and histone deacetylases, with the most well-known vitamin B-12 and folate providing methyl groups for DNA methylation reaction, and therefore can lead to different clinical phenotypes of obesity.[26,172–177]

A dietary intervention based on nutrigenetics or nutrigenomics could be helpful in prevention as a potential instrument that can complement dietary advice. However, application of nutrigenetics is currently limited by the high cost of the genetic analyses and the lack of studies analyzing the effects of common polymorphisms and polymorphisms with different ethnic background on obesity. More generally, compliance with nutrient-based recommendations, such as reducing intake of fat and sugar, has been very poor.[178–181] The mechanisms of person-to-person and population differences in response to food, and the ways in which food variably impacts the host, for example, nutrient-related disease outcomes are the major focus of nutrigenomics. While the nutrigenomics and related nutrition sciences are established, the efforts are also emerging to integrate the four major Big Data domains (agrigenomics, nutrigenomics, nutriproteomics, and nutrimetabolomics) that address complementary variability questions pertaining to individual differences in response to food-related environmental exposures.[182] The convergence of nutrigenomics with nutriproteomics, nutrimetabolomics, and agrigenomics may provide a robust basis for a trustworthy and sustainable precision nutrition 4.0 agenda for the guidance of dietary intervention helpful in prevention of food- and weight-related diseases.[182] Toward the precision medicine Pavlidis et al. recently presented the new concept of "Nutrigenomics 2.0," so as to cultivate and catalyze the next-generation research and funding priorities for responsible and sustainable knowledge-based innovations.[183] Through a study of the 38 genes included in commercially available nutrigenomics tests, they offered additional context in relation to the 2014 American Academy of Nutrition and Dietetics position. In the best interest of the nutrigenomics science community, governments, global society, and commercial nutrigenomics test providers, new evidence evaluation and synthesis platforms should be created with nutrigenomics tests before they become commercially available. The proposed assessment and synthesis of nutrigenomics data should be carried out on an ongoing dynamic basis with periodic intervals and/or when there is a specific demand for evidence synthesis, and importantly, in ways that are transparent where conflict of interests are disclosed fully by the involved parties, be they scientists, industry, governments, citizens, social scientists, or ethicists.[182,183]

4.2 Drug–Genotype Interactions in Obesity

The use of antiobesity drugs as a treatment option for obesity could be indicated for individuals with a BMI of 30 or greater, especially those with existing comorbidities such as diabetes, dyslipidemia, or hypertension.[12,184–186] As described earlier, common genetic variation is associated with increased risk of common metabolic diseases such as obesity. It is therefore not surprising that common polymorphisms also alter the response to pharmacotherapy affecting drug metabolism, drug transport, or drug targets.[12,185,187,188] At least 35 *loci* were validated as being associated with BMI and the advent of GWAS and NGS will likely lead to the identification of additional genetic biomarkers. In the future, it may be possible to determine which subpopulations will respond optimally to particular doses of drugs, allowing more effective personalized pharmacologic intervention. To achieve this end, it would be ideal if pharmacogenetic studies could identify differences in drug response and tolerability, and investigate gene regulation, epigenetic modifications, and DNA–protein interactions that could explain individual differences in responses to drugs beyond genetic variation. Ultimately, it will also be necessary for clinical trials to evaluate pharmacologic interventions that are guided by genetic tests. This will require large sample sizes to detect a weak to moderate genetic predisposition to disease, the need to reproduce such associations in independent cohorts, and the statistical criteria required to detect a true association.[12]

4.3 Physical Activity–Genotype Interactions in Obesity

Physical activity is another important component involved in the heterogeneous set of factors influencing obesity. Regular exercise is one of the most promising behavioral candidates for preventing and reducing weight gain, with other health and psychological benefits.[189] The most extensively studied example of a gene interaction with physical activity in obesity is the *FTO locus*; evidencing that physical activity attenuates the association of *FTO* variants with obesity-related traits.[189–192] Higher levels of physical activity may attenuate the influence of obesity susceptibility polymorphisms on BMI during adolescence. However, several studies have provided evidence that the propensity to be physically active also has a strong genetic component in both animals and humans.[193,194] In humans, physical activity has been shown to aggregate in families; more active parents have more active children relative to inactive parents. It appears that some variation in our DNA could contribute to the variation in the physical activity level. Thus, new studies and the identification of new loci implicated in this interaction could better enlighten and help to understand the causes contributing to the development of obesity.[195]

4.4 Epigenetics and Epigenome of Obesity

Epigenetics studies the heritable changes in gene expression that are caused not by the underlying DNA sequences but by the environmental impacts. These epigenetic gene regulation processes include DNA methylation, covalent histone modifications, chromatin folding, and the regulatory action of miRNAs and polycomb group complexes.[196,197] Over the last decade, there has been increasing interest in the role of epigenetics in the development of complex conditions such as obesity.[196–201] Epigenetic regulation of gene expression emerged as a potential factor that might explain individual differences in obesity risk. In contrast to genetic modifications, epigenetics refer to mitotically heritable modifications that regulate gene activity and expression but do not involve changes in DNA sequence.[201–204] At the molecular level, epigenetic marks including genomic DNA methylation, changes in chromatic organization by histone modifications, the noncoding miRNAs, genomic imprinting, noncovalent mechanisms, and other nuclear proteins can be programmed already in the intrauterine environment or can be modulated by environmental influences including diet.[201,203,205] Thus the epigenome is seen as a malleable interface between the environment and the genome. The changes in epigenome at critical developmental stages can be shaped by the environment and affects health and susceptibility to disease in later life, including obesity and metabolic syndrome.[201,203,205–207]

Methylation, a widespread feature of the genome, is the most well-known epigenetic marker, which has been proposed as a new generation of biomarkers. It is a biologic process that consists of the addition of a methyl group at the carbon-5 position of cytosine, in the context of the CpG dinucleotides, and usually associated with gene silencing in the promoter regions.[208] Methylation in a promoter region leads to the repression of gene expression, which may be achieved by a number of mechanisms including: obstructing access to transcription factors/activators and recruitment of corepressors. The universal methyl donor is DNA methyltransferases (Dnmts) that maintain the cellular DNA methylation patterns.[202]

Many "obesity genes" critical to energy balance are regulated by epigenetic mechanisms depending on nutritional clues. Using a genome-wide approach, obesity has been related to changes in DNA methylation status in peripheral blood leukocytes of lean and obese adolescents for two genes. In the ubiquitin-associated and SH3 domain-containing protein A (*UBASH3A*) gene, a CpG site showed higher methylation levels in obese cases, and one CpG site in the promoter region of Tripartite motif-containing 3 (*TRIM3*) gene, showed lower methylation levels in the obese cases.[209] Very recently, Huang et al. performed genome-wide methylation analysis and identified differentially methylated CpG loci associated with severe obesity in childhood. They provided convincing evidence that childhood obesity is associated with specific DNA methylation changes in whole blood, which may have utility as biomarkers of obesity risk.[210] The obesity risk allele of *FTO* has been associated with higher methylation of sites within the first intron of the *FTO* gene, suggesting an interaction between genetic and epigenetic factors.[210,211] Global DNA methylation changes were found in 17,975 individual CpG sites altering the levels of DNA methylation in response to physical activity.[212] The majority of studies examining the relationship between site-specific DNA methylation and obesity are cross-sectional, thus most of these DNA methylation sites need to be confirmed as being associated with obesity. Both methylation levels and the phenotype are measured at the same time point. Hence, it cannot be established whether the association between a specific DNA methylation mark and obesity is a cause or a consequence of the obese phenotype. However, the high number of new studies concerning obesity epigenetics will undoubtedly permit the confirmation of some of these associations, thereby establishing an epigenetic basis for human obesity.

Despite the high number of DNA methylation candidate genes identified in epigenetics, epigenomics, and recently, epigenome-wide association studies (EWAS), most of the associations have not yet been replicated in other samples to further confirm and establish whether those *loci* are reliably associated with obesity.[213–215] To what extent epigenetic modifications contribute to the total heritability and phenotype of obesity is presently unknown. Continuous advances in research show promising results about the implication of epigenetic mechanisms in the etiology of obesity. Epigenetics has shown that our genes are not the only factor to determine our phenotypes and that our behaviors can alter the expression of our genotypes.

5. PHENOME-WIDE ASSOCIATION STUDY OF OBESITY[216]

As described earlier, human adiposity is highly heritable with the estimated 20–90% genetic contribution to individual differences in relative body weight as estimated by BMI.[23–25] The identification of genetic variants associated with human adiposity and BMI in "omic" scale especially GWAS,[21,30,33–37] however, has revealed that multigenetic variants do not have direct correlations with the current clinical classification of obesity, which is not based on its etiology.[46,47] For example, while the most strongly associated variant at FTO only explains 0.34% of the phenotypic variance for

BMI in the general population the sum of 32 variants from GWAS increases the explained phenotypic variances to 1.45%, with each additional risk allele increasing BMI by 0.17 kg/m^2. Individuals carrying the lowest number of risk alleles have only an average BMI of 2.73 kg/m^2 lower than those carrying the highest number of risk alleles.[139] Furthermore, it is also well-known that, essential to the functionality of the human genome, a single genetic variant can be associated with multiple phenotypes, that is, the pleiotropy.[217,218] Through comparing multiple GWAS and candidate gene studies, pleiotropy has been noted in many SNPs and genes, potentially providing greater insight into putative biological mechanisms for pleiotropy.[218–221] In addition, multiple sociocultural, socioeconomic, lifestyle, behavioral, and biological factors also play important roles in the establishment and perpetuation of obese phenotypes or traits.[46–51] Unfortunately, our current clinical classification of human adiposity tends to ignore the diverse causes of obesity with measurement of BMI, WC, or WHR, and total cardiovascular risk factors but no considerations of etiology or genetic variants (Section 2). The gap between the narrowly defined obese phenotypes and the broader omic-scaled genotypes of obesity-associated genetic variants is getting wider and deeper. We should clearly recognize that a full understanding of the relationship between the diverse clinical forms of obesity and their different etiologies, including genetic and environmental mechanisms, is essential for improving risk prediction and facilitating more specific and personalized therapy for obesity.[222,223] Obesity is among the recognized public health relevant risk factors such as smoking, air pollution, and physical inactivity that are common to many noncommunicable diseases (NCDs). GWAS has identified pleiotropic genes and genetic variants linking NCD entities hitherto thought to be distant or unrelated in etiology. Accumulating evidence suggest that NCD disease mechanisms are in part shared. Therefore, in the era of precision medicine, the clinical phenotypes of human obesity need to be redefined in the "omic" scale with much greater details (or deep phenotyping) than using the underlying molecular causes as revealed from GWAS and other factors in addition to traditional signs and symptoms such as increased body weight and BMI.[53,54]

The availability of ever-increasing DNA biobanks linked to rich resources of physiological and pathological phenotypes and large epidemiological databases of body weight and chronic diseases including diabetes, cancer, and CVD has enabled the development of PheWAS as an additional approach and tools to investigate pleiotropy in obesity and related diseases.[46,224,225] As a counterpart of GWAS, PheWAS will validate genotype–phenotype associations identified not only by traditional GWAS but also through the generation of new hypotheses for potentially novel associations and putative instances of pleiotropy.[46,224–226] PheWAS has recently been used to enhance our understanding of the genetic determinants of complex traits discovered through GWAS.

5.1 Phenome and Phenomics of Obesity

A phenome is the sum of complete phenotypic characteristics (phenomic traits, Fig. 1) that signify the expression of the whole genome, proteome, and metabolome under specific environmental influence.[227–230] Phenomics is a

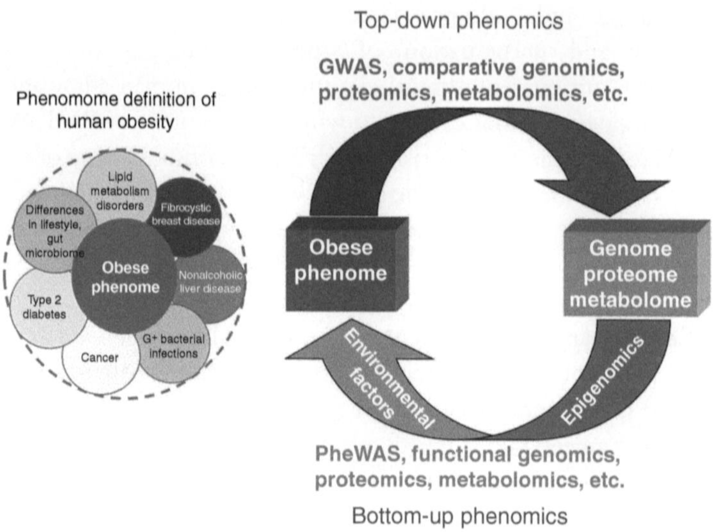

Figure 1 Illustration of phenomics approach to the new taxonomy for diagnosis and treatment of human obesity. The new taxonomy of human obesity is based on the clinical phenome, which is the sum of complete phenotypic characteristics (phenomic traits) that signify the expression of the whole genome, proteome, and metabolome under specific environmental influence (epigenome). Obese phenotypes will be systematically integrated and defined as a new obese phenome according to the association with other related (clustered) clinical phenotypes such as T2D, fibrocystic breast disease, nonalcoholic liver disease, gram-positive (G+) bacterial infections, lifestyle-related risk factors and gut microbiome, lipid metabolism disorders, cancer, etc. Multiple disease genes, SNPs, or proteins are identified under the scope of the whole genome, proteome, metabolome, using either the top-down (comparative genomics, proteomics, metabolomics, GWAS, etc.) or bottom-up (functional genomics, proteomics, metabolomics, PheWAS, etc.) or both strategies. Therefore, the obese phenome will now include not only clinical symptoms and signs of elevated BMI, WC, BF% but also a series of other systematically defined phenotypic characteristics at different levels, including metabolites, proteins, and genes.

recently developed new transdiscipline that studies phenome in order to correlate complex traits to variability not only in genome, but also in proteome, metabolome, interactome, and environmental impacts.[227,229,231–237] Phenomics is a multiphenotyping approach that requires strategic and comprehensive collection of a wide breadth of phenotypes with fine resolution and phenomic analysis and evaluation of patterns and relationships between individuals with related phenotypes and between phenotype–genotype associations.[238–240] It provides a suite of new technologies and platforms for the transition from focused phenotype–genotype study to a systematic phenome–genome approach (Fig. 2). Clinical phenomics is the systematic measurement and analysis of qualitative and quantitative clinical phenome (traits), including clinical signs and symptoms, and laboratory results obtained by biochemical, genomic, proteomic, metabolomic, and imaging methods, for the refinement and characterization of a clinical phenome, therefore can be used to redefine the clinical phenotypes of diseases.[238,241]

What will emerge from a phenomics (including clinical phenomics) approach is a more valid and etiologically based systematic definition of disease phenome (Fig. 1) that may be quite different from those of current

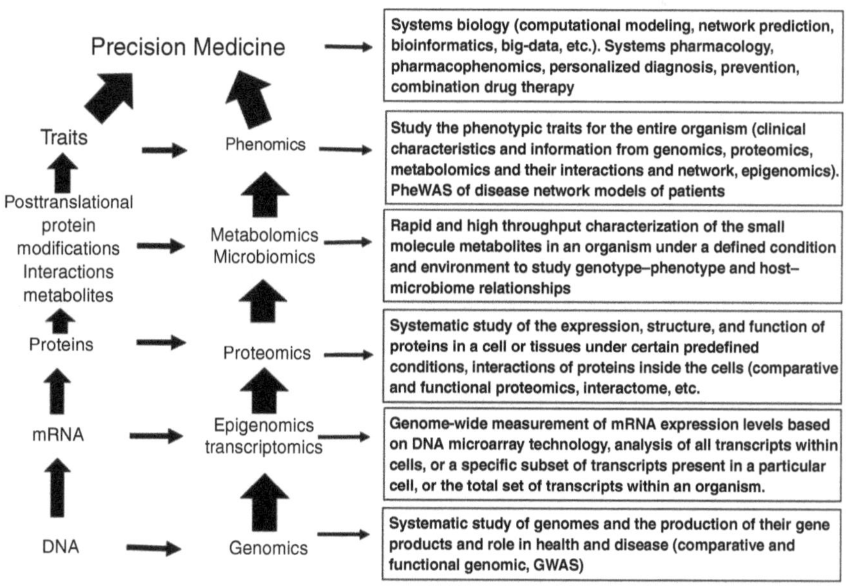

Figure 2 The platforms for GWAS and PheWAS of obesity. Phenomics is a novel transdiscipline as a complimentary approach to several other omics approaches, including genomics, transcriptomics, proteomics, and metabolomics in the study of human obesity.

disease phenotypes defined by using the clinical symptoms at an organ or system level alone, as has been the tradition in clinical practice of disease diagnosis and treatment of obesity for the past centuries. Accordingly, disease is now defined as a clinical phenome that is the sum total of a patient's clinical characteristics or phenomic traits that are systematically integrated (or clustered) to signify the expression of the whole genome, proteome, and metabolome under specific environmental influence (Fig. 1). Identification of multiple disease genes or proteins at the genomic and proteomic levels provide a solid platform for novel definition of the clinical phenomes. Therefore, the obese phenome will now include not only clinical symptoms of increased BMI or BF% but also a series of other systematically defined phenotypic characteristics at different levels, including genetic variants, proteomic and metabolomic profiles, lifestyles, complications in cardiovascular and other systems (Fig. 1). With well-defined obese phenomes and a suite of new phenomics technologies and platforms (Fig. 2), the phenomics approach may be used to characterize the clinical phenomes of human obesity and also to identify the corresponding therapeutic targets for combination therapy at the level of systems biology. Therefore, as a counterpart of genomics, proteomics, and metabolomics approaches, phenomics will not only refine the definition and diagnosis of human disease phenome with a new concept of wholism but also reform clinical treatment of disease with systematically defined therapeutic targets and improve predictive validity for outcomes of drug treatment.

Understanding the complexity and interrelation of risk factors and networks of disease phenomes requires the establishment of cohorts and biobanks for the collection of biologic samples, detailed and comprehensive phenotyping and genotyping, and broad risk-factor data. Many clinical cohort and DNA biobanks with rich resources of physiological and pathological phenotypes and large epidemiological databases of body weight and chronic diseases such as diabetes, cancer, and CVD are now available for GWAS and PheWAS to investigate pleiotropy in obesity and related diseases.[46,224,225] In 2007, the Electronic Medical Records and Genomics (eMERGE) Network was formed to use phenotypes derived from electronic health record (EHR) data to perform GWAS and other genomic investigations.[242–246] eMERGE investigators have also used EHR-based PheWAS methods to evaluate multiple phenotypes associated with specific genetic variants.[46,245,247] In 2001, Jackson Laboratory launched a Mouse Phenome Database (MPD) as the data coordination center for the international Mouse Phenome Project (phenome.jax.org).[248] MPD integrates quantitative phenotypes (grouped by mouse strains, subject area, ontology, targeted gene/protein etc.) with gene expression and genotype data into a common annotated framework. MPD now contains over 1330 strains of mice with >3500 phenotype measurements or traits relevant to human health and disease, including obesity, CVDs, cancer, infectious disease susceptibility, blood disorders, aging, neurosensory disorders, drug addiction, and toxicity. In addition, the MPD gene expression sector contains 12 million mean data points from 13 projects, representing 125,000 probe sets, 1.8 + billion SNPs, 18 + million indels and 600,000 structural variants consolidated from 18 community sources. MPD also provides an extensive library of detailed validated protocols for users to compare their own experimental results. Most data sets are directly associated with a peer-reviewed publication, providing an important layer of data validation. Several web-enabled tools for data analysis and visualization tools, public ontology annotated phenotype data and browser, and SNP query functionality with high-density coverage are also available for phenomic studies.[248] Therefore, MPD provides a very good source in the selection of mouse strains for phenomic study of obesity.

5.2 PheWAS of Obesity

PheWAS utilizes phenomics and big-data technologies to analyze all genetic/proteomic variants and all available phenotypic information from electronic medical records (EMRs), electronic health records (EHRs), or observational cohort containing all types of diagnosis of clinical phenotypes such as data from biobanks, the Clinical Data Warehouse (CDW), in the estimation of genome–phenome association and detection of pleiotropy.[247,249,250] With PheWAS, the genome-phenome associations between SNPs in a genome and a wide range of physiological and/or clinical phenotypes in a phenome can be explored by using algorithms to analyze the data collected either from EHRs, EMRs, CDW, or from observational cohort studies. In 2010, Denny et al. proposed to use PheWAS, as a mimic to the GWAS, to screen phenomic data for disease–gene associations in validating genetic associations derived from traditional genetic studies as well as identifying novel genetic associations.[247] PheWAS has now been used to investigate whether SNPs associated with one phenotype are also associated with other phenotypes.[46,247,249–252] For example, Denny and coworkers used phenotypic data from the EMRs of 13,835 individuals to look for associations between 1,358 phenotypes and 3,144 SNPs that had previously been found to show association with one or more traits in GWAS.[253] They found that 66% of associations from GWAS were replicated in cases for which the PheWAS was sufficiently powered. In addition, they also uncovered 63 cases of previously unknown associations, potentially as pleiotropic effects.[253]

Robust test of the EMR/EHR-based PheWAS allows unbiased interrogation across all domains of disease (cancers, diabetes, hypertension, stroke, brain diseases, heart diseases, etc.), and can not only replicate what is known about individual genotype–phenotype associations with various SNPs but also uncover novel associations with a wide range of phenotypes in the EMR/EHR-based cohorts. Larger EMR/EHR-based PheWASs may reveal more pleiotropy than has been estimated from GWAS and have the potential to significantly improve our understanding of the molecular etiologies of diseases. With the fast advance in big-data technology and phenomics, the application of the EMR/EHR data-based PheWAS provides important avenues to enhance systematically integrated analysis of the genomic basis of human disease. The EMR-based PheWAS provides a much simpler approach to pleiotropy analysis than the current GWAS-based approach, which requires complex integration of data from multiple studies. A recent application of PheWAS to 3,144 GWAS-identified SNPs (as mediators of human traits) and 1,358 EMR-derived phenotypes in 13,835 individuals of European ancestry, replicated 66% (51/77) of sufficiently powered prior GWAS associations, 210 known associations, and revealed 63 new pleiotropic associations.[253] These findings validate PheWAS as a tool to allow unbiased interrogation across multiple phenotypes in EMR-based cohorts and to enhance analysis of the genomic basis of human disease.

The predisposition to weight gain and clinical phenotypes of obesity vary significantly from person to person due to differences in the personal genetic profile, lifestyle, and environmental impacts. Although extensive genetics and GWAS of obesity will continue to identify and characterize obesity-associated SNPs and genetic variants, firm validation of these genes or combinations in clinics is still very limited. It has been noted that some of the *FTO* variants, including SNPs rs9939609 and rs8050136, are associated not only with obesity but also with T2D.[254–256] The SNP rs8050136 is located in an intronic region where the transcription factor cut-like homeobox (CUTL1) protein is predicted to bind.[257] This variant has been associated with T2D and obesity in Han Chinese and European populations[133,256,258] but other studies found no association between this variant and T2D or obesity in the Chinese Han population.[259,260] These differences in associations of SNPs with phenotypes have been further analyzed through fine mapping of BMI loci.[261] This study reported that GWAS studies primarily reformed in European populations of numerous loci associated with BMI are not generalizable to other ethnic groups, for example African Americans.[261] A more recent study noted that the mechanism of action for common variants in *FTO* may be through regulation of IRX3 expression, which is highly expressed in the brain.[262] There is also evidence of other putative disease associations with *FTO* variants that have not achieved genome-wide significance, such as endometrial cancer, Alzheimer's disease, and alcoholism.[263–265] These varied disease–SNP associations suggest that SNPs in *FTO* may have pleiotropic effects and cause distinct clinical phenomes. In a recent EHR-based PheWAS, data sets from the eMERGE network[266–269] were used to explore pleiotropy of genetic variants in *FTO*, some of which have been previously associated with obesity and T2D.[46] A metaanalysis of two study populations, one of 10,487 individuals of European ancestry with genome-wide genotyping from the eMERGE Network and the other of 13,711 individuals of European ancestry from the BioVU DNA biobank at Vanderbilt genotyped using Illumina Human-Exome BeadChip, replicated the well-described associations between *FTO* variants and obesity and T2D, implicating these two clinical phenotypes may be closely associated in the same phenome. This PheWAS also demonstrated that *FTO* variant rs8050136 was significantly associated with sleep apnea although the association was attenuated after adjustment for BMI. Other novel phenotype associations with the *FTO* variants associated with obesity included fibrocystic breast disease (rs9941349), nonalcoholic liver disease, and gram-positive bacterial infections. *FTO* variants not associated with obesity demonstrated other potential disease associations including noninflammatory disorders of the cervix and chronic periodontitis. These results suggest that genetic variants in *FTO* may have pleiotropic associations, some of which are not mediated by obesity.[46] Further characterization in larger populations and more carefully defined phenotypes are needed to determine whether these associations are real and where they belong to the same or different phenomes.

5.3 Challenges and Paradigm Shift in Obesity Research

A thorough understanding of the determinants and the mechanisms accounting for obesity is the central challenge of obesity prevention and management. Obesity is a highly heterogeneous disorder with substantial interindividual differences in terms of the body composition and degree of adiposity and the accompanying complications. The development and prevalence of obesity are determined by a combination of multiple factors, such as genetic population substructure, environmental factors, economic disadvantages, psychosocial stress, cultural context and diversity, social status, and access to medical care, etc. While the evidence for a genetic component of obesity has been well-established in recent years it remains largely unknown about the evolution of human body-weight regulation and obesity development. The search for underlying genotypes that cause obesity has been challenging due to the complex interactions involved in the regulation of adiposity and body composition. Indeed, many of the individual genotypes (especially those obtained with a lower odds ratio) that have been associated with elevated BMI have not been replicated in a reliable fashion. With the advent of GWAS, the field of obesity research has taken huge steps forward in the understanding of the genetic underpinnings and their associations with different obese phenotypes. We now know that common forms of obesity are highly polygenic although each variant may contribute to the complicated phenotypes with very small effects. While many regions in the genome have been identified to be associated with obesity by GWAS, the causal genes and their effects on the obese phenome remain to be identified, characterized, and determined. Complete understanding of identified associations in candidate genes or GWAS is often hampered by the lack of data on specific functional significance of the polymorphisms. The vast majority of obesity susceptibility variants identified in GWAS lie in noncoding regions.[81] Additional challenges come from the fact that the contribution of a given SNP to obesity could be modulated by the presence of other SNPs in the same gene or other genes. Therefore, the influence of haplotypes and gene–gene interactions need to be considered. Despite the advances in the genetics of obesity, the combined effect of all loci identified so far account for only about 2–4% of the total heritability of common forms of obesity. These numbers confirm the complex nature of obesity and the challenge to identify additional factors that may unravel some of the missing or hidden heritability of obesity. Such factors may include interactions between multiple genes and environmental factors and the contribution of other types of variants not covered by current GWAS design, including low-frequency and rare variants, copy number variations, and epigenetic modifications.[81,91,270] Numerous studies have suggested that disruptions in the relative proportions of gut microbial populations may contribute to weight gain and insulin resistance through roles in polysaccharide breakdown, nutrient absorption, inflammatory responses, gut permeability, and bile acid modification. However, the majority of studies are performed with stool or colonic samples and may not be representative of the metabolically active small intestine. Elucidating the mechanisms by which gut microbes contribute to obesity remains a challenge in the near future.

The current BMI, WHR, and BF% measurement-based clinical classification of obesity does not account for the wide interindividual variations in body composition, fat distribution, degree of adiposity or health risks, and genetic variants identified in the GWAS. This has created major challenges for a complete understanding of the relationship between the genomic/proteomic/metabonomic variability and the heterogeneous obese phenotypes in individual patients and the translation of the knowledge of GWAS, epigenomics, nutrigenomics, and microbiomics into the clinical practice of diagnosis, prevention, and therapy of obesity. The newly developed phenomics and PheWAS that assembles coherent sets of phenotypic features across individual measurements and diagnostic boundaries provide new and powerful approaches for identification and characterization of the genome–phenome relationship of obesity. The field of obesity research and clinical management is witnessing an ongoing paradigm shift from GWAS to PheWAS for the redefinition of clinical obesity classification according to the new category of clinical phenomes and may relieve the bottleneck of personalized medicine or precision medicine.

6. SUMMARY

Obesity represents a complex clinical health problem resulting from the interaction of multiple internal and external factors. Although most scientists and clinicians now acknowledge that genetic variants contribute to obesity, the specific *loci* involved and the mechanism by which they lead to the expression of obesity remain incompletely defined. Although remarkable advances in our understanding of the factors that give rise to obesity have occurred especially with GWAS of obesity, further research on the etiology, genomic variants, and other factors affecting obesity and the association of these to the complicated clinical obese phenome is still needed and probably will be a hot topic in obesity research in the years to come. The ongoing paradigm shift from GWAS to PheWAS of obesity for the redefinition/reclassification of clinical obesity according to the new category of clinical phenomes may relieve the bottleneck of obesity research for personalized medicine or precision medicine.

ACKNOWLEDGMENTS

Research in the Laboratory of Cardiovascular Phenomics in the Center for Molecular Medicine and Department of Pharmacology, School of Medicine at the University of Nevada is supported by NIH grants #HL106252 and #HL113598.

REFERENCES

1. Ajani UA, Lotufo PA, Gaziano JM, Lee IM, Spelsberg A, Buring JE, Willett WC, Manson JE. Body mass index and mortality among US male physicians. *Ann Epidemiol.* 2004;14:731–739.
2. Behn A, Ur E. The obesity epidemic and its cardiovascular consequences. *Curr Opin Cardiol.* 2006;21:353–360.
3. DiBaise JK, Foxx-Orenstein AE. Role of the gastroenterologist in managing obesity. *Expert Rev Gastroenterol Hepatol.* 2013;7:439–451.
4. Forouhi NG, Wareham NJ. Epidemiology of diabetes. *Medicine (Abingdon).* 2014;42:698–702.
5. James WP. WHO recognition of the global obesity epidemic. *Int J Obes (Lond).* 2008;32 (suppl 7):S120–S126.
6. Wang Y, Chen X, Klag MJ, Caballero B. Epidemic of childhood obesity: implications for kidney disease. *Adv Chronic Kidney Dis.* 2006;13:336–351.
7. Ng MC, Shriner D, Chen BH, Li J, Chen WM, Guo X, Liu J, Bielinski SJ, Yanek LR, Nalls MA, Comeau ME, Rasmussen-Torvik LJ, Jensen RA, Evans DS, Sun YV, An P, Patel SR, Lu Y, Long J, Armstrong LL, Wagenknecht L, Yang L, Snively BM, Palmer ND, Mudgal P, Langefeld CD, Keene KL, Freedman BI, Mychaleckyj JC, Nayak U, Raffel LJ, Goodarzi MO, Chen YD, Taylor Jr HA, Correa A, Sims M, Couper D, Pankow JS, Boerwinkle E, Adeyemo A, Doumatey A, Chen G, Mathias RA, Vaidya D, Singleton AB, Zonderman AB, Igo Jr RP, Sedor JR, Kabagambe EK, Siscovick DS, McKnight B, Rice K, Liu Y, Hsueh WC, Zhao W, Bielak LF, Kraja A, Province MA, Bottinger EP, Gottesman O, Cai Q, Zheng W, Blot WJ, Lowe WL, Pacheco JA, Crawford DC, Grundberg E, Rich SS, Hayes MG, Shu XO, Loos RJ, Borecki IB, Peyser PA, Cummings SR, Psaty BM, Fornage M, Iyengar SK, Evans MK, Becker DM, Kao WH, Wilson JG, Rotter JI, Sale MM, Liu S, Rotimi CN, Bowden DW. Meta-analysis of genome-wide association studies in African Americans provides insights into the genetic architecture of type 2 diabetes. *PLoS Genet.* 2014;10:e1004517.
8. James PT. Obesity: the worldwide epidemic. *Clin Dermatol.* 2004;22:276–280.
9. Sha BY, Yang TL, Zhao LJ, Chen XD, Guo Y, Chen Y, Pan F, Zhang ZX, Dong SS, Xu XH, Deng HW. Genome-wide association study suggested copy number variation may be associated with body mass index in the Chinese population. *J Hum Genet.* 2009;54:199–202.
10. Pradhan AD, Skerrett PJ, Manson JE. Obesity, diabetes, and coronary risk in women. *J Cardiovasc Risk.* 2002;9:323–330.
11. Burgio E, Lopomo A, Migliore L. Obesity and diabetes: from genetics to epigenetics. *Mol Biol Rep.* 2015;42:799–818.
12. Guzman AK, Ding M, Xie Y, Martin KA. Pharmacogenetics of obesity drug therapy. *Curr Mol Med.* 2014;14:891–908.
13. Faulkner G, Cohn TA. Pharmacologic and nonpharmacologic strategies for weight gain and metabolic disturbance in patients treated with antipsychotic medications. *Can J Psychiatry.* 2006;51:502–511.
14. Wang Y, Beydoun MA. The obesity epidemic in the United States—gender, age, socioeconomic, racial/ethnic, and geographic characteristics: a systematic review and meta-regression analysis. *Epidemiol Rev.* 2007;29:6–28.

15. Levian C, Ruiz E, Yang X. The pathogenesis of obesity from a genomic and systems biology perspective. *Yale J Biol Med.* 2014;87:113–126.
16. Rey-Lopez JP, de Rezende LF, Pastor-Valero M, Tess BH. The prevalence of metabolically healthy obesity: a systematic review and critical evaluation of the definitions used. *Obes Rev.* 2014;15:781–790.
17. Plourde G. Treating obesity. Lost cause or new opportunity? *Can Fam Physician.* 2000;46:1806–1813.
18. Williams EP, Mesidor M, Winters K, Dubbert PM, Wyatt SB. Overweight and obesity: prevalence, consequences, and causes of a growing public health problem. *Curr Obes Rep.* 2015;4:363–370.
19. Swinburn BA, Sacks G, Hall KD, McPherson K, Finegood DT, Moodie ML, Gortmaker SL. The global obesity pandemic: shaped by global drivers and local environments. *Lancet.* 2011;378:804–814.
20. Symonds ME, Sebert S, Budge H. The obesity epidemic: from the environment to epigenetics—not simply a response to dietary manipulation in a thermoneutral environment. *Front Genet.* 2011;2:24.
21. Almen MS, Jacobsson JA, Moschonis G, Benedict C, Chrousos GP, Fredriksson R, Schioth HB. Genome wide analysis reveals association of a FTO gene variant with epigenetic changes. *Genomics.* 2012;99:132–137.
22. Xu X, Zeng H, Xiao D, Zhou H, Liu Z. Genome wide association study of obesity. *Zhong Nan Da Xue Xue Bao Yi Xue Ban.* 2013;38:95–100.
23. Maes HH, Neale MC, Eaves LJ. Genetic and environmental factors in relative body weight and human adiposity. *Behav Genet.* 1997;27:325–351.
24. Nan C, Guo B, Warner C, Fowler T, Barrett T, Boomsma D, Nelson T, Whitfield K, Beunen G, Thomis M, Maes HH, Derom C, Ordonana J, Deeks J, Zeegers M. Heritability of body mass index in pre-adolescence, young adulthood and late adulthood. *Eur J Epidemiol.* 2012;27:247–253.
25. Willyard C. Heritability: the family roots of obesity. *Nature.* 2014;508:S58–S60.
26. Desai M, Jellyman JK, Ross MG. Epigenomics, gestational programming and risk of metabolic syndrome. *Int J Obes (Lond).* 2015;39:633–641.
27. Almen MS, Nilsson EK, Jacobsson JA, Kalnina I, Klovins J, Fredriksson R, Schioth HB. Genome-wide analysis reveals DNA methylation markers that vary with both age and obesity. *Gene.* 2014;548:61–67.
28. Lindgren CM, Heid IM, Randall JC, Lamina C, Steinthorsdottir V, Qi L, Speliotes EK, Thorleifsson G, Willer CJ, Herrera BM, Jackson AU, Lim N, Scheet P, Soranzo N, Amin N, Aulchenko YS, Chambers JC, Drong A, Luan J, Lyon HN, Rivadeneira F, Sanna S, Timpson NJ, Zillikens MC, Zhao JH, Almgren P, Bandinelli S, Bennett AJ, Bergman RN, Bonnycastle LL, Bumpstead SJ, Chanock SJ, Cherkas L, Chines P, Coin L, Cooper C, Crawford G, Doering A, Dominiczak A, Doney AS, Ebrahim S, Elliott P, Erdos MR, Estrada K, Ferrucci L, Fischer G, Forouhi NG, Gieger C, Grallert H, Groves CJ, Grundy S, Guiducci C, Hadley D, Hamsten A, Havulinna AS, Hofman A, Holle R, Holloway JW, Illig T, Isomaa B, Jacobs LC, Jameson K, Jousilahti P, Karpe F, Kuusisto J, Laitinen J, Lathrop GM, Lawlor DA, Mangino M, McArdle WL, Meitinger T, Morken MA, Morris AP, Munroe P, Narisu N, Nordstrom A, Nordstrom P, Oostra BA, Palmer CN, Payne F, Peden JF, Prokopenko I, Renstrom F, Ruokonen A, Salomaa V, Sandhu MS, Scott LJ, Scuteri A, Silander K, Song K, Yuan X, Stringham HM, Swift AJ, Tuomi T, Uda M, Vollenweider P, Waeber G, Wallace C, Walters GB, Weedon MN, Witteman JC, Zhang C, Zhang W, Caulfield MJ, Collins FS, Davey SG, Day IN, Franks PW, Hattersley AT, Hu FB, Jarvelin MR, Kong A, Kooner JS, Laakso M, Lakatta E, Mooser V, Morris AD, Peltonen L, Samani NJ, Spector TD, Strachan DP, Tanaka T, Tuomilehto J, Uitterlinden AG, van Duijn CM, Wareham NJ, Hugh W, Waterworth DM, Boehnke M, Deloukas P, Groop L, Hunter DJ, Thorsteinsdottir U, Schlessinger D, Wichmann

HE, Frayling TM, Abecasis GR, Hirschhorn JN, Loos RJ, Stefansson K, Mohlke KL, Barroso I, McCarthy MI. Genome-wide association scan meta-analysis identifies three Loci influencing adiposity and fat distribution. *PLoS Genet.* 2009;5:e1000508.
29. Scuteri A, Sanna S, Chen WM, Uda M, Albai G, Strait J, Najjar S, Nagaraja R, Orru M, Usala G, Dei M, Lai S, Maschio A, Busonero F, Mulas A, Ehret GB, Fink AA, Weder AB, Cooper RS, Galan P, Chakravarti A, Schlessinger D, Cao A, Lakatta E, Abecasis GR. Genome-wide association scan shows genetic variants in the FTO gene are associated with obesity-related traits. *PLoS Genet.* 2007;3:e115.
30. Fall T, Ingelsson E. Genome-wide association studies of obesity and metabolic syndrome. *Mol Cell Endocrinol.* 2014;382:740–757.
31. Thorleifsson G, Walters GB, Gudbjartsson DF, Steinthorsdottir V, Sulem P, Helgadottir A, Styrkarsdottir U, Gretarsdottir S, Thorlacius S, Jonsdottir I, Jonsdottir T, Olafsdottir EJ, Olafsdottir GH, Jonsson T, Jonsson F, Borch-Johnsen K, Hansen T, Andersen G, Jorgensen T, Lauritzen T, Aben KK, Verbeek AL, Roeleveld N, Kampman E, Yanek LR, Becker LC, Tryggvadottir L, Rafnar T, Becker DM, Gulcher J, Kiemeney LA, Pedersen O, Kong A, Thorsteinsdottir U, Stefansson K. Genome-wide association yields new sequence variants at seven loci that associate with measures of obesity. *Nat Genet.* 2009;41:18–24.
32. Maekawa R, Sato S, Yamagata Y, Asada H, Tamura I, Lee L, Okada M, Tamura H, Takaki E, Nakai A, Sugino N. Genome-wide DNA methylation analysis reveals a potential mechanism for the pathogenesis and development of uterine leiomyomas. *PLoS One.* 2013;8:e66632.
33. Ling H, Waterworth DM, Stirnadel HA, Pollin TI, Barter PJ, Kesaniemi YA, Mahley RW, McPherson R, Waeber G, Bersot TP, Cohen JC, Grundy SM, Mooser VE, Mitchell BD. Genome-wide linkage and association analyses to identify genes influencing adiponectin levels: the GEMS Study. *Obesity (Silver Spring).* 2009;17:737–744.
34. Dong C, Beecham A, Slifer S, Wang L, McClendon MS, Blanton SH, Rundek T, Sacco RL. Genome-wide linkage and peak-wide association study of obesity-related quantitative traits in Caribbean Hispanics. *Hum Genet.* 2011;129:209–219.
35. Liu AY, Gu D, Hixson JE, Rao DC, Shimmin LC, Jaquish CE, Liu DP, He J, Kelly TN. Genome-wide linkage and regional association study of obesity-related phenotypes: the GenSalt study. *Obesity (Silver Spring).* 2014;22:545–556.
36. Wheeler E, Huang N, Bochukova EG, Keogh JM, Lindsay S, Garg S, Henning E, Blackburn H, Loos RJ, Wareham NJ, O'Rahilly S, Hurles ME, Barroso I, Farooqi IS. Genome-wide SNP and CNV analysis identifies common and low-frequency variants associated with severe early-onset obesity. *Nat Genet.* 2013;45:513–517.
37. Jiao H, Arner P, Hoffstedt J, Brodin D, Dubern B, Czernichow S, van't Hooft F, Axelsson T, Pedersen O, Hansen T, Sorensen TI, Hebebrand J, Kere J, Dahlman-Wright K, Hamsten A, Clement K, Dahlman I. Genome wide association study identifies KCNMA1 contributing to human obesity. *BMC Med Genomics.* 2011;4:51.
38. Meyre D, Delplanque J, Chevre JC, Lecoeur C, Lobbens S, Gallina S, Durand E, Vatin V, Degraeve F, Proenca C, Gaget S, Korner A, Kovacs P, Kiess W, Tichet J, Marre M, Hartikainen AL, Horber F, Potoczna N, Hercberg S, Levy-Marchal C, Pattou F, Heude B, Tauber M, McCarthy MI, Blakemore AI, Montpetit A, Polychronakos C, Weill J, Coin LJ, Asher J, Elliott P, Jarvelin MR, Visvikis-Siest S, Balkau B, Sladek R, Balding D, Walley A, Dina C, Froguel P. Genome-wide association study for early-onset and morbid adult obesity identifies three new risk loci in European populations. *Nat Genet.* 2009;41:157–159.
39. Li J, Gui L, Wu C, He Y, Zhou L, Guo H, Yuan J, Yang B, Dai X, Deng Q, Huang S, Guan L, Hu D, Deng S, Wang T, Zhu J, Min X, Lang M, Li D, Yang H, Hu FB, Lin D, Wu T, He M. Genome-wide association study on serum alkaline phosphatase levels in a Chinese population. *BMC Genomics.* 2013;14:684.

40. Boraska V, Day-Williams A, Franklin CS, Elliott KS, Panoutsopoulou K, Tachmazidou I, Albrecht E, Bandinelli S, Beilin LJ, Bochud M, Cadby G, Ernst F, Evans DM, Hayward C, Hicks AA, Huffman J, Huth C, James AL, Klopp N, Kolcic I, Kutalik Z, Lawlor DA, Musk AW, Pehlic M, Pennell CE, Perry JR, Peters A, Polasek O, St Pourcain B, Ring SM, Salvi E, Schipf S, Staessen JA, Teumer A, Timpson N, Vitart V, Warrington NM, Yaghootkar H, Zemunik T, Zgaga L, An P, Anttila V, Borecki IB, Holmen J, Ntalla I, Palotie A, Pietilainen KH, Wedenoja J, Winsvold BS, Dedoussis GV, Kaprio J, Province MA, Zwart JA, Burnier M, Campbell H, Cusi D, Smith GD, Frayling TM, Gieger C, Palmer LJ, Pramstaller PP, Rudan I, Volzke H, Wichmann HE, Wright AF, Zeggini E. Genome-wide association study to identify common variants associated with brachial circumference: a meta-analysis of 14 cohorts. *PLoS One.* 2012;7:e31369.
41. Ollikainen M, Ismail K, Gervin K, Kyllonen A, Hakkarainen A, Lundbom J, Jarvinen EA, Harris JR, Lundbom N, Rissanen A, Lyle R, Pietilainen KH, Kaprio J. Genome-wide blood DNA methylation alterations at regulatory elements and heterochromatic regions in monozygotic twins discordant for obesity and liver fat. *Clin Epigenetics.* 2015;7:39.
42. Rahmioglu N, Macgregor S, Drong AW, Hedman AK, Harris HR, Randall JC, Prokopenko I, Nyholt DR, Morris AP, Montgomery GW, Missmer SA, Lindgren CM, Zondervan KT. Genome-wide enrichment analysis between endometriosis and obesity-related traits reveals novel susceptibility loci. *Hum Mol Genet.* 2015;24: 1185–1199.
43. Devuyst O. Genome-wide methylation and body-mass index. *Perit Dial Int.* 2014;34: 477.
44. den HM, Luan J, Langenberg C, Cooper C, Sayer AA, Jameson K, Kumari M, Kivimaki M, Hingorani AD, Grontved A, Khaw KT, Ekelund U, Wareham NJ, Loos RJ. Evaluation of common genetic variants identified by GWAS for early onset and morbid obesity in population-based samples. *Int J Obes (Lond).* 2013;37:191–196.
45. Di CA, Portincasa P. Fat, epigenome and pancreatic diseases. Interplay and common pathways from a toxic and obesogenic environment. *Eur J Intern Med.* 2014;25:865–873.
46. Cronin RM, Field JR, Bradford Y, Shaffer CM, Carroll RJ, Mosley JD, Bastarache L, Edwards TL, Hebbring SJ, Lin S, Hindorff LA, Crane PK, Pendergrass SA, Ritchie MD, Crawford DC, Pathak J, Bielinski SJ, Carrell DS, Crosslin DR, Ledbetter DH, Carey DJ, Tromp G, Williams MS, Larson EB, Jarvik GP, Peissig PL, Brilliant MH, McCarty CA, Chute CG, Kullo IJ, Bottinger E, Chisholm R, Smith ME, Roden DM, Denny JC. Phenome-wide association studies demonstrating pleiotropy of genetic variants within FTO with and without adjustment for body mass index. *Front Genet.* 2014;5:250.
47. Shabana. Hasnain S. Obesity, more than a 'cosmetic' problem. Current knowledge and future prospects of human obesity genetics. *Biochem Genet.* 2015;54:1–28.
48. Stenvinkel P. Obesity—a disease with many aetiologies disguised in the same oversized phenotype: has the overeating theory failed? *Nephrol Dial Transplant.* 2015;30: 1656–1664.
49. Ronn T, Volkov P, Gillberg L, Kokosar M, Perfilyev A, Jacobsen AL, Jorgensen SW, Brons C, Jansson PA, Eriksson KF, Pedersen O, Hansen T, Groop L, Stener-Victorin E, Vaag A, Nilsson E, Ling C. Impact of age, BMI and HbA1c levels on the genome-wide DNA methylation and mRNA expression patterns in human adipose tissue and identification of epigenetic biomarkers in blood. *Hum Mol Genet.* 2015;24:3792–3813.
50. Storlien L, Huang XF, Tapsell LC, Karlsson AC, Niklasson M, Pears J, Carlsson BC. Lifestyle-gene-drug interactions in relation to the metabolic syndrome. *World Rev Nutr Diet.* 2005;94:84–95.
51. Barua S, Junaid MA. Lifestyle, pregnancy and epigenetic effects. *Epigenomics.* 2015;7: 85–102.
52. Willard MD. Obesity: types and treatments. *Am Fam Physician.* 1991;43:2099–2108.

53. Duarte CW, Klimentidis YC, Harris JJ, Cardel M, Fernandez JR. Discovery of phenotypic networks from genotypic association studies with application to obesity. *Int J Data Min Bioinform.* 2015;12:129–143.
54. Herman MA, Rosen ED. Making biological sense of GWAS data: lessons from the FTO locus. *Cell Metab.* 2015;22:538–539.
55. Bray GA, Ryan DH. Clinical evaluation of the overweight patient. *Endocrine.* 2000;13:167–186.
56. Luca AC, Iordache C. Obesity—a risk factor for cardiovascular diseases. *Rev Med Chir Soc Med Nat Iasi.* 2013;117:65–71.
57. Tahrani A, Boelaert K, Barnes R, Palin S, Field A, Redwayne H, Aytok L, Rahim A. Body volume index: time to replace body mass index? *Endocr Abstr.* 2008;15:104.
58. Kyle UG, Piccoli A, Pichard C. Body composition measurements: interpretation finally made easy for clinical use. *Curr Opin Clin Nutr Metab Care.* 2003;6:387–393.
59. Alemany M. The etiologic basis for the classification of obesity. *Prog Food Nutr Sci.* 1989;13:45–66.
60. Keke LM, Samouda H, Jacobs J, di Pompeo C, Lemdani M, Hubert H, Zitouni D, Guinhouya BC. Body mass index and childhood obesity classification systems: a comparison of the French, International Obesity Task Force (IOTF) and World Health Organization (WHO) references. *Rev Epidemiol Sante Publique.* 2015;63: 173–182.
61. Freedman DS, Ogden CL, Berenson GS, Horlick M. Body mass index and body fatness in childhood. *Curr Opin Clin Nutr Metab Care.* 2005;8:618–623.
62. Simmonds M, Burch J, Llewellyn A, Griffiths C, Yang H, Owen C, Duffy S, Woolacott N. The use of measures of obesity in childhood for predicting obesity and the development of obesity-related diseases in adulthood: a systematic review and meta-analysis. *Health Technol Assess.* 2015;19:1–336.
63. Kim SH, Despres JP, Koh KK. Obesity and cardiovascular disease: friend or foe? *Eur Heart J.* 2015.
64. de Koning L, Gerstein HC, Bosch J, Diaz R, Mohan V, Dagenais G, Yusuf S, Anand SS. Anthropometric measures and glucose levels in a large multi-ethnic cohort of individuals at risk of developing type 2 diabetes. *Diabetologia.* 2010;53:1322–1330.
65. de Koning L, Denhoff E, Kellogg MD, de Ferranti SD. Associations of total and abdominal adiposity with risk marker patterns in children at high-risk for cardiovascular disease. *BMC Obes.* 2015;2:15.
66. Ashwell M, Gunn P, Gibson S. Waist-to-height ratio is a better screening tool than waist circumference and BMI for adult cardiometabolic risk factors: systematic review and meta-analysis. *Obes Rev.* 2012;13:275–286.
67. Kitagawa T, Yamamoto H, Sentani K, Takahashi S, Tsushima H, Senoo A, Yasui W, Sueda T, Kihara Y. The relationship between inflammation and neoangiogenesis of epicardial adipose tissue and coronary atherosclerosis based on computed tomography analysis. *Atherosclerosis.* 2015;243:293–299.
68. Abraham TM, Pedley A, Massaro JM, Hoffmann U, Fox CS. Association between visceral and subcutaneous adipose depots and incident cardiovascular disease risk factors. *Circulation.* 2015;132:1639–1647.
69. Scheuer SH, Faerch K, Philipsen A, Jorgensen ME, Johansen NB, Carstensen B, Witte DR, Andersen I, Lauritzen T, Andersen GS. Abdominal fat distribution and cardiovascular risk in men and women with different levels of glucose tolerance. *J Clin Endocrinol Metab.* 2015;100:3340–3347.
70. Luna-Luna M, Medina-Urrutia A, Vargas-Alarcon G, Coss-Rovirosa F, Vargas-Barron J, Perez-Mendez O. Adipose tissue in metabolic syndrome: onset and progression of atherosclerosis. *Arch Med Res.* 2015;46:392–407.

71. Jackson AS, Stanforth PR, Gagnon J, Rankinen T, Leon AS, Rao DC, Skinner JS, Bouchard C, Wilmore JH. The effect of sex, age and race on estimating percentage body fat from body mass index: the Heritage Family Study. *Int J Obes Relat Metab Disord.* 2002;26:789–796.
72. Romero-Corral A, Somers VK, Sierra-Johnson J, Korenfeld Y, Boarin S, Korinek J, Jensen MD, Parati G, Lopez-Jimenez F. Normal weight obesity: a risk factor for cardiometabolic dysregulation and cardiovascular mortality. *Eur Heart J.* 2010;31:737–746.
73. Jean N, Somers VK, Sochor O, Medina-Inojosa J, Llano EM, Lopez-Jimenez F. Normal-weight obesity: implications for cardiovascular health. *Curr Atheroscler Rep.* 2014;16:464.
74. Lobstein T, Jackson-Leach R, Moodie ML, Hall KD, Gortmaker SL, Swinburn BA, James WP, Wang Y, McPherson K. Child and adolescent obesity: part of a bigger picture. *Lancet.* 2015;385:2510–2520.
75. Chinn S. Definitions of childhood obesity: current practice. *Eur J Clin Nutr.* 2006;60: 1189–1194.
76. Doco-Fenzy M, Leroy C, Schneider A, Petit F, Delrue MA, Andrieux J, Perrin-Sabourin L, Landais E, Aboura A, Puechberty J, Girard M, Tournaire M, Sanchez E, Rooryck C, Ameil A, Goossens M, Jonveaux P, Lefort G, Taine L, Cailley D, Gaillard D, Leheup B, Sarda P, Genevieve D. Early-onset obesity and paternal 2pter deletion encompassing the ACP1, TMEM18, and MYT1L genes. *Eur J Hum Genet.* 2014;22:471–479.
77. Rhee KE, Phelan S, McCaffery J. Early determinants of obesity: genetic, epigenetic, and in utero influences. *Int J Pediatr.* 2012;2012:463850.
78. Pietrobelli A, Malavolti M, Battistini NC, Fuiano N. Metabolic syndrome: a child is not a small adult. *Int J Pediatr Obes.* 2008;3(suppl 1):67–71.
79. Simmonds M, Llewellyn A, Owen CG, Woolacott N. Predicting adult obesity from childhood obesity: a systematic review and meta-analysis. *Obes Rev.* 2015;17:95–107.
80. Choh AC, Lee M, Kent JW, Diego VP, Johnson W, Curran JE, Dyer TD, Bellis C, Blangero J, Siervogel RM, Towne B, Demerath EW, Czerwinski SA. Gene-by-age effects on BMI from birth to adulthood: the Fels Longitudinal Study. *Obesity (Silver Spring).* 2014;22:875–881.
81. Sandholt CH, Hansen T, Pedersen O. Beyond the fourth wave of genome-wide obesity association studies. *Nutr Diab.* 2012;2:e37.
82. Apalasamy YD, Mohamed Z. Obesity and genomics: role of technology in unraveling the complex genetic architecture of obesity. *Hum Genet.* 2015;134:361–374.
83. Rao KR, Lal N, Giridharan NV. Genetic & epigenetic approach to human obesity. *Indian J Med Res.* 2014;140:589–603.
84. Farooqi IS. Genetic, molecular and physiological insights into human obesity. *Eur J Clin Invest.* 2011;41:451–455.
85. Herrera BM, Keildson S, Lindgren CM. Genetics and epigenetics of obesity. *Maturitas.* 2011;69:41–49.
86. Park MH, Kwak SH, Kim KJ, Go MJ, Lee HJ, Kim KS, Hwang JY, Kimm K, Cho YM, Lee HK, Park KS, Lee JY. Identification of a genetic locus on chromosome 4q34-35 for type 2 diabetes with overweight. *Exp Mol Med.* 2013;45:e7.
87. Peters U, North KE, Sethupathy P, Buyske S, Haessler J, Jiao S, Fesinmeyer MD, Jackson RD, Kuller LH, Rajkovic A, Lim U, Cheng I, Schumacher F, Wilkens L, Li R, Monda K, Ehret G, Nguyen KD, Cooper R, Lewis CE, Leppert M, Irvin MR, Gu CC, Houston D, Buzkova P, Ritchie M, Matise TC, Le ML, Hindorff LA, Crawford DC, Haiman CA, Kooperberg C. A systematic mapping approach of 16q12.2/FTO and BMI in more than 20,000 African Americans narrows in on the underlying functional variation: results from the Population Architecture using Genomics and Epidemiology (PAGE) study. *PLoS Genet.* 2013;9:e1003171.

88. Fontanesi L, Calo DG, Galimberti G, Negrini R, Marino R, Nardone A, Ajmone-Marsan P, Russo V. A candidate gene association study for nine economically important traits in Italian Holstein cattle. *Anim Genet.* 2014;45:576–580.
89. Do DN, Strathe AB, Ostersen T, Jensen J, Mark T, Kadarmideen HN. Genome-wide association study reveals genetic architecture of eating behavior in pigs and its implications for humans obesity by comparative mapping. *PLoS One.* 2013;8:e71509.
90. Urbanek M, Hayes MG, Armstrong LL, Morrison J, Lowe LP, Badon SE, Scheftner D, Pluzhnikov A, Levine D, Laurie CC, McHugh C, Ackerman CM, Mirel DB, Doheny KF, Guo C, Scholtens DM, Dyer AR, Metzger BE, Reddy TE, Cox NJ, Lowe Jr WL. The chromosome 3q25 genomic region is associated with measures of adiposity in newborns in a multi-ethnic genome-wide association study. *Hum Mol Genet.* 2013;22: 3583–3596.
91. Sandholt CH, Grarup N, Pedersen O, Hansen T. Genome-wide association studies of human adiposity: zooming in on synapses. *Mol Cell Endocrinol.* 2015;418:90–100.
92. Graff M, Ngwa JS, Workalemahu T, Homuth G, Schipf S, Teumer A, Volzke H, Wallaschofski H, Abecasis GR, Edward L, Francesco C, Sanna S, Scheet P, Schlessinger D, Sidore C, Xiao X, Wang Z, Chanock SJ, Jacobs KB, Hayes RB, Hu F, Van Dam RM, Crout RJ, Marazita ML, Shaffer JR, Atwood LD, Fox CS, Heard-Costa NL, White C, Choh AC, Czerwinski SA, Demerath EW, Dyer TD, Towne B, Amin N, Oostra BA, van Duijn CM, Zillikens MC, Esko T, Nelis M, Nikopensius T, Metspalu A, Strachan DP, Monda K, Qi L, North KE, Cupples LA, Gordon-Larsen P, Berndt SI. Genome-wide analysis of BMI in adolescents and young adults reveals additional insight into the effects of genetic loci over the life course. *Hum Mol Genet.* 2013;22:3597–3607.
93. Garver WS. Gene–diet interactions in childhood obesity. *Curr Genomics.* 2011;12: 180–189.
94. van Vliet-Ostaptchouk JV, Snieder H, Lagou V. Gene–lifestyle interactions in obesity. *Curr Nutr Rep.* 2012;1:184–196.
95. Warden CH, Fisler JS. Gene-nutrient and gene-physical activity summary—genetics viewpoint. *Obesity (Silver Spring).* 2008;16(suppl 3):S55–S59.
96. Knoll N, Jarick I, Volckmar AL, Klingenspor M, Illig T, Grallert H, Gieger C, Wichmann HE, Peters A, Hebebrand J, Scherag A, Hinney A. Gene set of nuclear-encoded mitochondrial regulators is enriched for common inherited variation in obesity. *PLoS One.* 2013;8:e55884.
97. Ahmad S, Rukh G, Varga TV, Ali A, Kurbasic A, Shungin D, Ericson U, Koivula RW, Chu AY, Rose LM, Ganna A, Qi Q, Stancakova A, Sandholt CH, Elks CE, Curhan G, Jensen MK, Tamimi RM, Allin KH, Jorgensen T, Brage S, Langenberg C, Aadahl M, Grarup N, Linneberg A, Pare G, Magnusson PK, Pedersen NL, Boehnke M, Hamsten A, Mohlke KL, Pasquale LT, Pedersen O, Scott RA, Ridker PM, Ingelsson E, Laakso M, Hansen T, Qi L, Wareham NJ, Chasman DI, Hallmans G, Hu FB, Renstrom F, Orho-Melander M, Franks PW. Gene physical activity interactions in obesity: combined analysis of 111,421 individuals of European ancestry. *PLoS Genet.* 2013;9:e1003607.
98. Farooqi IS. Monogenic human obesity. *Front Horm Res.* 2008;36:1–11.
99. Rankinen T, Perusse L, Weisnagel SJ, Snyder EE, Chagnon YC, Bouchard C. The human obesity gene map: the 2001 update. *Obes Res.* 2002;10:196–243.
100. Farooqi IS, O'Rahilly S. Monogenic obesity in humans. *Annu Rev Med.* 2005;56: 443–458.
101. Xia Q, Grant SF. The genetics of human obesity. *Ann NY Acad Sci.* 2013;1281:178–190.
102. Hinney A, Vogel CI, Hebebrand J. From monogenic to polygenic obesity: recent advances. *Eur Child Adolesc Psychiatry.* 2010;19:297–310.
103. Farooqi IS, O'Rahilly S. 20 years of leptin: human disorders of leptin action. *J Endocrinol.* 2014;223:T63–T70.

104. Ichihara S, Yamada Y. Genetic factors for human obesity. *Cell Mol Life Sci.* 2008;65: 1086–1098.
105. Hinney A, Hebebrand J. Polygenic obesity in humans. *Obes Facts.* 2008;1:35–42.
106. Rankinen T, Sarzynski MA, Ghosh S, Bouchard C. Are there genetic paths common to obesity, cardiovascular disease outcomes, and cardiovascular risk factors? *Circ Res.* 2015;116:909–922.
107. Razquin C, Marti A, Martinez JA. Evidences on three relevant obesogenes: MC4R, FTO and PPARgamma. Approaches for personalized nutrition. *Mol Nutr Food Res.* 2011;55:136–149.
108. Saeed S, Bonnefond A, Manzoor J, Philippe J, Durand E, Arshad M, Sand O, Butt TA, Falchi M, Arslan M, Froguel P. Novel LEPR mutations in obese Pakistani children identified by PCR-based enrichment and next generation sequencing. *Obesity (Silver Spring).* 2014;22:1112–1117.
109. Sallman AM, Rask-Andersen M, Jacobsson JA, Ameur A, Kalnina I, Moschonis G, Juhlin S, Bringeland N, Hedberg LA, Ignatovica V, Chrousos GP, Manios Y, Klovins J, Marcus C, Gyllensten U, Fredriksson R, Schioth HB. Determination of the obesity-associated gene variants within the entire FTO gene by ultra-deep targeted sequencing in obese and lean children. *Int J Obes (Lond).* 2013;37:424–431.
110. Bonnefond A, Philippe J, Durand E, Muller J, Saeed S, Arslan M, Martinez R, De GF, Dhennin V, Rabearivelo I, Polak M, Cave H, Castano L, Vaxillaire M, Mandel JL, Sand O, Froguel P. Highly sensitive diagnosis of 43 monogenic forms of diabetes or obesity through one-step PCR-based enrichment in combination with next-generation sequencing. *Diab Care.* 2014;37:460–467.
111. Abente EJ, Subramanian M, Ramachandran V, Najafi-Shoushtari SH. MicroRNAs in obesity-associated disorders. *Arch Biochem Biophys.* 2016;589:108–119.
112. Kloting N, Berthold S, Kovacs P, Schon MR, Fasshauer M, Ruschke K, Stumvoll M, Bluher M. MicroRNA expression in human omental and subcutaneous adipose tissue. *PLoS One.* 2009;4:e4699.
113. Ono K. MicroRNA links obesity and impaired glucose metabolism. *Cell Res.* 2011;21:864–866.
114. Arner P, Kulyte A. MicroRNA regulatory networks in human adipose tissue and obesity. *Nat Rev Endocrinol.* 2015;11:276–288.
115. Dehwah MA, Xu A, Huang Q. MicroRNAs and type 2 diabetes/obesity. *J Genet Genomics.* 2012;39:11–18.
116. Jinwei Z, Yi L, Yuhao W, Liujun H, Mingzhou L, Xun W. MicroRNA regulates animal adipocyte differentiation. *Yi Chuan.* 2015;37:1175–1184.
117. Rottiers V, Naar AM. MicroRNAs in metabolism and metabolic disorders. *Nat Rev Mol Cell Biol.* 2012;13:239–250.
118. Hilton C, Neville MJ, Karpe F. MicroRNAs in adipose tissue: their role in adipogenesis and obesity. *Int J Obes (Lond).* 2013;37:325–332.
119. Williams MD, Mitchell GM. MicroRNAs in insulin resistance and obesity. *Exp Diab Res.* 2012;2012:484696.
120. Berndt SI, Gustafsson S, Magi R, Ganna A, Wheeler E, Feitosa MF, Justice AE, Monda KL, Croteau-Chonka DC, Day FR, Esko T, Fall T, Ferreira T, Gentilini D, Jackson AU, Luan J, Randall JC, Vedantam S, Willer CJ, Winkler TW, Wood AR, Workalemahu T, Hu YJ, Lee SH, Liang L, Lin DY, Min JL, Neale BM, Thorleifsson G, Yang J, Albrecht E, Amin N, Bragg-Gresham JL, Cadby G, den HM, Eklund N, Fischer K, Goel A, Hottenga JJ, Huffman JE, Jarick I, Johansson A, Johnson T, Kanoni S, Kleber ME, Konig IR, Kristiansson K, Kutalik Z, Lamina C, Lecoeur C, Li G, Mangino M, McArdle WL, Medina-Gomez C, Muller-Nurasyid M, Ngwa JS, Nolte IM, Paternoster L, Pechlivanis S, Perola M, Peters MJ, Preuss M, Rose LM, Shi J, Shungin D, Smith AV, Strawbridge RJ, Surakka I, Teumer A, Trip MD, Tyrer J, van

Vliet-Ostaptchouk JV, Vandenput L, Waite LL, Zhao JH, Absher D, Asselbergs FW, Atalay M, Attwood AP, Balmforth AJ, Basart H, Beilby J, Bonnycastle LL, Brambilla P, Bruinenberg M, Campbell H, Chasman DI, Chines PS, Collins FS, Connell JM, Cookson WO, de Faire U, de Vegt F, Dei M, Dimitriou M, Edkins S, Estrada K, Evans DM, Farrall M, Ferrario MM, Ferrieres J, Franke L, Frau F, Gejman PV, Grallert H, Gronberg H, Gudnason V, Hall AS, Hall P, Hartikainen AL, Hayward C, Heard-Costa NL, Heath AC, Hebebrand J, Homuth G, Hu FB, Hunt SE, Hypponen E, Iribarren C, Jacobs KB, Jansson JO, Jula A, Kahonen M, Kathiresan S, Kee F, Khaw KT, Kivimaki M, Koenig W, Kraja AT, Kumari M, Kuulasmaa K, Kuusisto J, Laitinen JH, Lakka TA, Langenberg C, Launer LJ, Lind L, Lindstrom J, Liu J, Liuzzi A, Lokki ML, Lorentzon M, Madden PA, Magnusson PK, Manunta P, Marek D, Marz W, Mateo LI, McKnight B, Medland SE, Mihailov E, Milani L, Montgomery GW, Mooser V, Muhleisen TW, Munroe PB, Musk AW, Narisu N, Navis G, Nicholson G, Nohr EA, Ong KK, Oostra BA, Palmer CN, Palotie A, Peden JF, Pedersen N, Peters A, Polasek O, Pouta A, Pramstaller PP, Prokopenko I, Putter C, Radhakrishnan A, Raitakari O, Rendon A, Rivadeneira F, Rudan I, Saaristo TE, Sambrook JG, Sanders AR, Sanna S, Saramies J, Schipf S, Schreiber S, Schunkert H, Shin SY, Signorini S, Sinisalo J, Skrobek B, Soranzo N, Stancakova A, Stark K, Stephens JC, Stirrups K, Stolk RP, Stumvoll M, Swift AJ, Theodoraki EV, Thorand B, Tregouet DA, Tremoli E, van der Klauw MM, van Meurs JB, Vermeulen SH, Viikari J, Virtamo J, Vitart V, Waeber G, Wang Z, Widen E, Wild SH, Willemsen G, Winkelmann BR, Witteman JC, Wolffenbuttel BH, Wong A, Wright AF, Zillikens MC, Amouyel P, Boehm BO, Boerwinkle E, Boomsma DI, Caulfield MJ, Chanock SJ, Cupples LA, Cusi D, Dedoussis GV, Erdmann J, Eriksson JG, Franks PW, Froguel P, Gieger C, Gyllensten U, Hamsten A, Harris TB, Hengstenberg C, Hicks AA, Hingorani A, Hinney A, Hofman A, Hovingh KG, Hveem K, Illig T, Jarvelin MR, Jockel KH, Keinanen-Kiukaanniemi SM, Kiemeney LA, Kuh D, Laakso M, Lehtimaki T, Levinson DF, Martin NG, Metspalu A, Morris AD. Genome-wide meta-analysis identifies 11 new loci for anthropometric traits and provides insights into genetic architecture. *Nat Genet.* 2013;45:501–512.

121. Heneghan HM, Miller N, Kerin MJ. Role of microRNAs in obesity and the metabolic syndrome. *Obes Rev.* 2010;11:354–361.
122. Keller P, Gburcik V, Petrovic N, Gallagher IJ, Nedergaard J, Cannon B, Timmons JA. Gene-chip studies of adipogenesis-regulated microRNAs in mouse primary adipocytes and human obesity. *BMC Endocr Disord.* 2011;11:7.
123. Ortega FJ, Moreno-Navarrete JM, Pardo G, Sabater M, Hummel M, Ferrer A, Rodriguez-Hermosa JI, Ruiz B, Ricart W, Peral B, Fernandez-Real JM. MiRNA expression profile of human subcutaneous adipose and during adipocyte differentiation. *PLoS One.* 2010;5:e9022.
124. Ortega FJ, Mercader JM, Catalan V, Moreno-Navarrete JM, Pueyo N, Sabater M, Gomez-Ambrosi J, Anglada R, Fernandez-Formoso JA, Ricart W, Fruhbeck G, Fernandez-Real JM. Targeting the circulating microRNA signature of obesity. *Clin Chem.* 2013;59:781–792.
125. Ortega FJ, Moreno M, Mercader JM, Moreno-Navarrete JM, Fuentes-Batllevell N, Sabater M, Ricart W, Fernandez-Real JM. Inflammation triggers specific microRNA profiles in human adipocytes and macrophages and in their supernatants. *Clin Epigenetics.* 2015;7:49.
126. Osmai M, Osmai Y, Bang-Berthelsen CH, Pallesen EM, Vestergaard AL, Novotny GW, Pociot F, Mandrup-Poulsen T. microRNAs as regulators of beta-cell function and dysfunction. *Diab Metab Res Rev.* 2015. doi:10.1002/dmrr.2719.
127. Hinney A, Nguyen TT, Scherag A, Friedel S, Bronner G, Muller TD, Grallert H, Illig T, Wichmann HE, Rief W, Schafer H, Hebebrand J. Genome wide association (GWA)

study for early onset extreme obesity supports the role of fat mass and obesity associated gene (FTO) variants. *PLoS One.* 2007;2:e1361.
128. Frayling TM, Timpson NJ, Weedon MN, Zeggini E, Freathy RM, Lindgren CM, Perry JR, Elliott KS, Lango H, Rayner NW, Shields B, Harries LW, Barrett JC, Ellard S, Groves CJ, Knight B, Patch AM, Ness AR, Ebrahim S, Lawlor DA, Ring SM, Ben-Shlomo Y, Jarvelin MR, Sovio U, Bennett AJ, Melzer D, Ferrucci L, Loos RJ, Barroso I, Wareham NJ, Karpe F, Owen KR, Cardon LR, Walker M, Hitman GA, Palmer CN, Doney AS, Morris AD, Smith GD, Hattersley AT, McCarthy MI. A common variant in the FTO gene is associated with body mass index and predisposes to childhood and adult obesity. *Science.* 2007;316:889–894.
129. Farooqi IS. FTO and obesity: the missing link. *Cell Metab.* 2011;13:7–8.
130. Yajnik CS, Janipalli CS, Bhaskar S, Kulkarni SR, Freathy RM, Prakash S, Mani KR, Weedon MN, Kale SD, Deshpande J, Krishnaveni GV, Veena SR, Fall CH, McCarthy MI, Frayling TM, Hattersley AT, Chandak GR. FTO gene variants are strongly associated with type 2 diabetes in South Asian Indians. *Diabetologia.* 2009;52:247–252.
131. Yang J, Loos RJ, Powell JE, Medland SE, Speliotes EK, Chasman DI, Rose LM, Thorleifsson G, Steinthorsdottir V, Magi R, Waite L, Smith AV, Yerges-Armstrong LM, Monda KL, Hadley D, Mahajan A, Li G, Kapur K, Vitart V, Huffman JE, Wang SR, Palmer C, Esko T, Fischer K, Zhao JH, Demirkan A, Isaacs A, Feitosa MF, Luan J, Heard-Costa NL, White C, Jackson AU, Preuss M, Ziegler A, Eriksson J, Kutalik Z, Frau F, Nolte IM, van Vliet-Ostaptchouk JV, Hottenga JJ, Jacobs KB, Verweij N, Goel A, Medina-Gomez C, Estrada K, Bragg-Gresham JL, Sanna S, Sidore C, Tyrer J, Teumer A, Prokopenko I, Mangino M, Lindgren CM, Assimes TL, Shuldiner AR, Hui J, Beilby JP, McArdle WL, Hall P, Haritunians T, Zgaga L, Kolcic I, Polasek O, Zemunik T, Oostra BA, Junttila MJ, Gronberg H, Schreiber S, Peters A, Hicks AA, Stephens J, Foad NS, Laitinen J, Pouta A, Kaakinen M, Willemsen G, Vink JM, Wild SH, Navis G, Asselbergs FW, Homuth G, John U, Iribarren C, Harris T, Launer L, Gudnason V, O'Connell JR, Boerwinkle E, Cadby G, Palmer LJ, James AL, Musk AW, Ingelsson E, Psaty BM, Beckmann JS, Waeber G, Vollenweider P, Hayward C, Wright AF, Rudan I, Groop LC, Metspalu A, Khaw KT, van Duijn CM, Borecki IB, Province MA, Wareham NJ, Tardif JC, Huikuri HV, Cupples LA, Atwood LD, Fox CS, Boehnke M, Collins FS, Mohlke KL, Erdmann J, Schunkert H, Hengstenberg C, Stark K, Lorentzon M, Ohlsson C, Cusi D, Staessen JA, van der Klauw MM, Pramstaller PP, Kathiresan S, Jolley JD, Ripatti S, Jarvelin MR, de Geus EJ, Boomsma DI, Penninx B, Wilson JF, Campbell H, Chanock SJ, van der Harst P, Hamsten A, Watkins H, Hofman A, Witteman JC, Zillikens MC, Uitterlinden AG, Rivadeneira F, Zillikens MC, Kiemeney LA, Vermeulen SH, Abecasis GR, Schlessinger D, Schipf S, Stumvoll M, Tonjes A, Spector TD, North KE, Lettre G, McCarthy MI, Berndt SI, Heath AC, Madden PA, Nyholt DR, Montgomery GW, Martin NG, McKnight B, Strachan DP, Hill WG, Snieder H, Ridker PM, Thorsteinsdottir U, Stefansson K, Frayling TM, Hirschhorn JN, Goddard ME, Visscher PM. FTO genotype is associated with phenotypic variability of body mass index. *Nature.* 2012;490:267–272.
132. Mei H, Chen W, Srinivasan SR, Jiang F, Schork N, Murray S, Smith E, So JD, Berenson GS. FTO influences on longitudinal BMI over childhood and adulthood and modulation on relationship between birth weight and longitudinal BMI. *Hum Genet.* 2010;128:589–596.
133. Liu G, Zhu H, Lagou V, Gutin B, Stallmann-Jorgensen IS, Treiber FA, Dong Y, Snieder H. FTO variant rs9939609 is associated with body mass index and waist circumference, but not with energy intake or physical activity in European- and African-American youth. *BMC Med Genet.* 2010;11:57.
134. Heid IM, Huth C, Loos RJ, Kronenberg F, Adamkova V, Anand SS, Ardlie K, Biebermann H, Bjerregaard P, Boeing H, Bouchard C, Ciullo M, Cooper JA,

Corella D, Dina C, Engert JC, Fisher E, Frances F, Froguel P, Hebebrand J, Hegele RA, Hinney A, Hoehe MR, Hu FB, Hubacek JA, Humphries SE, Hunt SC, Illig T, Jarvelin MR, Kaakinen M, Kollerits B, Krude H, Kumar J, Lange LA, Langer B, Li S, Luchner A, Lyon HN, Meyre D, Mohlke KL, Mooser V, Nebel A, Nguyen TT, Paulweber B, Perusse L, Qi L, Rankinen T, Rosskopf D, Schreiber S, Sengupta S, Sorice R, Suk A, Thorleifsson G, Thorsteinsdottir U, Volzke H, Vimaleswaran KS, Wareham NJ, Waterworth D, Yusuf S, Lindgren C, McCarthy MI, Lange C, Hirschhorn JN, Laird N, Wichmann HE. Meta-analysis of the INSIG2 association with obesity including 74,345 individuals: does heterogeneity of estimates relate to study design? *PLoS Genet.* 2009;5:e1000694.

135. Loos RJ, Lindgren CM, Li S, Wheeler E, Zhao JH, Prokopenko I, Inouye M, Freathy RM, Attwood AP, Beckmann JS, Berndt SI, Jacobs KB, Chanock SJ, Hayes RB, Bergmann S, Bennett AJ, Bingham SA, Bochud M, Brown M, Cauchi S, Connell JM, Cooper C, Smith GD, Day I, Dina C, De S, Dermitzakis ET, Doney AS, Elliott KS, Elliott P, Evans DM, Sadaf F, Froguel I, Ghori P, Groves J, Gwilliam CJ, Hadley R, Hall D, Hattersley AS, Hebebrand AT, Heid J, Lamina IM, Gieger C, Illig C, Meitinger T, Wichmann T, Herrera HE, Hinney B, Hunt A, Jarvelin SE, Johnson MR, Jolley T, Karpe JD, Keniry F, Khaw A, Luben KT, Mangino RN, Marchini M, McArdle J, McGinnis WL, Meyre R, Munroe D, Morris PB, Ness AD, Neville AR, Nica MJ, Ong AC, O'Rahilly KK, Owen S, Palmer KR, Papadakis CN, Potter K, Pouta S, Qi A, Randall L, Rayner JC, Ring NW, Sandhu SM, Scherag MS, Sims A, Song MA, Soranzo K, Speliotes N, Syddall EK, Teichmann HE, Timpson SA, Tobias NJ, Uda JH, Vogel M, Wallace CI, Waterworth C, Weedon DM, Willer MN, Wraight CJ, Yuan X, Zeggini E, Hirschhorn JN, Strachan DP, Ouwehand WH, Caulfield MJ, Samani NJ, Frayling TM, Vollenweider P, Waeber G, Mooser V, Deloukas P, McCarthy MI, Wareham NJ, Barroso I, Jacobs KB, Chanock SJ, Hayes RB, Lamina C, Gieger C, Illig T, Meitinger T, Wichmann HE, Kraft P, Hankinson SE, Hunter DJ, Hu FB, Lyon HN, Voight BF, Ridderstrale M, Groop L, Scheet P, Sanna S, Abecasis GR, Albai G, Nagaraja R, Schlessinger D, Jackson AU, Tuomilehto J, Collins FS, Boehnke M, Mohlke KL. Common variants near MC4R are associated with fat mass, weight and risk of obesity. *Nat Genet.* 2008;40:768–775.

136. Chambers JC, Elliott P, Zabaneh D, Zhang W, Li Y, Froguel P, Balding D, Scott J, Kooner JS. Common genetic variation near MC4R is associated with waist circumference and insulin resistance. *Nat Genet.* 2008;40:716–718.

137. Willer CJ, Speliotes EK, Loos RJ, Li S, Lindgren CM, Heid IM, Berndt SI, Elliott AL, Jackson AU, Lamina C, Lettre G, Lim N, Lyon HN, McCarroll SA, Papadakis K, Qi L, Randall JC, Roccasecca RM, Sanna S, Scheet P, Weedon MN, Wheeler E, Zhao JH, Jacobs LC, Prokopenko I, Soranzo N, Tanaka T, Timpson NJ, Almgren P, Bennett A, Bergman RN, Bingham SA, Bonnycastle LL, Brown M, Burtt NP, Chines P, Coin L, Collins FS, Connell JM, Cooper C, Smith GD, Dennison EM, Deodhar P, Elliott P, Erdos MR, Estrada K, Evans DM, Gianniny L, Gieger C, Gillson CJ, Guiducci C, Hackett R, Hadley D, Hall AS, Havulinna AS, Hebebrand J, Hofman A, Isomaa B, Jacobs KB, Johnson T, Jousilahti P, Jovanovic Z, Khaw KT, Kraft P, Kuokkanen M, Kuusisto J, Laitinen J, Lakatta EG, Luan J, Luben RN, Mangino M, McArdle WL, Meitinger T, Mulas A, Munroe PB, Narisu N, Ness AR, Northstone K, O'Rahilly S, Purmann C, Rees MG, Ridderstrale M, Ring SM, Rivadeneira F, Ruokonen A, Sandhu MS, Saramies J, Scott LJ, Scuteri A, Silander K, Sims MA, Song K, Stephens J, Stevens S, Stringham HM, Tung YC, Valle TT, van Duijn CM, Vimaleswaran KS, Vollenweider P, Waeber G, Wallace C, Watanabe RM, Waterworth DM, Watkins N, Witteman JC, Zeggini E, Zhai G, Zillikens MC, Altshuler D, Caulfield MJ, Chanock SJ, Farooqi IS, Ferrucci L, Guralnik JM, Hattersley AT, Hu FB, Jarvelin MR, Laakso M, Mooser V, Ong KK, Ouwehand WH, Salomaa V, Samani NJ, Spector TD, Tuomi T,

Tuomilehto J, Uda M, Uitterlinden AG, Wareham NJ, Deloukas P, Frayling TM, Groop LC, Hayes RB, Hunter DJ, Mohlke KL, Peltonen L, Schlessinger D, Strachan DP, Wichmann HE, McCarthy MI, Boehnke M, Barroso I, Abecasis GR, Hirschhorn JN. Six new loci associated with body mass index highlight a neuronal influence on body weight regulation. *Nat Genet.* 2009;41:25–34.

138. Xi B, Takeuchi F, Chandak GR, Kato N, Pan HW, Zhou DH, Pan HY, Mi J. Common polymorphism near the MC4R gene is associated with type 2 diabetes: data from a meta-analysis of 123,373 individuals. *Diabetologia.* 2012;55:2660–2666.

139. Speliotes EK, Willer CJ, Berndt SI, Monda KL, Thorleifsson G, Jackson AU, Lango AH, Lindgren CM, Luan J, Magi R, Randall JC, Vedantam S, Winkler TW, Qi L, Workalemahu T, Heid IM, Steinthorsdottir V, Stringham HM, Weedon MN, Wheeler E, Wood AR, Ferreira T, Weyant RJ, Segre AV, Estrada K, Liang L, Nemesh J, Park JH, Gustafsson S, Kilpelainen TO, Yang J, Bouatia-Naji N, Esko T, Feitosa MF, Kutalik Z, Mangino M, Raychaudhuri S, Scherag A, Smith AV, Welch R, Zhao JH, Aben KK, Absher DM, Amin N, Dixon AL, Fisher E, Glazer NL, Goddard ME, Heard-Costa NL, Hoesel V, Hottenga JJ, Johansson A, Johnson T, Ketkar S, Lamina C, Li S, Moffatt MF, Myers RH, Narisu N, Perry JR, Peters MJ, Preuss M, Ripatti S, Rivadeneira F, Sandholt C, Scott LJ, Timpson NJ, Tyrer JP, van WS, Watanabe RM, White CC, Wiklund F, Barlassina C, Chasman DI, Cooper MN, Jansson JO, Lawrence RW, Pellikka N, Prokopenko I, Shi J, Thiering E, Alavere H, Alibrandi MT, Almgren P, Arnold AM, Aspelund T, Atwood LD, Balkau B, Balmforth AJ, Bennett AJ, Ben-Shlomo Y, Bergman RN, Bergmann S, Biebermann H, Blakemore AI, Boes T, Bonnycastle LL, Bornstein SR, Brown MJ, Buchanan TA, Busonero F, Campbell H, Cappuccio FP, Cavalcanti-Proenca C, Chen YD, Chen CM, Chines PS, Clarke R, Coin L, Connell J, Day IN, den HM, Duan J, Ebrahim S, Elliott P, Elosua R, Eiriksdottir G, Erdos MR, Eriksson JG, Facheris MF, Felix SB, Fischer-Posovszky P, Folsom AR, Friedrich N, Freimer NB, Fu M, Gaget S, Gejman PV, Geus EJ, Gieger C, Gjesing AP, Goel A, Goyette P, Grallert H, Grassler J, Greenawalt DM, Groves CJ, Gudnason V, Guiducci C, Hartikainen AL, Hassanali N, Hall AS, Havulinna AS, Hayward C, Heath AC, Hengstenberg C, Hicks AA, Hinney A, Hofman A, Homuth G, Hui J, Igl W, Iribarren C, Isomaa B, Jacobs KB, Jarick I, Jewell E, John U, Jorgensen T, Jousilahti P, Jula A, Kaakinen M, Kajantie E, Kaplan LM, Kathiresan S, Kettunen J, Kinnunen L, Knowles JW, Kolcic I, Konig IR, Koskinen S, Kovacs P, Kuusisto J, Kraft P, Kvaloy K, Laitinen J, Lantieri O, Lanzani C, Launer LJ, Lecoeur C, Lehtimaki T, Lettre G, Liu J, Lokki ML, Lorentzon M, Luben RN, Ludwig B, Manunta P, Marek D, Marre M, Martin NG, McArdle WL, McCarthy A, McKnight B, Meitinger T, Melander O, Meyre D, Midthjell K, Montgomery GW, Morken MA, Morris AP, Mulic R, Ngwa JS, Nelis M, Neville MJ, Nyholt DR, O'Donnell CJ, O'Rahilly S, Ong KK, Oostra B, Pare G, Parker AN, Perola M, Pichler I, Pietilainen KH, Platou CG, Polasek O, Pouta A, Rafelt S, Raitakari O, Rayner NW, Ridderstrale M, Rief W, Ruokonen A, Robertson NR, Rzehak P, Salomaa V, Sanders AR, Sandhu MS, Sanna S, Saramies J, Savolainen MJ, Scherag S, Schipf S, Schreiber S, Schunkert H, Silander K, Sinisalo J, Siscovick DS, Smit JH, Soranzo N, Sovio U, Stephens J, Surakka I, Swift AJ, Tammesoo ML, Tardif JC, Teder-Laving M, Teslovich TM, Thompson JR, Thomson B, Tonjes A, Tuomi T, van Meurs JB, van Ommen GJ. Association analyses of 249,796 individuals reveal 18 new loci associated with body mass index. *Nat Genet.* 2010;42:937–948.

140. Whitaker KL, Jarvis MJ, Beeken RJ, Boniface D, Wardle J. Comparing maternal and paternal intergenerational transmission of obesity risk in a large population-based sample. *Am J Clin Nutr.* 2010;91:1560–1567.

141. Khanolkar AR, Byberg L, Koupil I. Parental influences on cardiovascular risk factors in Swedish children aged 5–14 years. *Eur J Public Health.* 2012;22:840–847.

142. Cooper R, Pinto Pereira SM, Power C, Hypponen E. Parental obesity and risk factors for cardiovascular disease among their offspring in mid-life: findings from the 1958 British Birth Cohort Study. *Int J Obes (Lond).* 2013;37:1590–1596.
143. McCarthy K, Ye YL, Yuan S, He QQ. Parental weight status and offspring cardiovascular disease risks: a cross-sectional study of Chinese children. *Prev Chronic Dis.* 2015;12: E01.
144. Ge ZJ, Zhang CL, Schatten H, Sun QY. Maternal diabetes mellitus and the origin of non-communicable diseases in offspring: the role of epigenetics. *Biol Reprod.* 2014;90:139.
145. Blackmore HL, Ozanne SE. Maternal diet-induced obesity and offspring cardiovascular health. *J Dev Orig Health Dis.* 2013;4:338–347.
146. Alm PS, Krook A, de Castro BT. Maternal obesity legacy: exercise it away!. *Diabetologia.* 2016;59:5–8.
147. Vidal AC, Benjamin Neelon SE, Liu Y, Tuli AM, Fuemmeler BF, Hoyo C, Murtha AP, Huang Z, Schildkraut J, Overcash F, Kurtzberg J, Jirtle RL, Iversen ES, Murphy SK. Maternal stress, preterm birth, and DNA methylation at imprint regulatory sequences in humans. *Genet Epigenet.* 2014;6:37–44.
148. Sasaki A, de Vega W, Sivanathan S, St-Cyr S, McGowan PO. Maternal high-fat diet alters anxiety behavior and glucocorticoid signaling in adolescent offspring. *Neuroscience.* 2014;272:92–101.
149. Huang T, Hu FB. Gene-environment interactions and obesity: recent developments and future directions. *BMC Med Genomics.* 2015;8(suppl 1):S2.
150. Harrington JM, Phillips CM. Nutrigenetics: bridging two worlds to understand type 2 diabetes. *Curr Diab Rep.* 2014;14:477.
151. Wang J, Wu Z, Li D, Li N, Dindot SV, Satterfield MC, Bazer FW, Wu G. Nutrition, epigenetics, and metabolic syndrome. *Antioxid Redox Signal.* 2012;17:282–301.
152. Grayson M. Nutrigenomics. *Nature.* 2010;468:S1.
153. Neeha VS, Kinth P. Nutrigenomics research: a review. *J Food Sci Technol.* 2013;50: 415–428.
154. Garcia-Rios A, Perez-Martinez P, Delgado-Lista J, Lopez-Miranda J, Perez-Jimenez F. Nutrigenetics of the lipoprotein metabolism. *Mol Nutr Food Res.* 2012;56:171–183.
155. Waller-Evans H, Hue C, Fearnside J, Rothwell AR, Lockstone HE, Calderari S, Wilder SP, Cazier JB, Scott J, Gauguier D. Nutrigenomics of high fat diet induced obesity in mice suggests relationships between susceptibility to fatty liver disease and the proteasome. *PLoS One.* 2013;8:e82825.
156. Phillips CM. Nutrigenetics and metabolic disease: current status and implications for personalised nutrition. *Nutrients.* 2013;5:32–57.
157. Lau FC, Bagchi M, Sen CK, Bagchi D. Nutrigenomic basis of beneficial effects of chromium(III) on obesity and diabetes. *Mol Cell Biochem.* 2008;317:1–10.
158. Ardekani AM, Jabbari S. Nutrigenomics and cancer. *Avicenna J Med Biotechnol.* 2009;1:9–17.
159. Nicastro HL, Trujillo EB, Milner JA. Nutrigenomics and cancer prevention. *Curr Nutr Rep.* 2012;1:37–43.
160. Kang JX. Nutrigenomics and cancer therapy. *J Nutrigenet Nutrigenomics.* 2013;6:I–II.
161. Davis CD. Nutrigenomics and the prevention of colon cancer. *Pharmacogenomics.* 2007;8:121–124.
162. Riscuta G, Dumitrescu RG. Nutrigenomics: implications for breast and colon cancer prevention. *Methods Mol Biol.* 2012;863:343–358.
163. Ferguson LR. Nutrigenetics, nutrigenomics and inflammatory bowel diseases. *Expert Rev Clin Immunol.* 2013;9:717–726.
164. Bollati V, Favero C, Albetti B, Tarantini L, Moroni A, Byun HM, Motta V, Conti DM, Tirelli AS, Vigna L, Bertazzi PA, Pesatori AC. Nutrients intake is associated with DNA

methylation of candidate inflammatory genes in a population of obese subjects. *Nutrients.* 2014;6:4625–4639.
165. Lovegrove JA, Gitau R. Nutrigenetics and CVD: what does the future hold? *Proc Nutr Soc.* 2008;67:206–213.
166. Merched AJ, Chan L. Nutrigenetics and nutrigenomics of atherosclerosis. *Curr Atheroscler Rep.* 2013;15:328.
167. Engler MB. Nutrigenomics in cardiovascular disease: implications for the future. *Prog Cardiovasc Nurs.* 2009;24:190–195.
168. Iacoviello L, Santimone I, Latella MC, de Gaetano G, Donati MB. Nutrigenomics: a case for the common soil between cardiovascular disease and cancer. *Genes Nutr.* 2008;3:19–24.
169. Corella D, Ordovas JM. Nutrigenomics in cardiovascular medicine. *Circ Cardiovasc Genet.* 2009;2:637–651.
170. Godard B, Ozdemir V. Nutrigenomics and personalized diet: from molecule to intervention and nutri-ethics. *OMICS.* 2008;12:227–228.
171. Wittwer J, Rubio-Aliaga I, Hoeft B, Bendik I, Weber P, Daniel H. Nutrigenomics in human intervention studies: current status, lessons learned and future perspectives. *Mol Nutr Food Res.* 2011;55:341–358.
172. Milagro FI, Mansego ML, De Miguel C, Martinez JA. Dietary factors, epigenetic modifications and obesity outcomes: progresses and perspectives. *Mol Aspects Med.* 2013;34:782–812.
173. Busch C, Burkard M, Leischner C, Lauer UM, Frank J, Venturelli S. Epigenetic activities of flavonoids in the prevention and treatment of cancer. *Clin Epigenetics.* 2015;7:64.
174. Martinez-Jimenez CP, Sandoval J. Epigenetic crosstalk: a molecular language in human metabolic disorders. *Front Biosci (Schol Ed).* 2015;7:46–57.
175. Verduci E, Banderali G, Barberi S, Radaelli G, Lops A, Betti F, Riva E, Giovannini M. Epigenetic effects of human breast milk. *Nutrients.* 2014;6:1711–1724.
176. Lillycrop KA, Burdge GC. Epigenetic mechanisms linking early nutrition to long term health. *Best Pract Res Clin Endocrinol Metab.* 2012;26:667–676.
177. Lee JH, Friso S, Choi SW. Epigenetic mechanisms underlying the link between non-alcoholic fatty liver diseases and nutrition. *Nutrients.* 2014;6:3303–3325.
178. Suzuki MM, Bird A. DNA methylation landscapes: provocative insights from epigenomics. *Nat Rev Genet.* 2008;9:465–476.
179. Marti A, Goyenechea E, Martinez JA. Nutrigenetics: a tool to provide personalized nutritional therapy to the obese. *J Nutrigenet Nutrigenomics.* 2010;3:157–169.
180. Downs BW, Chen AL, Chen TJ, Waite RL, Braverman ER, Kerner M, Braverman D, Rhoades P, Prihoda TJ, Palomo T, Oscar-Berman M, Reinking J, Blum SH, DiNubile NA, Liu HH, Blum K. Nutrigenomic targeting of carbohydrate craving behavior: can we manage obesity and aberrant craving behaviors with neurochemical pathway manipulation by Immunological Compatible Substances (nutrients) using a Genetic Positioning System (GPS) Map? *Med Hypotheses.* 2009;73: 427–434.
181. Bird A. Perceptions of epigenetics. *Nature.* 2007;447:396–398.
182. Ozdemir V, Kolker E. Precision nutrition 4.0: a big data and ethics foresight analysis-convergence of agrigenomics, nutrigenomics, nutriproteomics, and nutrimetabolomics. *OMICS.* 2016;20:69–75.
183. Pavlidis C, Nebel JC, Katsila T, Patrinos GP. Nutrigenomics 2.0: the need for ongoing and independent evaluation and synthesis of commercial nutrigenomics tests' scientific knowledge base for responsible innovation. *OMICS.* 2015;20:65–68.
184. Ioannides-Demos LL, Proietto J, McNeil JJ. Pharmacotherapy for obesity. *Drugs.* 2005;65:1391–1418.

185. Vella A, Camilleri M. Pharmacogenetics: potential role in the treatment of diabetes and obesity. *Expert Opin Pharmacother.* 2008;9:1109–1119.
186. Freemark M. Pharmacotherapy of childhood obesity: an evidence-based, conceptual approach. *Diab Care.* 2007;30:395–402.
187. Lett TA, Wallace TJ, Chowdhury NI, Tiwari AK, Kennedy JL, Muller DJ. Pharmacogenetics of antipsychotic-induced weight gain: review and clinical implications. *Mol Psychiatry.* 2012;17:242–266.
188. McMurray F, Demetriades M, Aik W, Merkestein M, Kramer H, Andrew DS, Scudamore CL, Hough TA, Wells S, Ashcroft FM, McDonough MA, Schofield CJ, Cox RD. Pharmacological inhibition of FTO. *PLoS One.* 2015;10:e0121829.
189. Richardson AS, North KE, Graff M, Young KM, Mohlke KL, Lange LA, Lange EM, Harris KM, Gordon-Larsen P. Moderate to vigorous physical activity interactions with genetic variants and body mass index in a large US ethnically diverse cohort. *Pediatr Obes.* 2014;9:e35–e46.
190. Andreasen CH, Stender-Petersen KL, Mogensen MS, Torekov SS, Wegner L, Andersen G, Nielsen AL, Albrechtsen A, Borch-Johnsen K, Rasmussen SS, Clausen JO, Sandbaek A, Lauritzen T, Hansen L, Jorgensen T, Pedersen O, Hansen T. Low physical activity accentuates the effect of the FTO rs9939609 polymorphism on body fat accumulation. *Diabetes.* 2008;57:95–101.
191. Rampersaud E, Mitchell BD, Pollin TI, Fu M, Shen H, O'Connell JR, Ducharme JL, Hines S, Sack P, Naglieri R, Shuldiner AR, Snitker S. Physical activity and the association of common FTO gene variants with body mass index and obesity. *Arch Intern Med.* 2008;168:1791–1797.
192. Ruiz JR, Labayen I, Ortega FB, Legry V, Moreno LA, Dallongeville J, Martinez-Gomez D, Bokor S, Manios Y, Ciarapica D, Gottrand F, De Henauw S, Molnar D, Sjostrom M, Meirhaeghe A. Attenuation of the effect of the FTO rs9939609 polymorphism on total and central body fat by physical activity in adolescents: the HELENA study. *Arch Pediatr Adolesc Med.* 2010;164:328–333.
193. Reddon H, Gerstein HC, Engert JC, Mohan V, Bosch J, Desai D, Bailey SD, Diaz R, Yusuf S, Anand SS, Meyre D. Physical activity and genetic predisposition to obesity in a multiethnic longitudinal study. *Sci Rep.* 2016;6:18672.
194. Speakman JR. The 'Fat Mass and Obesity Related' (FTO) gene: mechanisms of impact on obesity and energy balance. *Curr Obes Rep.* 2015;4:73–91.
195. Rankinen T, Roth SM, Bray MS, Loos R, Perusse L, Wolfarth B, Hagberg JM, Bouchard C. Advances in exercise, fitness, and performance genomics. *Med Sci Sports Exerc.* 2010;42:835–846.
196. Dolinoy DC, Weidman JR, Jirtle RL. Epigenetic gene regulation: linking early developmental environment to adult disease. *Reprod Toxicol.* 2007;23:297–307.
197. Dolinoy DC, Jirtle RL. Environmental epigenomics in human health and disease. *Environ Mol Mutagen.* 2008;49:4–8.
198. Mussa A, Russo S, Larizza L, Riccio A, Ferrero GB. (Epi)genotype-phenotype correlations in Beckwith–Wiedemann syndrome: a paradigm for genomic medicine. *Clin Genet.* 2015;89:401–519.
199. Symonds ME, Budge H, Frazier-Wood AC. Epigenetics and obesity: a relationship waiting to be explained. *Hum Hered.* 2013;75:90–97.
200. Murphy TM, Mill J. Epigenetics in health and disease: heralding the EWAS era. *Lancet.* 2014;383:1952–1954.
201. Greenhill C. Epigenetics: obesity-induced hypermethylation of adiponectin gene. *Nat Rev Endocrinol.* 2015;11:504.
202. Campion J, Milagro FI, Martinez JA. Individuality and epigenetics in obesity. *Obes Rev.* 2009;10:383–392.

203. Cordero P, Li J, Oben JA. Epigenetics of obesity: beyond the genome sequence. *Curr Opin Clin Nutr Metab Care.* 2015;18:361–366.
204. Milagro FI, Martinez JA. Epigenetics of obesity and weight loss. *Endocrinol Nutr.* 2013;60(suppl 1):12–14.
205. Ahmed F. Epigenetics: tales of adversity. *Nature.* 2010;468:S20.
206. Gardner KR, Sapienza C, Fisher JO. Genetic and epigenetic associations to obesity-related appetite phenotypes among African-American children. *Pediatr Obes.* 2015;10:476–482.
207. Schwenk RW, Vogel H, Schurmann A. Genetic and epigenetic control of metabolic health. *Mol Metab.* 2013;2:337–347.
208. Waterland RA. Is epigenetics an important link between early life events and adult disease? *Horm Res.* 2009;suppl 1(71):13–16.
209. Wang X, Zhu H, Snieder H, Su S, Munn D, Harshfield G, Maria BL, Dong Y, Treiber F, Gutin B, Shi H. Obesity related methylation changes in DNA of peripheral blood leukocytes. *BMC Med.* 2010;8:87.
210. Huang RC, Garratt ES, Pan H, Wu Y, Davis EA, Barton SJ, Burdge GC, Godfrey KM, Holbrook JD, Lillycrop KA. Genome-wide methylation analysis identifies differentially methylated CpG loci associated with severe obesity in childhood. *Epigenetics.* 2015;10:995–1005.
211. Gemma C, Sookoian S, Alvarinas J, Garcia SI, Quintana L, Kanevsky D, Gonzalez CD, Pirola CJ. Maternal pregestational BMI is associated with methylation of the PPARGC1A promoter in newborns. *Obesity (Silver Spring).* 2009;17:1032–1039.
212. Ronn T, Volkov P, Davegardh C, Dayeh T, Hall E, Olsson AH, Nilsson E, Tornberg A, Dekker NM, Eriksson KF, Jones HA, Groop L, Ling C. A six months exercise intervention influences the genome-wide DNA methylation pattern in human adipose tissue. *PLoS Genet.* 2013;9:e1003572.
213. Rozek LS, Dolinoy DC, Sartor MA, Omenn GS. Epigenetics: relevance and implications for public health. *Annu Rev Public Health.* 2014;35:105–122.
214. Demerath EW, Guan W, Grove ML, Aslibekyan S, Mendelson M, Zhou YH, Hedman AK, Sandling JK, Li LA, Irvin MR, Zhi D, Deloukas P, Liang L, Liu C, Bressler J, Spector TD, North K, Li Y, Absher DM, Levy D, Arnett DK, Fornage M, Pankow JS, Boerwinkle E. Epigenome-wide association study (EWAS) of BMI, BMI change and waist circumference in African American adults identifies multiple replicated loci. *Hum Mol Genet.* 2015;24:4464–4479.
215. Aslibekyan S, Demerath EW, Mendelson M, Zhi D, Guan W, Liang L, Sha J, Pankow JS, Liu C, Irvin MR, Fornage M, Hidalgo B, Lin LA, Thibeault KS, Bressler J, Tsai MY, Grove ML, Hopkins PN, Boerwinkle E, Borecki IB, Ordovas JM, Levy D, Tiwari HK, Absher DM, Arnett DK. Epigenome-wide study identifies novel methylation loci associated with body mass index and waist circumference. *Obesity (Silver Spring).* 2015;23:1493–1501.
216. Hegele RA. Phenomics, lipodystrophy, and the metabolic syndrome. *Trends Cardiovasc Med.* 2004;14:133–137.
217. Pendergrass SA, Brown-Gentry K, Dudek SM, Torstenson ES, Ambite JL, Avery CL, Buyske S, Cai C, Fesinmeyer MD, Haiman C, Heiss G, Hindorff LA, Hsu CN, Jackson RD, Kooperberg C, Le ML, Lin Y, Matise TC, Moreland L, Monroe K, Reiner AP, Wallace R, Wilkens LR, Crawford DC, Ritchie MD. The use of phenome-wide association studies (PheWAS) for exploration of novel genotype-phenotype relationships and pleiotropy discovery. *Genet Epidemiol.* 2011;35:410–422.
218. Sivakumaran S, Agakov F, Theodoratou E, Prendergast JG, Zgaga L, Manolio T, Rudan I, McKeigue P, Wilson JF, Campbell H. Abundant pleiotropy in human complex diseases and traits. *Am J Hum Genet.* 2011;89:607–618.

219. Solovieff N, Cotsapas C, Lee PH, Purcell SM, Smoller JW. Pleiotropy in complex traits: challenges and strategies. *Nat Rev Genet.* 2013;14:483–495.
220. Stranger BE, Stahl EA, Raj T. Progress and promise of genome-wide association studies for human complex trait genetics. *Genetics.* 2011;187:367–383.
221. Tyler AL, Crawford DC, Pendergrass SA. The detection and characterization of pleiotropy: discovery, progress, and promise. *Brief Bioinform.* 2016;17:13–22.
222. Ozdemir V, Dove ES, Gursoy UK, Sardas S, Yildirim A, Yilmaz SG, Omer BI, Gungor K, Mete A, Srivastava S. Personalized medicine beyond genomics: alternative futures in big data-proteomics, environtome and the social proteome. *J Neural Transm (Vienna).* 2015;1–8.
223. Martinez JA, Navas-Carretero S, Saris WH, Astrup A. Personalized weight loss strategies-the role of macronutrient distribution. *Nat Rev Endocrinol.* 2014;10:749–760.
224. Pendergrass SA, Ritchie MD. Phenome-wide association studies: leveraging comprehensive phenotypic and genotypic data for discovery. *Curr Genet Med Rep.* 2015;3: 92–100.
225. Blows MW, Allen SL, Collet JM, Chenoweth SF, McGuigan K. The phenome-wide distribution of genetic variance. *Am Nat.* 2015;186:15–30.
226. Pendergrass SA, Verma A, Okula A, Hall MA, Crawford DC, Ritchie MD. Phenome-wide association studies: embracing complexity for discovery. *Hum Hered.* 2015;79: 111–123.
227. Duan DD. Phenomics of cardiac chloride channels. *Compr Physiol.* 2013;3:667–692.
228. Duan DD, Han Y-S, Li L, Zhao J-Z, Wang Z. Pharmacophenomics: a new paradigm for pharmacology, toxicology, and personalized medicine. *Chin J Pharmacol Toxicol.* 2014;28:1–9.
229. Gerlai R. Phenomics: fiction or the future? *Trends Neurosci.* 2002;25:506–509.
230. Oti M, Huynen MA, Brunner HG. Phenome connections. *Trends Genet.* 2008;24: 103–106.
231. Bilder RM. Phenomics: building scaffolds for biological hypotheses in the post-genomic era. *Biol Psychiatry.* 2008;63:439–440.
232. Bilder RM, Sabb FW, Cannon TD, London ED, Jentsch JD, Parker DS, Poldrack RA, Evans C, Freimer NB. Phenomics: the systematic study of phenotypes on a genome-wide scale. *Neuroscience.* 2009;164:30–42.
233. Furbank RT, Tester M. Phenomics—technologies to relieve the phenotyping bottleneck. *Trends Plant Sci.* 2011;16:635–644.
234. Han Y, Li LI, Zhang Y, Yuan H, Ye L, Zhao J, Duan DD. Phenomics of vascular disease: the systematic approach to the combination therapy. *Curr Vasc Pharmacol.* 2015;13: 433–440.
235. Hegele RA, Pollex RL. Hypertriglyceridemia: phenomics and genomics. *Mol Cell Biochem.* 2009;326:35–43.
236. Houle D, Govindaraju DR, Omholt S. Phenomics: the next challenge. *Nat Rev Genet.* 2010;11:855–866.
237. Lanktree MB, Hassell RG, Lahiry P, Hegele RA. Phenomics: expanding the role of clinical evaluation in genomic studies. *J Investig Med.* 2010;58:700–706.
238. Freimer N, Sabatti C. The human phenome project. *Nat Genet.* 2003;34:15–21.
239. Joy T, Hegele RA. Genetics of metabolic syndrome: is there a role for phenomics? *Curr Atheroscler Rep.* 2008;10:201–208.
240. Tracy RP. 'Deep phenotyping': characterizing populations in the era of genomics and systems biology. *Curr Opin Lipidol.* 2008;19:151–157.
241. Plomin R, Haworth CM, Davis OS. Common disorders are quantitative traits. *Nat Rev Genet.* 2009;10:872–878.
242. Crosslin DR, McDavid A, Weston N, Zheng X, Hart E, de AM, Kullo IJ, McCarty CA, Doheny KF, Pugh E, Kho A, Hayes MG, Ritchie MD, Saip A, Crawford DC, Crane

PK, Newton K, Carrell DS, Gallego CJ, Nalls MA, Li R, Mirel DB, Crenshaw A, Couper DJ, Tanaka T, van Rooij FJ, Chen MH, Smith AV, Zakai NA, Yango Q, Garcia M, Liu Y, Lumley T, Folsom AR, Reiner AP, Felix JF, Dehghan A, Wilson JG, Bis JC, Fox CS, Glazer NL, Cupples LA, Coresh J, Eiriksdottir G, Gudnason V, Bandinelli S, Frayling TM, Chakravarti A, van Duijn CM, Melzer D, Levy D, Boerwinkle E, Singleton AB, Hernandez DG, Longo DL, Witteman JC, Psaty BM, Ferrucci L, Harris TB, O'Donnell CJ, Ganesh SK, Larson EB, Carlson CS, Jarvik GP. Genetic variation associated with circulating monocyte count in the eMERGE Network. *Hum Mol Genet.* 2013;22:2119–2127.

243. Ding K, de Andrade M, Manolio TA, Crawford DC, Rasmussen-Torvik LJ, Ritchie MD, Denny JC, Masys DR, Jouni H, Pachecho JA, Kho AN, Roden DM, Chisholm R, Kullo IJ. Genetic variants that confer resistance to malaria are associated with red blood cell traits in African-Americans: an electronic medical record-based genome-wide association study. *G3 (Bethesda).* 2013;3:1061–1068.
244. Kullo IJ, Haddad R, Prows CA, Holm I, Sanderson SC, Garrison NA, Sharp RR, Smith ME, Kuivaniemi H, Bottinger EP, Connolly JJ, Keating BJ, McCarty CA, Williams MS, Jarvik GP. Return of results in the genomic medicine projects of the eMERGE network. *Front Genet.* 2014;5:50.
245. Pathak J, Wang J, Kashyap S, Basford M, Li R, Masys DR, Chute CG. Mapping clinical phenotype data elements to standardized metadata repositories and controlled terminologies: the eMERGE Network experience. *J Am Med Inform Assoc.* 2011;18:376–386.
246. Zuvich RL, Armstrong LL, Bielinski SJ, Bradford Y, Carlson CS, Crawford DC, Crenshaw AT, de Andrade AM, Doheny KF, Haines JL, Hayes MG, Jarvik GP, Jiang L, Kullo IJ, Li R, Ling H, Manolio TA, Matsumoto ME, McCarty CA, McDavid AN, Mirel DB, Olson LM, Paschall JE, Pugh EW, Rasmussen LV, Rasmussen-Torvik LJ, Turner SD, Wilke RA, Ritchie MD. Pitfalls of merging GWAS data: lessons learned in the eMERGE network and quality control procedures to maintain high data quality. *Genet Epidemiol.* 2011;35:887–898.
247. Denny JC, Ritchie MD, Basford MA, Pulley JM, Bastarache L, Brown-Gentry K, Wang D, Masys DR, Roden DM, Crawford DC. PheWAS: demonstrating the feasibility of a phenome-wide scan to discover gene-disease associations. *Bioinformatics.* 2010;26: 1205–1210.
248. Grubb SC, Bult CJ, Bogue MA. Mouse phenome database. *Nucleic Acids Res.* 2014;42: D825–D834.
249. Neuraz A, Chouchana L, Malamut G, Le BC, Roche D, Beaune P, Degoulet P, Burgun A, Loriot MA, Avillach P. Phenome-wide association studies on a quantitative trait: application to TPMT enzyme activity and thiopurine therapy in pharmacogenomics. *PLoS Comput Biol.* 2013;9:e1003405.
250. Pendergrass SA, Brown-Gentry K, Dudek S, Frase A, Torstenson ES, Goodloe R, Ambite JL, Avery CL, Buyske S, Buzkova P, Deelman E, Fesinmeyer MD, Haiman CA, Heiss G, Hindorff LA, Hsu CN, Jackson RD, Kooperberg C, Le ML, Lin Y, Matise TC, Monroe KR, Moreland L, Park SL, Reiner A, Wallace R, Wilkens LR, Crawford DC, Ritchie MD. Phenome-wide association study (PheWAS) for detection of pleiotropy within the Population Architecture using Genomics and Epidemiology (PAGE) Network. *PLoS Genet.* 2013;9:e1003087.
251. Pendergrass SA, Verma A, Okula A, Hall MA, Crawford DC, Ritchie MD. Phenome-wide association studies: embracing complexity for discovery. *Hum Hered.* 2015;79: 111–123.
252. Pendergrass SA, Ritchie MD. Phenome-wide association studies: leveraging comprehensive phenotypic and genotypic data for discovery. *Curr Genet Med Rep.* 2015;3:92–100.
253. Denny JC, Bastarache L, Ritchie MD, Carroll RJ, Zink R, Mosley JD, Field JR, Pulley JM, Ramirez AH, Bowton E, Basford MA, Carrell DS, Peissig PL, Kho AN, Pacheco

JA, Rasmussen LV, Crosslin DR, Crane PK, Pathak J, Bielinski SJ, Pendergrass SA, Xu H, Hindorff LA, Li R, Manolio TA, Chute CG, Chisholm RL, Larson EB, Jarvik GP, Brilliant MH, McCarty CA, Kullo IJ, Haines JL, Crawford DC, Masys DR, Roden DM. Systematic comparison of phenome-wide association study of electronic medical record data and genome-wide association study data. *Nat Biotechnol.* 2013;31: 1102–1110.

254. Hertel JK, Johansson S, Raeder H, Midthjell K, Lyssenko V, Groop L, Molven A, Njolstad PR. Genetic analysis of recently identified type 2 diabetes loci in 1,638 unselected patients with type 2 diabetes and 1,858 control participants from a Norwegian population-based cohort (the HUNT study). *Diabetologia.* 2008;51:971–977.
255. Rees SD, Islam M, Hydrie MZ, Chaudhary B, Bellary S, Hashmi S, O'Hare JP, Kumar S, Sanghera DK, Chaturvedi N, Barnett AH, Shera AS, Weedon MN, Basit A, Frayling TM, Kelly MA, Jafar TH. An FTO variant is associated with Type 2 diabetes in South Asian populations after accounting for body mass index and waist circumference. *Diabet Med.* 2011;28:673–680.
256. Li X, Song F, Jiang H, Zhang M, Lin J, Bao W, Yao P, Yang X, Hao L, Liu L. A genetic variation in the fat mass- and obesity-associated gene is associated with obesity and newly diagnosed type 2 diabetes in a Chinese population. *Diab Metab Res Rev.* 2010;26:128–132.
257. Stratigopoulos G, Padilla SL, LeDuc CA, Watson E, Hattersley AT, McCarthy MI, Zeltser LM, Chung WK, Leibel RL. Regulation of Fto/Ftm gene expression in mice and humans. *Am J Physiol Regul Integr Comp Physiol.* 2008;294:R1185–R1196.
258. Hubacek JA, Bohuslavova R, Kuthanova L, Kubinova R, Peasey A, Pikhart H, Marmot MG, Bobak M. The FTO gene and obesity in a large Eastern European population sample: the HAPIEE study. *Obesity (Silver Spring).* 2008;16:2764–2766.
259. Li H, Wu Y, Loos RJ, Hu FB, Liu Y, Wang J, Yu Z, Lin X. Variants in the fat mass- and obesity-associated (FTO) gene are not associated with obesity in a Chinese Han population. *Diabetes.* 2008;57:264–268.
260. Xi B, Takeuchi F, Meirhaeghe A, Kato N, Chambers JC, Morris AP, Cho YS, Zhang W, Mohlke KL, Kooner JS, Shu XO, Pan H, Tai ES, Pan H, Wu JY, Zhou D, Chandak GR. Associations of genetic variants in/near body mass index-associated genes with type 2 diabetes: a systematic meta-analysis. *Clin Endocrinol (Oxf).* 2014;81:702–710.
261. Gong J, Schumacher F, Lim U, Hindorff LA, Haessler J, Buyske S, Carlson CS, Rosse S, Buzkova P, Fornage M, Gross M, Pankratz N, Pankow JS, Schreiner PJ, Cooper R, Ehret G, Gu CC, Houston D, Irvin MR, Jackson R, Kuller L, Henderson B, Cheng I, Wilkens L, Leppert M, Lewis CE, Li R, Nguyen KD, Goodloe R, Farber-Eger E, Boston J, Dilks HH, Ritchie MD, Fowke J, Pooler L, Graff M, Fernandez-Rhodes L, Cochrane B, Boerwinkle E, Kooperberg C, Matise TC, Le ML, Crawford DC, Haiman CA, North KE, Peters U. Fine mapping and identification of BMI Loci in African Americans. *Am J Hum Genet.* 2013;93:661–671.
262. Smemo S, Tena JJ, Kim KH, Gamazon ER, Sakabe NJ, Gomez-Marin C, Aneas I, Credidio FL, Sobreira DR, Wasserman NF, Lee JH, Puviindran V, Tam D, Shen M, Son JE, Vakili NA, Sung HK, Naranjo S, Acemel RD, Manzanares M, Nagy A, Cox NJ, Hui CC, Gomez-Skarmeta JL, Nobrega MA. Obesity-associated variants within FTO form long-range functional connections with IRX3. *Nature.* 2014;507:371–375.
263. Corella D, Ortega-Azorin C, Sorli JV, Covas MI, Carrasco P, Salas-Salvado J, Martinez-Gonzalez MA, Aros F, Lapetra J, Serra-Majem L, Lamuela-Raventos R, Gomez-Gracia E, Fiol M, Pinto X, Ros E, Marti A, Coltell O, Ordovas JM, Estruch R. Statistical and biological gene-lifestyle interactions of MC4R and FTO with diet and physical activity on obesity: new effects on alcohol consumption. *PLoS One.* 2012;7:e52344.
264. Lurie G, Gaudet MM, Spurdle AB, Carney ME, Wilkens LR, Yang HP, Weiss NS, Webb PM, Thompson PJ, Terada K, Setiawan VW, Rebbeck TR, Prescott J, Orlow I,

O'Mara T, Olson SH, Narod SA, Matsuno RK, Lissowska J, Liang X, Levine DA, Le Marchand L, Kolonel LN, Henderson BE, Garcia-Closas M, Doherty JA, De Vivo I, Chen C, Brinton LA, Akbari MR, Goodman MT. The obesity-associated polymorphisms FTO rs9939609 and MC4R rs17782313 and endometrial cancer risk in non-Hispanic white women. *PLoS One.* 2011;6:e16756.
265. Reitz C, Tosto G, Mayeux R, Luchsinger JA. Genetic variants in the Fat and Obesity Associated (FTO) gene and risk of Alzheimer's disease. *PLoS One.* 2012;7:e50354.
266. Gottesman O, Drill E, Lotay V, Bottinger E, Peter I. Can genetic pleiotropy replicate common clinical constellations of cardiovascular disease and risk? *PLoS One.* 2012;7: e46419.
267. Crosslin DR, Tromp G, Burt A, Kim DS, Verma SS, Lucas AM, Bradford Y, Crawford DC, Armasu SM, Heit JA, Hayes MG, Kuivaniemi H, Ritchie MD, Jarvik GP, de AM. Controlling for population structure and genotyping platform bias in the eMERGE multi-institutional biobank linked to electronic health records. *Front Genet.* 2014;5:352.
268. Gottesman O, Kuivaniemi H, Tromp G, Faucett WA, Li R, Manolio TA, Sanderson SC, Kannry J, Zinberg R, Basford MA, Brilliant M, Carey DJ, Chisholm RL, Chute CG, Connolly JJ, Crosslin D, Denny JC, Gallego CJ, Haines JL, Hakonarson H, Harley J, Jarvik GP, Kohane I, Kullo IJ, Larson EB, McCarty C, Ritchie MD, Roden DM, Smith ME, Bottinger EP, Williams MS. The Electronic Medical Records and Genomics (eMERGE) Network: past, present, and future. *Genet Med.* 2013;15:761–771.
269. McCarty CA, Chisholm RL, Chute CG, Kullo IJ, Jarvik GP, Larson EB, Li R, Masys DR, Ritchie MD, Roden DM, Struewing JP, Wolf WA. The eMERGE Network: a consortium of biorepositories linked to electronic medical records data for conducting genomic studies. *BMC Med Genomics.* 2011;4:13.
270. Palou A, Bonet ML. Challenges in obesity research. *Nutr Hosp.* 2013;28(suppl 5): 144–153.

CHAPTER EIGHT

Genetics of Adiposity in Large Animal Models for Human Obesity—Studies on Pigs and Dogs

M. Stachowiak[a], I. Szczerbal[a], M. Switonski*
Department of Genetics, Animal Breeding, Poznań University of Life Sciences, Poznań, Poland
* Corresponding author. E-mail address: switonsk@up.poznan.pl
[a] M. Stachowiak and I. Szczerbal these authors contributed equally.

Contents

Abstract

The role of domestic mammals in the development of human biomedical sciences has been widely documented. Among these model species the pig and dog are of special importance. Both are useful for studies on the etiology of human obesity. Genome sequences of both species are known and advanced genetic tools [eg, microarray SNP for genome wide association studies (GWAS), next generation sequencing (NGS), etc.] are commonly used in such studies. In the domestic pig the accumulation of

Progress in Molecular Biology and Translational Science, Volume 140
ISSN 1877-1173
http://dx.doi.org/10.1016/bs.pmbts.2016.01.001

adipose tissue is an important trait, which influences meat quality and fattening efficiency. Numerous quantitative trait loci (QTLs) for pig fatness traits were identified, while gene polymorphisms associated with these traits were also described. The situation is different in dog population. Generally, excessive accumulation of adipose tissue is considered, similar to humans, as a complex disease. However, research on the genetic background of canine obesity is still in its infancy. Between-breed differences in terms of adipose tissue accumulation are well known in both animal species. In this review we show recent advances of studies on adipose tissue accumulation in pigs and dogs, and their potential importance for studies on human obesity.

1. INTRODUCTION

Long-lasting breeding of domestic animals has resulted in the creation of a wide variety of breeds, differing in terms of numerous traits, related to morphology, behavior, coating, productivity, quality of animal products, adaptation to abiotic (eg, temperature) and biotic (eg, resistance to specific pathogens) environmental conditions, etc. The observed phenotypic variation between breeds of the same species reflects differences in their gene pools.

Breeding strategies, mainly based on selection, are quite different in livestock and companion species. In livestock the most important are production (meat, milk, wool, eggs, etc.) and the related functional traits (fertility, longevity, adaptation to specific environmental conditions, etc.). Thus, a directional selection is focused on continuous improvement of economically efficient production of high quality animal products by healthy animals managed under welfare conditions. A crucial step of selection in mammalian livestock (eg, cattle and pigs) is evaluation of breeding value of males, carried on the basis of their offspring performance (eg, milk yield, daily body mass gain, fatness traits, etc.). In this procedure, called progeny testing, sophisticated statistical methods are used. Recently a new approach, based on the use of SNP microarrays and called genome selection, has been implemented into this methodology. In contrast, a stabilizing selection in accordance with breed standards is applied in companion species, mainly dogs and cats. In this approach, called pedigree selection, a crucial role is played by pedigree information, concerning phenotypes of the ancestors, including prizes won on animal shows.

The global number of livestock and companion animal breeds is large and the exact figures remain unknown. It is estimated that several hundreds of pig breeds have been established worldwide. At the Pig website (http://www.thepigsite.com) 75 common breeds are specified, but this database does not include a wide variety of local breeds. For example, in the United Kindom there are 18 autochthonous breeds (the British Pig Association, http://www.britishpigs.org.uk/), while in China 48 native breeds are recognized (http://www.Pigprogress.net). In the case of dogs this knowledge is more precise due to an international database run by Fédération Cynologique Internationale (FCI) (http://www.fci.be/en/). According to this database, approx. 340 dog breeds are internationally recognized. It is also important to point out that among domestic mammals an exceptional phenotypic variability is observed in dogs.[1]

Due to the extensive use of selected sires in artificial insemination (eg, in cattle and pigs) or a limited number of founders of a breed (numerous dog breeds) a random increase of undesired mutations in a gene pool may occur. Among domestic mammals the highest number of monogenic diseases with a known causative mutation was described in the dog.[2] Importantly, a majority of these diseases have counterparts in humans, and thus the dog has become an important large animal model for human biomedical science.[3] Knowledge on complex diseases (eg, diabetes, metabolic syndrome, obesity, cardiovascular diseases, neoplasms, etc.) in livestock and companion species is different, due to the difference of their life span. In addition, attentive veterinary care, common in the case of dogs and cats, facilitates a precise diagnosis of such diseases.

Among domestic mammals two species, namely the pig and the dog, are crucial models for human obesity. The pig is an omnivorous species showing a wide range of anatomical, histological, and physiological similarities to humans,[4] while the dog is a carnivorous species sharing with the human the same environment, including diet and life style, for example, physical activity.[5] Deposition of adipose tissue in livestock species is an important production trait, influencing meat quality and its dietetic value, as well as fattening effectiveness.[6] Thus, mechanisms responsible for nonpathogenic accumulation of subcutaneous, visceral, or intramuscular fat tissue in these species, including pigs, have been extensively studied. The pig is a model of special interest because similarly to humans, an excessive accumulation of adipose tissue is responsible for the development of obesity-related diseases. This knowledge may help us to understand a known relationship between specific fat depots (eg, abdominal) and human obesity-related diseases, that is, diabetes.[7] Recently, it was shown that diet-induced obesity in pigs can be a model in studies on the effect of obesity on the development of metabolic syndrome.[8] Also physiological indices related with obesity are similar in both species, that is, a high serum concentration of low density lipoproteins (LDL) and a low high density lipoprotein (HDL) concentration.[9]

Studies on adiposity of companion animals (dogs and cats) have recently received special attention because obesity is considered as an emerging health problem in these species.[10] The major factors affecting canine and human obesity are similar: nutrition, physical activity, and hereditary predisposition.[11] An analysis of over 2000 dogs living in Beijing (China), aimed on the identification of risk factors for obesity indicated the following ones: feeding (the use of noncommercial food type and increased feeding frequency), age, physical activity, neutering, breed, and sex.[12] Another similarity between human and canine obesity was also endocrinologically documented in terms of the relationship between adiposity and the level of adipokines.[13–15]

The hereditary background of adipose tissue accumulation may be analyzed in terms of the effect of breed, revealing chromosomal regions [quantitative trait locus (QTL)] harboring the predisposing DNA variants, identification of the associated gene polymorphisms and deciphering variation of gene expression, including epigenetic mechanisms. Such studies have been extensively carried out with regard to porcine fatness traits (reviewed by Switonski et al.[16]), while in dogs they are in infancy.[17,18] This review presents recent advances, focused on the importance of these studies to provide insight to the genetic background of human obesity.

2. BREED SPECIFIC DIFFERENCES OF ADIPOSE TISSUE ACCUMULATION

Variability of fat tissue deposition between pig breeds, as well as predisposition to obesity in some dog breeds are well-known phenomena. Similarly to the human, heritability of pig adiposity fluctuates around 0.5.[16] It may be assumed that also in dogs this coefficient has a similar value.

Commercial pig breeds are classified as a meat type, since accumulation of fat tissue is accepted on a relatively low level, essential for high quality of meat. However, there are numerous autochthonous breeds, representing a fat-type pig, which accumulate an excessive amount of adipose tissue. These include well-known European breeds—Mangalica[19] and the Iberian pig,[20] as well as several Chinese breeds, including the Erhualian breed.[21] One can speculate that predisposition to adipose tissue accumulation in these breeds, also referred to as the thrifty genotype, is an adaptive mechanism to uneven availability of feed. A classic example of this genotype is a miniature Ossabaw breed, which develops morbid obesity and metabolic syndrome under a calorie-rich diet.[22] This breed is considered to be a very useful large animal model in studies on human obesity and obesity-related diseases, for example, metabolic syndrome.[23–25]

There are several adipose traits, which have been extensively studied in pig breeds due to their importance for quality of meat and efficiency of fattening. These traits are usually studied *postmortem*, when carcass dissection is carried out. Among them the most crucial include thickness of subcutaneous fat tissue on the back at different locations [back fat thickness (BFT), expressed in cm abdominal fat weight (AFW, expressed in kg), intramuscular fat content (IMF, expressed in percentage)] and fatty acid composition of fat tissues. These analyses provided detailed insights into mechanisms related with adipose tissue accumulation. It is well known that some breeds have a very thin subcutaneous fat tissue (eg, the Pietrain breed), in a majority of breeds this tissue is moderately thick (eg, breeds belonging to Large White breeds), while in some breeds this depot is very thick (eg, the Mangalica breed). For example, a comparison of BFT of Mangalica and Hungarian Large White, slaughtered at the same body weight (approx. 130 kg), showed a significant difference of the average thickness, of 5.0 and 2.5 cm, respectively.[26] Another example concerns the influence of breed and production system on performance traits, including BFT and mass of perineal fat. Pigs of two breeds, the commercial Large White (meat-type breed) and Basque (autochthonous, fat-type breed), were fed the same growing diet and slaughtered at the same body weight (on average 145 kg). It was found that the accumulation of adipose tissue was significantly higher in Basque pigs. The average BFT and perineal fat mass in Basque pigs were 4.7 cm and 4.8 kg (in the conventional indoor housing system), or 5.0 cm and 4.8 kg (indoor housing system with free access to outdoor area), respectively. These values for Large White pigs were 2.3 cm and 1.9 kg, or 2.4 cm and 1.9 kg, respectively.[27] Differences of porcine fatness is also manifested on the cellular level. It has been found that the size of adipocytes is larger in fat breeds than in leaner ones. For example, the commercial Landrace breed with a low backfat thickness has smaller adipocytes in the subcutaneous adipose tissues when compared with the fat-type Meishan breed.[28] Moreover, tissue-specific differences in adipocyte size are also observed. Adipocytes derived from intramuscular fat tissue are smaller than those from subcutaneous and visceral fat tissues. The aforementioned examples clearly show that gene pools involved in adipose tissue accumulation are different in the compared breeds.

Obesity in dogs is an emerging health problem, which has received serious attention of veterinarians, breeders, and owners. Evaluation of adipose tissue accumulation in dogs is usually performed with the use of a

Table 1 Prevalence (%) of Obesity and Overweight in Dogs in the USA (http://www.petobesityprevention.org/pet-obesity-fact-risks/).

Year	Obese Dogs, BCS = 5	Overweight Dogs, BCS = 4	Total
2011	21.5	31.5	53.0
2012	15.7	36.8	52.5
2013	16.7	35.9	52.6
2014	17.6	35.1	52.7

subjective 5- or 9-level body condition score (BCS) scale. In the 5-level scale the following BCS values mean: 3, normal body condition, 4, overweight, and 5, obesity. It should be pointed out that more objective methods are also available, for example, dual-energy X-ray absorptiometry (DEXA) and deuterium oxide (D_2O) dilution.[29] Interestingly, the results of the DEXA approach very well correlate with the subjective BCS scale.[30] The incidence of overweight and obesity is constantly increasing. It was estimated by the Association for Pet Obesity Prevention that in the USA the prevalence of obesity and overweight in dogs exceeded 52% (Table 1). It is also well known that some breeds are predisposed to obesity. A wide survey of 21,754 dogs in the USA showed the highest prevalence of obesity in Beagles (47.7%), Cocker Spaniels (46.3%), Shetland Sheep Dog (45%), Golden Retrievers (44.3%), and Labrador Retrievers (41.1%).[11] Another study, comprising 2391 dogs from Beijing (China), showed a very high overall prevalence of adiposity (44.4%), with the highest rates noted in Pugs (70.7%), Cocker Spaniels (69.4%), Pomeranians (54.6%), Pekingese dogs (51.9%), Golden Retrievers (51.9%), Chihuahuas (46.9%), and Labrador Retrievers (46%), while the lowest rate was observed in Huskies (25%), Miniature Poodles (23.9%), and Poodles (20.3%).[12]

These observations indicate that the presence of genetic variants predisposing to an excessive accumulation of adipose tissue in gene pools of the aforementioned breeds are different. Thus, breeds differing significantly in terms of their predisposition to obesity are of special interest when searching for the predisposing and protecting genetic variants.

3. EPIGENETIC MARKERS OF ADIPOGENESIS AND ADIPOSE TISSUE

Development of adipose tissue is a consequence of two processes, namely generation of new adipocytes—hyperplasia, and an increase in their

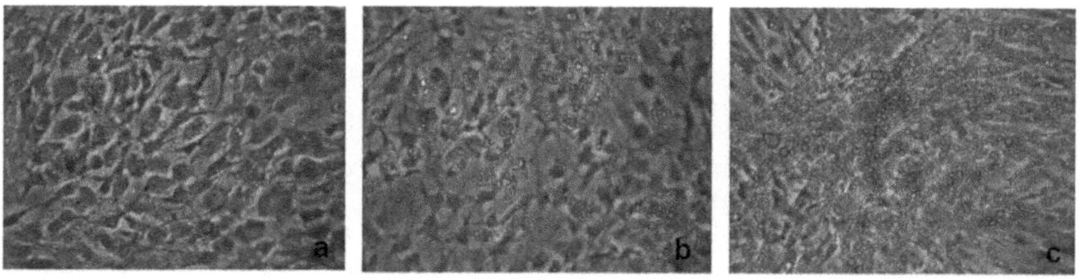

Figure 1 Accumulation of lipid droplets during porcine in vitro adipogenesis. Oil Red O staining was performed at days 3 (A), 7 (B), and 14 (C) of the differentiation. *Photo: I. Szczerbal.*

volume—hypertrophy.[31] Thus, genes involved in adipocyte differentiation or lipid metabolism are considered to be functional candidates for porcine fatness traits.[32,33] A majority of model studies on adipogenesis were carried out on mice, while knowledge about this process in large animal models is limited in the pig and very scarce in the dog. Since the pig is considered as a valuable model organism for studies on human obesity, understanding genetic and epigenetic mechanisms governing porcine adipocyte differentiation is highly required.

Adipogenesis in the pig (Fig. 1), similar to other species, is regulated by a complex network of transcription factors, among which PPARγ and members of the C/EBP family play a major role.[34] A number of other positive and negative regulators of porcine adipogenesis have been described.[35,36] Interestingly, there are transcription factors (eg, KLF13), which showed no effect on mouse adipogenesis, but have been recognized as proadipogenic factors in the pig.[37] In recent years increasing attention has been addressed to epigenetic factors involved in porcine adipogenesis, including DNA methylation, the role of noncoding RNA (ncRNA) and nuclear architecture. It is assumed that epigenetics is an important contributor of "the missing heritability" in complex traits.[38]

3.1 DNA Methylation

DNA methylation has gained special interest among epigenetic mechanisms involved in the development of human obesity. Different approaches concerning global, gene-specific, and genome-wide DNA methylation levels were applied to search for its role in obesity. New methodologies such as epigenome-wide association studies (EWAS) allowed researchers to expand these studies in recent years. Changes in DNA methylation were observed in a number of candidate genes in the human, which functionally are linked mainly with adiposity or appetite control, but also other cellular processes. In addition, obesity associated differentially methylated (DM) sites have been identified with the use of genome-wide methods.[39] There are examples showing that perinatal nutritional exposures can cause epigenetic consequences in the offspring.[40] Moreover, the DNA methylation profile is not stable throughout adult life, but may be changed by different interventions, such as exercise,[41] diet,[42] and weight loss surgery.[43]

A comprehensive study of three pig breeds differing in terms of their fat accumulation, that is, the Landrace (a lean, meat type), Tibetan (a feral type) and Rongchang (a fat type), produced a DNA methylation map for adipose and muscle tissues.[44] Altogether, 10 samples from various anatomic locations were analyzed: 8 adipose tissues (subcutaneous, 3; visceral, 4; intermuscular, 1) and two skeletal muscle tissues (the longissimus dorsi muscle and psoas major muscle). Differentially methylated regions (DMRs) were identified and their number varied between breed, tissues, and sexes. Analyses of breed specific DMRs in promoters of adipose tissues revealed that the Rongchang and Tibetan breeds are phylogenetically closer to each other than the Landrace, while the Landrace is closer to Tibetan than the Rongchang breeds when muscle tissues were evaluated. This indicates that breed-specific differences are found not only on the genetic, but also the epigenetic level. DMRs in promoters were also tissue-specific and methylation in these genomic elements correlated with adipose tissue from different locations. For example, intermuscular adipose tissue was more similar to visceral adipose tissue in terms of the methylation pattern. Among 282 porcine gene orthologs to human obesity-related genes, 80% has shown location within the identified DMRs. Also more than 70% of porcine genes from the QTL regions affecting fatness and pork quality overlapped with the identified DMRs.

A detailed analysis of methylated genes from different adipose depots (the same samples as in Li et al.[44]) has shown that comethylated genes from visceral and intermuscular adipose tissues are associated with inflammatory and immune processes, while comethylated genes in subcutaneous adipose tissue are mainly associated with metabolic processes.[45] These observations were also confirmed by a comprehensive genome-wide comparison of gene expression profiling.[46] In addition, different layers of porcine backfat tissue (superficial vs. deep) were studied in terms of DNA methylation and DMRs associated with differentially expressed genes involved in lipid metabolism and regulation of immune-related cytokines have been reported.[47] Distinct features of adipose tissues from various locations identified on molecular levels reflect their functional and metabolic differences. With the use of transcriptomic studies it has been recognized that both visceral and intramuscular adipose tissues are associated mainly with immune and inflammation responses and these tissues have been identified as a metabolic risk factor for obesity.[48,49]

3.2 ncRNA

Two categories of noncoding RNAs, that is, microRNA (miRNA) and long noncoding RNA (lncRNA), have been recognized as important regulators of gene expression. The role of miRNA in mammalian adipogenesis has been recently reviewed by Peng et al.[50] A number of proadipogenic and antiadipogenic miRNAs have been identified. A comprehensive study concerning porcine miRNAs involved in adipogenesis and a comparison of miRNAomes during differentiation of intramuscular (IVSC) and subcutaneous vascular stem cells (SVSC) were performed by Guo et al.[51] The authors identified 224 known and 280 potential porcine miRNAs and showed expression similarities and differences during differentiation of IVSC and SVSC. The miRNome of back fat in two adult Italian Large White pigs was also analyzed by Gaffo et al.[52] Unfortunately, in the aforementioned studies no specific miRNAs associated with the accumulation of adipose tissue were indicated.

In recent years it has been shown that also long-noncoding RNAs (lncRNA) may play an important role in murine adipogenesis.[53] So far, lncRNAs have not been extensively studied in the pig. There is only a single study concerning antisense long noncoding RNAs (AS lncRNAs), which revealed that a novel AS lncRNA (PU.1) has a positive effect on porcine adipogenesis.[54] The authors hypothesized that modulation of lncRNA may provide a new target for the control of fat accumulation.

3.3 Nucleus Architecture

Regulation of transcription is also controlled by nuclear architecture. It was shown that the arrangement of key adipogenesis genes and chromosomes harboring these genes within a nuclear space have specific patterns during different stages of in vitro adipogenesis in pigs (Fig. 2). The radial nuclear position of selected genes (eg, *PPARG*, *GATA2*, *SREBF1*) has changed during differentiation and these changes correlated with the transcriptional status, as upregulated genes were more internally located within nuclei.[55] Further studies revealed spatial coassociation of genes activated during

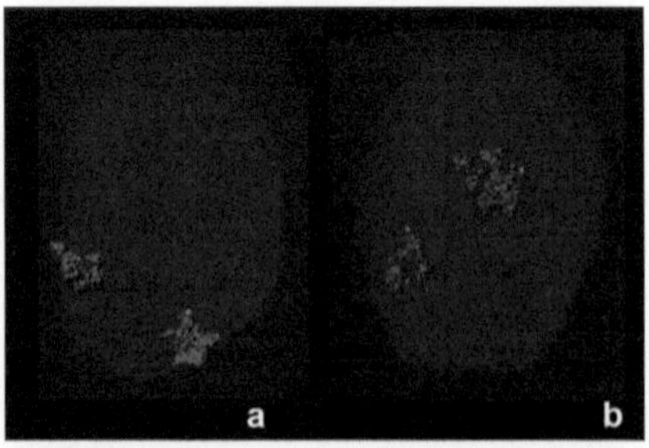

Figure 2 Changes of nuclear architecture during porcine in vitro adipogenesis. Distribution of the *FABP4* gene and SSC4 chromosome territories in a mesenchymal stem cell (A) and an adipocyte from day 14 of differentiation (B) are shown. *Photo: I. Szczerbal.*

porcine adipogenesis with nuclear speckles (SC-35). It was suggested that it can be a mechanism which enhances gene expression.[56] On the other hand, analysis of the nuclear architecture of matured adipocytes derived from fat tissue has shown that correlation between nuclear position and transcription activity is not common in case of genes involved in lipid metabolism.[57] It has been suggested that the relationship between transcriptional status of adipogenic genes and their nuclear positioning is characteristic for differentiation process, when genes switch from the silent to the active state.

3.4 Nutrigenomics and Epigenetic Modifications

There is a growing body of evidence that epigenetic modifications may be caused by environmental factors, including nutrition.[58] Many studies have shown that dietary interventions during gestation induce changes in DNA methylation and histone modification patterns of offspring. Different types of diets (eg, high-fat, high-protein, or undernutrition) as well as supplementation or restriction with a range of dietary factors, including those which affect the methionine cycle (eg, folate, vitamin B6 and B12), were examined. Moreover, nutrition during pregnancy has not only the direct effect on a fetus, but may also show a transgenerational effect.[59] Rodent models, but also farm animals have been widely used in such experiments.[60]

An example of a transgenerational epigenetic effect in pigs was reported by Braunschweig et al.[61] Three generations of Large White pigs were studied in terms of heritable epigenetic changes caused by a methyl-enriched diet. Males from the F0 generation were fed with a diet supplemented with high amounts of methylating micronutrients and the control group received a standard diet. The feeding effect was analyzed in the F2 generation. It was found that F2 animals had a lower fat percentage and a higher shoulder muscle percentage as compared to the controls. Moreover, differences in transcript levels of genes from the liver and muscle tissue as well as DNA methylation changes in the *IYD* (iodotyrosine deiodinase) gene were found in that experiment.

The effects of maternal diet on transcriptional regulation of the *MSTN* (myostatin) gene, a negative regulator of skeletal muscle growth, was studied in Meishan pigs.[62] This indigenous breed is traditionally raised on a low-protein diet. Sows were fed with low-protein or standard-protein diets throughout gestation and lactation periods. The *MSTN* transcript level was analyzed in the longissimus dorsi muscle of their offspring at weaning (approx. body weight 20 kg) and finishing stages (approx. body weight 40–70 kg). The *MSTN* expression was affected in pigs exposed to a maternal low-protein diet at the finishing stage through regulatory mechanisms, including histone modification and microRNA activity. It has been shown that the immediate effect is not related with epigenetic regulations, while the long-term effects were epigenetically controlled. The impact of the maternal low-protein diet during gestation and lactation was also investigated in Meishan pigs in relation to hepatic cholesterol metabolism.[63] Piglets at weaning had a lower body weight and liver weight. The phenotypic changes were associated with upregulation of hepatic genes through epigenetic modifications.

Nutrigenomics studies were also carried out in dogs, but they were focused on the impact of different diets and supplementations on transcript levels of key genes related with obesity.[18] However, to date no reports on the influence of dog nutrition on epigenetic modifications have been presented.

4. QTL FOR ADIPOSITY TRAITS

Discovery of common DNA markers, mainly microsatellites [short tandem repeats (STR), short simple repeats (SS)] has facilitated development of high density marker genome maps. In the early 1990s international marker genome map projects were launched and they were focused on mapping microsatellite markers. The most important porcine projects were established in Europe [the European Pig Gene Mapping Project (PiGMaP)] and the USA [the porcine linkage map supported by the United States Department of Agriculture (USDA)]. In 1993 an international consortium DogMap was launched for the development of the dog genome marker map. Within a relatively short time advanced maps, comprising thousands of mapped STR markers, were presented. The porcine maps were extensively used in the so-called genome scanning, aimed at the identification of chromosomal regions [quantitative trait loci (QTLs)] harboring markers linked with unknown DNA variants influencing variability of numerous production traits, including fatness. For this purpose several reference families were established, in which founders differed phenotypically. Some of these families were created by crossing commercial meat-type breeds (eg, Large White, Duroc, Landrace) and local fat-type breeds, for example, Mangalica.[21,64,65]

The next step in deciphering genome organization was its sequencing. The genome sequence of the dog was described in 2005 by Lindblad-Toh et al.,[66] while the porcine one in 2012 by Groenen et al.[67] The genome sequencing projects revealed the presence of numerous single nucleotide polymorphisms (SNPs) in the studied genomes and it facilitated development of SNP microarrays, comprising up to 170k SNPs for the dog and 60k SNPs for the pig. It is expected that soon a high density porcine microarray (510k) will be commercialized. These powerful tools were extensively used in the so-called genome-wide association study (GWAS) for QTL studies in pigs and the identification of gene mutations responsible for canine monogenic diseases, as well as phenotypic variation between breeds.

Genome scanning resulted in the detection of numerous QTLs for adipose tissue accumulation in the pig. In the PigQTL database (http://www.animalgenome.org/cgi-bin/QTLdb/SS/index), among the reported 13,030 QTLs for 663 pig traits there are 243 regions for an average backfat thickness, 180 for IMF and 34 for AFW. This summary clearly shows that the genetic background of adipose tissue accumulation in pigs have been extensively studied.

The first QTL for pig fatness traits was published over 20 years ago.[68] This region, later called *FAT1*, was assigned to pig chromosome 4 [Sus scrofa chromosome 4 (SSC4)], but the search for a candidate gene in this region is still not completed. Recently, a novel candidate gene—*PLAG1* (pleiomorphic adenoma gene 1) has been proposed.[69] Interestingly, this gene has been under a strong selection during pig domestication, since distinct sequence differences, related with body size, were found between the domestic pig and the wild boar.[70]

According to the PigQTL database, QTL regions for fatness traits are distributed on all chromosomes, but their significance is not similar. During the last 5 years a detailed study on QTLs for porcine fatness traits confirmed the importance of regions residing on the following chromosomes: SSC2,[21,69] SSC4,[21,69] SSC7,[21,69,71] and SSCX.[21,69,72]

An interesting comparative whole genome search for candidate genes involved in lipid metabolism, performed with the use of the next generation sequencing (NGS) approach, was presented by Molnar et al.[73] The authors compared the genomes of three males of Mangalica, representing three color variants (blond, red, and swallow-belly), with the genome of the commercial meat-type Duroc pig and the reference pig genome. The comparison revealed nonsynonymous SNPs that were found in Mangalicas, but were not present in the Duroc pig in 52 genes involved in lipid-related metabolic processes and 49 of them were localized within fat-related QTLs. Interestingly, among the 52 highlighted genes, 42 were not previously associated with fatness traits. A comparison of this gene list with the set of 97 human genes associated with predisposition to obesity, presented by Locke et al.,[74] revealed that none of the 52 pig candidate genes was indicated in the human.

A vast majority of QTL studies were carried out with the aim to identify chromosomal regions for production traits; however, there are also reports on comparative genome analyses of human and porcine genomes focused on the hereditary background of adiposity. A whole genome comparison of obesity related traits in the human [subscapular skinfold thickness (SUB)] and the pig (BFT), based on genotypes obtained with the use of SNP microarrays—porcine 60k and human 500k, was reported by Kim et al.[75] This analysis showed that human chromosome 2 (HSA2), which is orthologous to pig chromosomes SSC3 and SSC15, harbors loci (*MRPL33*, *STK39*, *ZNF385B*, *PARD3B*, and *ERDB4*), which variants probably predispose to obesity. However, until now no reports confirming the role of these genes have been published. Moreover, these genes were not indicated as predisposing to human obesity.[74]

The GWAS methodology has been widely applied in studies on behavioral, morphological traits as well as complex diseases, including cancers.[76] Several studies concerned the identification of QTLs for body size. Using the GWAS approach the phenotypic variation between dog breeds was studied by Vaysse et al.[77] The authors identified 44 genomic regions responsible for extreme phenotypic differentiation across 46 breeds. One of the analyzed traits was body weight of normal (standard) dogs. In the identified region on chromosome CFA15 the *IGF1* gene is present. This finding confirmed an earlier study of Sutter et al.,[78] who showed that different gene variants of this gene are characteristic of large and small breeds. Unfortunately, until now there have been no reports on the use of GWAS to study canine obesity.

5. POLYMORPHISMS ASSOCIATED WITH FAT TISSUE ACCUMULATION

Extensive studies with the use of GWAS technology, as well as sequencing of functional and positional candidate genes resulted in the identification of polymorphisms associated with pig fatness, while in dogs such analyses are not so advanced. The associated SNPs are predominant; however, in recent years also some copy number variations (CNV) appeared to be associated with pig fatness and dog obesity.

5.1 SNPs and Indels in Pigs

Identification of genetic variants associated with lipid metabolism and fat tissue accumulation is an important goal of pig breeding, presently focused on fast growing and lean, meat-type pigs. The most favorable fatness-related traits in pig breeding include low back fat thickness (BFT, cm) and abdominal fat weight (AFW, kg), while a moderate intramuscular fat content (IMF, %) or marbling score (subjectively evaluated deposition of intramuscular fat) improve sensory parameters of meat. Studies of candidate genes, known as predisposing humans to obesity, often did not confirm similar evidence in pigs or other domestic mammals.

The candidate gene approach, based on the function of the gene, preferably supported by data from a GWAS or QTL mapping, is the most common strategy in searching for genes with a potential phenotypic effect. An extensive study of human obesity in multiple populations showed 97 loci, significantly associated with predisposition to obesity.[74] Among genes located nearest five most significant markers there are loci of *FTO* (fat mass and obesity associated), *MC4R* (melanocortin 4 receptor), *TMEM18* (transmembrane protein 18), *GNPDA2* (glucosamine-6-phosphate deaminase 2), *GABRG1* (gamma-aminobutyric acid A receptor, gamma 1) and *SEC16B* (SEC16 homolog B, endoplasmic reticulum export factor) genes. Two of these genes, namely *FTO* and *MC4R*, have also been studied in pigs (Table 2).

The *FTO* gene encodes a nuclear protein alpha-ketoglutarate-dependent dioxygenase, which physiological function remains unknown. The most extensively studied variant of this gene, SNP in intron 1, has been repeatedly indicated as associated with an increased risk of obesity (elevated BMI) in multiple human populations. The molecular mechanism of this phenomenon, an interaction between the enhancer region harboring the SNP and

Table 2 Recent Reports (Not Included in the Review by Switonski et al.[16]) on the Association of Polymorphism in Selected Pig Candidate Genes With Fat Tissue Accumulation.

Gene	Polymorphism (Localization)	Breed	Traits Showing an Association	References
Genes, which human orthologs are associated with predisposition to obesity[74]				
FTO	g.276T > G	Italian Duroc	BFT	[79]
		Commercial pigs	BFT IMF	[79]
	c.594C > G FM244720:g.400C > G (exon3)	Meishan × Pietrain	AFW BFT	[80]
		Commercial crossbred pigs	BFT Fat in the belly IMF	[80]
	g.−167T > G (5′ flanking region)	Polish Landrace	BFT AFW	[81]
		Polish Large White	ns	[81]
		Synthetic line 990	ns	[81]
	A227G	Suzhong	Marbling score	[82]
MC4R	c.1426A > G, p.Asp298Asn (exon 1)	Duroc × Iberian	ns	[83]
		Italian Large White	BFT	[84]
		Italian Duroc	ns	[84]
		Synthetic line DIV_2	BFT	[85]
		Commercial pigs	IMF	[86]
		Duroc	BFT	[87]

(*Continued*)

Table 2 Recent Reports (Not Included in the Review by Switonski et al.[16]) on the Association of Polymorphism in Selected Pig Candidate Genes With Fat Tissue Accumulation.—cont'd.

Gene	Polymorphism (Localization)	Breed	Traits Showing an Association	References
Genes encoding adipokines and their receptors				
ADIPOQ	c.−67G > A c.−106_−91delGCCAGGGGTGTGAGCC (promoter)	Polish Landrace Line 990	ns	[88]
LEP	c.3469T > C (exon 3)	Synthetic line DIV_2	ns	[85]
		Duroc	ns	[87]
	g.1387C > T (intron 2)	Iberian × Landrace cross	BFT	[89]
	g.+846C > T (3′UTR)	Polish Landrace	AFW	[90]
LEPR	c.2002C > T (described previously as c.1987C > T; p.Leu663Phe) (exon 14)	Duroc × Iberian cross	BFT IMF	[83]
		Iberian × Landrace cross	BFT IMF	[89]
		Duroc	BFT	[91]
		Duroc	BFT	[87]
RETN	AM:157180:g.1473A > G (p.Ala36Thr)	Landrace × Chinese-European	IMF	[92]
Other candidate genes				
ACACA	c.*99A > T (3′UTR)	Polish Large White	AFW BFT	[93]
		Polish Landrace	BFT	[93]
	c.*195C > A	Polish Landrace	BFT	[93]
ACSL4	G2645A (3′UTR)	(Landrace × Yorkshire) × Duroc	IMF	[94]

BMP5	c.*131C > T (3′UTR)	Large White × Meishan	BFT	[95]
CART	T415 C (intron 1)	Landrace × Lantang F_2 population	BFT IMF	[96]
	C640 T (intron 2)	Landrace × Lantang F_2 population	BFT	[96]
	T847 C (intron 2)	Landrace × Lantang F_2 population	BFT	[96]
GNAS	g.324C > T (exon 10) and rs196952953 haplotype	Landrace	IMF	[97]
HNF1A	c.327 – 13A > G (intron 1)	Berkshire × Yorkshire	BFT	[98]
	c.1768 + 40_23del (intron 9)	Berkshire × Yorkshire	BFT	[98]
ME1	c.*488A > G (3′UTR)	Polish Large White	BFT	[99]
		Polish Landrace	BFT	[99]
MSTN	g.435G > A and g.447A > G haplotype (promoter)	Duroc and Landrace	BFT	[100]
NAMPT	AM999341:g.669T > C (intron 9)	Wild boar × Meishan	BFT	[101]
		Landrace × Chinese-European synthetic population	ns	[101]
SCD	c.*2041T > C (3′UTR)	Berkshire	IMF	[102]
	c.−353C > T (5′flanking) c.−233T > C (5′flanking) c.*164A > G (3′UTR) c.*928G > C (3′UTR) c.*2574_257delGTC (3′UTR)	Polish Large White, Line 990	ns	[103]

(*Continued*)

Table 2 Recent Reports (Not Included in the Review by Switonski et al.[16]) on the Association of Polymorphism in Selected Pig Candidate Genes With Fat Tissue Accumulation.—cont'd.

Gene	Polymorphism (Localization)	Breed	Traits Showing an Association	References
PCSK1	g.1696C > A (intron 4)	Italian Duroc	IMF	[104]
		Italian Large White	ns	[104]
	g.5182A > T (intron 11)	Italian Duroc	BFT IMF	[104]
		Italian Large White	ns	[104]
PPARA	c.*636A > G (3′UTR)	Polish Landrace	BFT IMF	[105]
		Polish Large White	ns	[105]
PPARG	c.−1633C > T and c.−1572G > A haplotype (promoter)	Erhualian	IMF	[106]
PRKAG3	I199V	Large White	IMF	[107]
		Pietrain	IMF	[107]
		Duroc	ns	[107]
		Duroc × Landrace-Large White cross	ns	[108]
RYR1	c.1843C > T	Line 990	BFT AFW	[109]
SREBF1	c.1006T > C (exon 6) c.1033C > T (exon 6) c.1045C > T (exon 6) c.2911 + 130G > C (intron 17)	Polish Landrace	ns	[93]
	c.2911 + 320A > G (intron 17)	Polish Landrace	BFT, AFW	[93]

BFT, back fat thickness; *AFW*, abdominal fat weight; *IMF*, intramuscular fat; *ns*, nonsignificant.

expression of the *IRX3* gene, has recently been unraveled.[110] In contrast to studies on humans, there are no causative mutations detected in the pig *FTO* gene, which could be used as a universal marker of adipose tissue deposition. However, in recent years several reports have been focused on the relationship between *FTO* polymorphism and porcine fatness traits (Table 2). A g.276T > G SNP in intron 3 was associated with BFT both in Italian Duroc and commercial pig populations, while a weak association with IMF was observed in commercial pigs.[79] Moreover, it was also found that the frequency of the g.276G allele changed significantly during 20 years of selection toward leaner meat and lower BFT in Italian Large White pigs.[111] Another SNP, localized in the 5′flanking region of the *FTO* (g.−167T > G), was associated with fatness traits in Polish Landrace pigs, but not in the Polish Large White or a synthetic line 990.[81] A synonymous SNP (c.594C > G, alternatively described as FM244720: g.400C > G) also showed an association with some fatness traits, but the effects were again dependent on the population.[80] An extensive analysis of the *FTO* gene sequence in eight breeds revealed more than 30 polymorphisms arranged in 20 haplotypes and it was suggested that such a heterogeneity may hinder identification of variants predisposing to excessive lipid accumulation.[112] Interestingly, the *FTO* transcript level showed an inconsistent relationship with IMF in different breeds in a study performed by Tao et al.[113] It was reported that the *FTO* transcript level varied significantly between breeds, but it was not associated with IMF content in the muscle of various pig breeds. This may suggest that not the *FTO* itself, but an unknown regulatory element in this gene, being in a linkage disequilibrium with some SNPs in the *FTO* gene, exerts a significant phenotypic effect on fatness traits, similarly as it was described in humans by Smemo et al.[110]

The human *MC4R* gene, encoding a G protein-coupled transmembrane receptor involved in the control of appetite, energy homeostasis and body weight regulation, is highly polymorphic.[114] In the porcine ortholog numerous polymorphic variants have also been identified, but the results concerning their association with fatness traits are not conclusive.[115] During the last years several studies have been performed, supporting the hypothesis that the effect of the most extensively studied SNP (c.1426A > G, p.Asp298Asn) on fatness traits should be regarded as breed-specific (Table 2). For example, this polymorphism showed no significant effect on fatness traits in Duroc × Iberian crossbred pigs,[83] while in a study performed by Davoli et al.[84] a significant association with BFT in the Italian Large White was observed, whereas no effect was found in Italian Duroc pigs. The association with IMF in commercial pigs, slaughtered at three local abattoirs, was reported by Rohrer et al.[86] and with BFT in Duroc pigs bred in Japan.[87] It was found that this SNP was the second most significant marker for BFT in Large White sows,[116] while the most significant was the g.3072G > A substitution in intron 3 of the *IGF2* gene, described earlier as a major marker for porcine meatiness and fatness by Van Laere et al.[117] The role of *MC4R* polymorphism was confirmed by a retrospective analysis of frequency changes in major genes, caused by 20-year long selection in the Italian Large White pig breed.[111] It was found that the frequency of the c.1426A allele increased significantly in that period.

Adipokines play a crucial role in energy homeostasis and thus the encoding genes are functional candidates for fat tissue accumulation traits. Polymorphisms of adipokine genes and their receptors have been extensively studied in the recent 5 years in relation to fatness phenotypes. The following ones were analyzed: leptin (*LEP*) and its receptor (*LEPR*), adiponectin (*ADIPOQ*), and resistin (*RETN*) (Table 2). Earlier studies of these genes were reviewed by Switonski et al.[16]

To date over 100 SNPs have been identified in the pig *LEP* gene.[90] Among them an effect of synonymous c.3469C > T substitution was the most extensively studied, but its association with fatness traits seems to be doubtful (Table 2). Other SNPs located in intron 2 of this gene were analyzed in the Iberian × Landrace cross and one of them (g.1387C > T) was associated with several traits, including BFT.[89] Two SNPs located in the 3′untranslated region (3′UTR), namely g.+747A > G and g.+846C > T, were tested for an association with fatness traits in the Polish Landrace breed, but only one of them, g.+846C > T substitution, showed a weak association with AFW (Table 2).[90]

The *LEPR* gene seems to be a more promising candidate gene. In a study of Munoz et al.[83] a missense SNP (c.2002C > T, p.Leu663Phe) showed significant effects on BFT and IMF (Table 2) and the T allele was associated with higher fatness. The same polymorphism, described by Perez-Montarelo et al.[89] as c.1987C > T, was analyzed in the Iberian × Landrace cross and the results for BFT and IMF were consistent with these reported by Munoz et al.[83] Further association analyses, performed in the Duroc breed, confirmed that the T allele has a strong effect on the increase of BFT.[87,91] The c.1987 T allele is fixed in the gene pool of a fat type Iberian breed and no haplotypic variability in this gene was observed when compared to other pig populations and wild boars from diverse European and Asian locations.[118] It is speculated that the *LEPR* gene region could have been under selection in this breed, which resulted in increased fatness and other traits related to leptin resistance. The in silico analysis of RNA folding, performed by Ovilo et al.[119] to predict the effect of this SNP on mRNA, revealed an alteration in the RNA structure suggesting possible changes in the transcript stability. Concluding, the *LEPR* missense substitution c.2002C > T (c.1987C > T) appears to be a promising marker for economically important fatness traits in pigs. It should be mentioned here that based on the metaanalyses on the association of various *LEPR* gene variants with BMI and a GWAS, this gene should not be considered as an important candidate for excessive lipid accumulation in humans, except for rare monogenic forms of obesity.[74,120,121]

Polymorphisms of other adipokine genes have also been studied in relation to fat deposition in pigs, but there are fewer reports regarding their phenotypic effects. Two polymorphisms (c.−67G > A and 16 bp indel c.−106_−91delGCCAGGGGTGTGAGCC) in the promoter region of the *ADIPOQ* gene have been analyzed, but no association with fatness traits was found in the Polish Landrace and a synthetic line 990.[88] In the *RETN* gene a missense g.1473A > G SNP (p.Ala36Thr) showed a significant effect on IMF in the Landrace × Chinese–European crossbred population.[92] None of the aforementioned genes encoding adipokines and their receptors were among candidate genes showing significant associations with obesity in the large GWAS metaanalysis for BMI in humans.[74]

In studies focused on the identification of genetic markers for porcine fatness traits other genes have also been analyzed, including *ACACA*, *ACSL4*, *BMP5*, *CART*, *GNAS*, *HNF1A*, *ME1*, *MSTN*, *NAMPT*, *SCD*, *PCSK1*, *PPARA*, *PPARG*, *PRKAG3*, *RYR1*, and *SREBF1*. An overview of the association effects is presented in Table 2. Unfortunately, a majority of the identified polymorphic variants were not extensively analyzed in independent studies and quite often their effects were not consistent across breeds. Therefore their possible effect on fatness traits should be verified in further studies using various pig populations. Again, the aforementioned genes were not considered as candidates for predisposition to human obesity, listed by Locke et al.[74]

The association analyses quite often are accompanied by functional studies to unravel the possible molecular mechanism of the observed phenotypic effects. Such studies were undertaken when polymorphic variants were identified in regulatory regions (mainly promoter and 3′UTR) of the candidate genes. For example, the relevance of polymorphisms in the upstream regulatory regions of the *ADIPOQ*, *PPARG*, and *KDR* genes and 3′UTR (near a putative miRNA target sequence) of the *PPARA* gene, were comprehensively analyzed using the in vitro luciferase assay and real time PCR.[88,105,106,122] A c.−1316A > G substitution in the *KDR* gene, encoding a vascular endothelial growth factor receptor, showed a significant effect on promoter activity (in vitro luciferase assay), the transcript level and was associated with IMF in the fat type Erhualian breed.[122] Similar results were obtained in the case of the *PPARG* gene, where two completely linked SNPs (c.−1633C > T and c.−1572G > A) affected the promoter activity and its transcript level was correlated with IMF in Erhualian pigs.[106] On the other hand, it was demonstrated that a 16 bp deletion in the 5′regulatory region (c.−106_−91delGCCAGGGGTGTGAGCC) of the *ADIPOQ* gene significantly changed the promoter activity in two cell lines, studied with the use of the luciferase assay approach, but no effect of the genotype was observed at the transcript level for any fatness traits.[88] Another in vitro study showed that c.*636A > G substitution in a putative microRNA target sequence in 3′UTR of the *PPARA* gene did not affect the interaction with miR-224, despite significant changes in the transcript level and an association with fatness traits in Polish Landrace pigs.[105] Such evidence coming from extensive molecular analyses facilitates verification of functionality of tested polymorphisms, as it was demonstrated in the case of the human *FTO* gene.[110]

Among candidate genes searched for association with pig fatness only *FTO* and *MC4R* showed a significant relationship with human obesity. Due to a strong selection to obtain lean and fast-growing pigs, a different set of genes may be involved in the regulation of lipid metabolism and adipose tissue accumulation in pigs in comparison to humans. Thus, knowledge concerning porcine genes with relevant functions in fat tissue accumulation may be useful in searching for novel candidate genes involved in hereditary predisposition to obesity in humans.

5.2 SNPs in Dogs

Importance of nongenetic factors predisposing dogs to obesity has been widely elucidated, while knowledge on DNA polymorphism associated with predisposition to this disease is poorly advanced.[17] Until now polymorphism was analyzed in the following candidate genes: *FTO*, *GPR120*, *INSIG2*, *IL6*, *MC3R*, *MC4R*, *PPARG*, *RETN*, and *TNF* (Table 3). Unfortunately, very few reports concerned their association with predisposition to obesity.

Table 3 Polymorphism and Association Studies of Candidate Genes for Dog Obesity.

Gene	Polymorphism (Localization)	Breed	Association Study	References
Genes, which human orthologs are associated with predisposition to obesity[74]				
FTO	23C > T Thr > Met (exon 1), 192A > T (intron 1), 223T > C (intron 1), 378053G > A (3′flanking), 378284T > C (3′flanking), 378318G > C (3′flanking)	Various breeds	Not tested	[123]
MC4R	c.637G > T p.Val213Phe (exon 1), c. 777 T > C (exon 1), c.*33C > G (3′UTR)	Various breeds	Not tested	[124]
	c.637G > T p.Val213Phe (exon 1), c. 777 T > C (exon 1), c.868C > T (exon 1), c.*33C > G (3′UTR)	Golden Retriever	No association with morphological measures	[125]
	A420 C (Asp101Thr) (exon 1)	Beagle	Significant association with body weight	[126]
	C895 T (exon 1)	Beagle	Significant association with female body weight	[126]
Other candidate genes				
GPR120	c.252C > G, c.282C > G, c.287T > G p.Leu96Arg, c.307G > A p.Ala103Thr, c.446G > C p. Gly149Ala, c.595C > A p.Pro199Thr, c.702A > G, c.726G > A, c.984T > C	Various breeds	Frequency of variants at c.595C > A significantly differed between lean and overweight/obese dogs	[127]

(*Continued*)

Table 3 Polymorphism and Association Studies of Candidate Genes for Dog Obesity.—cont'd.

Gene	Polymorphism (Localization)	Breed	Association Study	References
INSIG2	−91G > A (5′flanking), −1C > T (5′flanking), 40C > A Arg > Ser (exon 1) 1483A > T (intron 1) 1637C > T (intron 1) 2169G > A (intron 1) 10820T > A (intron 4)	Various breeds	Not tested	[123]
MC3R	c.−90T(11_13) (5′ flanking region), c.142C > T (exon 1)	Various breeds	Not tested	[128]
PPARG	C1362 T (exon 7)	Mongrels, Miniature Dachshund	Not tested	[129]
TNF	c.−40A > C (5′UTR), c.249C > T (exon 1), c.548A > T p. Glu183Val (exon 4), c.627C > T (exon 4), c.186 + 16A > G (intron 1), c.186 + 174GAAT$_{[N]}$ (intron 1), c.186 + 211C > T (intron 1), c.187 − 47T > C (intron 1), c.233 − 54T > C (intron 2), c.233 + 14G > A (intron 3), c.233 + 17G > T (intron 3), c.233 + 108A > G (intron 3), c.*107G > A (3′UTR)	Various breeds and mongrels	Frequency of variants at c.−40A > C and c.233 + 14G > A significantly differed between lean, overweight and obese Labradors. Other polymorphisms not analyzed.	[130]

IL6	c.102T > C (exon 2), c.572G > A (exon 5), c.309 + 215T > C (intron 3), c.*283G > A (3′UTR)	Various breeds and mongrels	No significant differences of allele frequencies at c.309 + 215T > C between lean, overweight and obese dogs of various breeds. Other SNPs not analyzed.	[130]
RETN	c.19C > T (p.Leu7Phe), c.75G > A, c.115 + 29G > C, c.115 + 143T > G, c.116 − 179G > A, c.141C > T, c.194–69T > A, c.236C > G	Various breeds and mongrels	No significant differences of allele frequencies at c.19C > T, c.75G > A and c.194 − 69T > A between lean, overweight and obese dogs of various breeds. Other SNPs not analyzed.	[130]

The *MC4R* was the most frequently investigated gene. Altogether six polymorphic variants, including two missense ones, have been reported (Table 3). In comparison with the human *MC4R* gene,[114] the canine counterpart seems to be much less polymorphic, however, it should be pointed out that this suggestion may be biased due to a much smaller number of the analyzed dogs, when compared with studies in humans. A missense A420 C (Asp101Thr) substitution showed an effect on body weight in a small cohort of Beagles (males and females), while a silent C895 T SNP was associated with body weight of females only.[126] Other SNPs in this gene (missense c.637G > T and two silent substitutions: c. 777 T> C and c.868C > T, and c.*33C > G in 3′UTR) showed no association with morphological measures, including weight and body index score [weight/(length × height)] of Golden Retrievers.[125]

Among numerous polymorphisms found in the canine *MC3R*, *FTO*, *INSIG2*, and *PPARG* genes there were two nonsynonymous SNPs. Unfortunately, none of the identified variants was tested for an association with canine obesity.

Nine SNPs were detected in the coding sequence of the *GPR120* (G-protein-coupled receptor 120) gene, including four missense substitutions (Table 3). The frequencies of a nonsynonymous substitution (c.595C > A, p. Pro199Thr) were analyzed across 20 breeds and a group of mongrel dogs with different BCSs, with a significant difference found between BCS3 (lean) and BCS4/BCS5 (overweight/obese) dogs.[127]

Recently three adipokine genes (*TNF*, *RETN*, and *IL6*) were studied in lean, overweight, and obese dogs.[130] In a screen comprising 17 breeds and mongrels, a total of 25 polymorphisms were found in these genes, including nonsynonymous substitutions in *TNF* and *RETN*. Distribution of six SNPs (two in *TNF*, one in *IL6*, and three in *RETN*) was analyzed in dogs in relation to obesity (Table 3). It was found that frequencies of two SNPs in noncoding regions, c.−40A > C in 5′UTR and c.233 + 14G > A in intron 3 of the *TNF* showed significant differences in Labrador dogs with different BCS.

It is rather surprising that genetic studies of canine obesity, an emerging and serious health problem, are so scarce, even though the dog genome sequence has been available for a decade.[66] It seems that the *MC4R*, *GPR120*, and *TNF* genes are potential markers for hereditary predisposition to obesity at least in some breeds (Labrador Retriever and Beagle). It is expected that in the near future more genetic studies will be focused on the association of polymorphic variants of candidate genes, selected on the basis of knowledge concerning genetic markers predisposing humans to obesity (eg, *FTO, MC4R, TMEM18, GNPDA2, GABRG1, SEC16B*, etc.).

5.3 CNVs in Pigs

CNV is a class of structural variation of the genome and is defined as a DNA segment (from 1 kb up to several Mb) that is present at a variable copy number when compared to a reference genome.[131] The recent human CNV map revealed that CNVs represent about 4.8–9.5% of the genome.[132] It was shown that this type of genomic variations is responsible for human diseases.[133] There are several reports concerning an association of CNVs with human obesity. In these CNV regions several candidate genes were indicated, for example, *SH2B1* on chromosome HSA16,[134,135] *PPYR1* on HSA10,[136] and *AMY1* on HSA1.[137]

Many efforts have been made to detect CNVs in the pig genome with the use of different technologies, such as the CGH array, SNP genotyping array and genome resequencing. The latest study on the characterization of CNVs in the pig genome was performed with the use of a custom-designed 1 M CGH array in 12 pig samples from diverse pig breeds, including the Asian wild boar, Chinese indigenous, and European commercial breeds.[138] The authors identified 758 CNV regions (CNVRs), covering 47.43 Mb of the pig genome sequence, which corresponds to 1.69% of the genome. More than 1200 genes were completely or partially overlapped with the identified CNVRs. So far, only a few attempts were made to identify CNVs in pigs in relation to production traits, including fatness. Trait-related CNVs were studied in 18 diverse pig populations with the use of the porcine 60k SNP microarray (Porcine SNP60 BeadChip) and 538 CNVs were identified in a White Duroc × Erhualian F2 population.[139] Integrating previous QTL mapping data with the detected CNV regions facilitated the identification of seven candidate genes (*ANP32B, BSCL2, LTBP3, GDF3, GYS1, KIT*, and *CAV1*) for several traits, including fatness parameters: BFT, AFW, and IMF. In 2013 Fowler et al.[140] using the same Porcine SNP60 BeadChip compared fat and lean samples derived from 3 commercial breeds (the Large White, Duroc, and a white Pietrain composite line) and identified 12 CNV regions, which harbored 4 functional candidate genes (*MCHR1, PPARA, SLC5A1*, and *SLC5A4*) for fatness traits.

Analysis of CNVs specific to indigenous pig breeds may be useful in the identification of the genetic background responsible for phenotypic variation, including fatness. Resequencing of genomes for three variants of the

Mangalica breed (Blond, Red, and Swallow-belly) led to the identification of approx. 1000 CNV gains (CNV losses were not analyzed due to a low coverage of sequencing).[73] However, no CNVs associated with fatness traits were detected. In another study, carried out in Chinese breeds, it was shown that Meishan pigs have a high number of copies of *AADAT* and *ZNF622* genes,[141] which may be related with a lower growth rate, but again no CNVs for fatness traits were identified. A comparison of CNVs in four Chinese normal size breeds (Luchuan, Tongcheng and Laiwu pigs) and two minipig breeds (Bama and Wuzhishan) did not identify CNVs influencing growth of these breeds as well as fatness.[142]

Searching for CNVs in commercial, meat-type breeds, were also carried out. Such a study, concerning Italian Large White pigs, was reported by Schiavo et al.[143] They analyzed two groups of animals (approx. 150 animals each) with extreme BFT values using the 60k SNP microarray (Illumina Porcine SNP60 BeadChip). Sixteen CNV regions (CNVR) located on six autosomes (SSC8, 11, 12, 13, 14, and 15) showed an association with BFT; however, this result appeared to be nonsignificant after the Bonferroni correction. Among the detected CNVRs, 14 represented low-frequency CNV events, while 9 high-frequent were observed in fatter pigs. One of the CNVRs encompassed the *ZPLD1* gene, which was also found in the human CNV associated with childhood obesity.[144] Thus, further studies of this chromosomal region are advisable. The lack of CNVs significantly associated with BFT may be explained by the power of the experiment, since the identification of low frequent markers may require a larger cohort of investigated animals. It is also possible that this type of polymorphism may have a limited impact on adiposity in pigs. A previous analysis of CNVs in both domestic and wild pigs,[145] which allowed to identify 3118 CNVRs, revealed that CNVRs reflect demographic history rather than phenotypic diversity.

5.4 CNVs in Dogs

So far, there have been no reports on the discovery of CNVs associated with obesity in dogs. However, there is an extensive study on dog domestication involving genotyping 1611 dog CNVs in 23 wolf-like canids.[146] This approach facilitated the identification of 25 CNVRs that showed the largest differentiation between dogs and wild canids. Within CNV regions there are 12 candidate genes, which function is related with growth and neurological function. One of the promising candidate genes for body weight is the *PDE4D* gene. The region containing this gene presented gain in all wild canids, whereas losses were observed in a majority of the studied dogs (Boxer, Beagle, and Basenji). It is known from studies on mice that PDE4D-deficient young mice exhibited delayed growth and a decreased level of circulating IGF-I.[147] The second gene identified in the study of Ramirez et al.[146] was *CRTC3*, for which a higher copy number was observed in dogs than in gray wolves. Interestingly, variants of this gene have been associated with adiposity in humans.[148,149]

An interesting discovery concerns copy numbers of the canine amylase gene. It should be mentioned that most mammals, including the human, express amylase in saliva, but dogs produce amylase in the pancreas only. The gene encoding the pancreatic type of amylase (*AMY2B*) is found only in two copies in wolves, while in dogs the copy number ranges from 4 to 30.[150] The increase in the number of copies is associated with a high activity of this enzyme and it is believed that it allows more efficient digestion of starch. A further study of Arendt et al.[151] revealed that copy numbers of the *AMY2B* vary between breeds, as well as individuals. Since amylase activity may predispose to obesity and diabetes mellitus in humans,[152] the *AMY2B* copy number and susceptibility to diabetes mellitus in dogs was analyzed, but no association was found.[151] However, the authors indicated that a larger collection of cases should be evaluated in future studies. Studies on pigs have shown that conversely to humans and dogs, this species has a universally high number of copies—from 8 to 21 copies of amylase genes (*AMY1*, *AMY2A*, *AMY2B*).[145]

6. CONCLUSIONS AND PERSPECTIVES

The pig and the dog are known as important large animal models for human diseases, including obesity. However, an excessive accumulation of adipose tissue has different significance in these model species. In the pig it is considered as an important production trait and thus extensive studies on its genetic background have been carried out. In the dog obesity is an emerging health problem, but until now genetic studies of this disease are not advanced.

A comparison of candidate gene sets, which polymorphisms are associated with adipose tissue accumulation, in the human (97 loci listed by Locke et al.[74]) and pigs (22 loci presented in Table 2) revealed distinct differences. Only two porcine genes, associated with adiposity (*FTO* and *MC4R*), were indicated in the human and the pig, while others (eg, *LEP*, *LEPR*, *SCD*, etc.) were specific to pigs only. It may be caused by different factors, including the size of the studied cohorts (very large in humans, especially when metaanalyses were carried out), number of the studied traits—usually body mass index (BMI) in humans and several traits in pigs (BFT at several points, ABW and IMF), and different breeding strategies, which modified the gene pool of pig breeds representing a specific type (fat, meat, autochthonous, experimental strains—minipigs, etc.).

Phenotypic evaluation of dog adiposity is commonly based on a subjective BCS scale. Unfortunately, it may bias the association analysis. On the other hand, the dog seems to be a very suitable model for human obesity because both species share the same environment, including diet and physical activity and a quite long life span of dogs. Moreover, it is important to point out that there are dog breeds predisposed to obesity (eg, Labrador and Golden Retrievers, the Cocker Spaniel, Beagle, etc.), as well as those characterized by limited depots of adipose tissue (eg, the Greyhound, Whippet, etc.).

Clearly, in the near future more studies will be carried out on the genetic background of obesity, especially in dogs. Since the impact of common polymorphisms (SNPs and CNVs) on human obesity is limited, it is suggested to focus on epigenetic markers in further studies on model organisms, including the pig and the dog.

ACKNOWLEDGMENTS

This study was supported by the National Science Centre in Poland—grants UMO-2012/07/E/NZ9/02573 and 2013/09/B/NZ2/02208.

REFERENCES

1. Rimbault M, Ostrander EA. So many doggone traits: mapping genetics of multiple phenotypes in the domestic dog. *Hum Mol Genet.* 2012;21(R1):R52–R57.
2. Slutsky J, Raj K, Yuhnke S, et al. A web resource on DNA tests for canine and feline hereditary diseases. *Vet J.* 2013;197(2):182–187.
3. Switonski M. Dog as a model in studies on human hereditary diseases and their gene therapy. *Reprod Biol.* 2014;14(1):44–50.
4. Bassols A, Costa C, Eckersall PD, Osada J, Sabria J, Tibau J. The pig as an animal model for human pathologies: a proteomics perspective. *Proteomics Clin Appl.* 2014;8(9–10):715–731.
5. German AJ. Style over substance: what can parenting styles tell us about ownership styles and obesity in companion animals? *Br J Nutr.* 2015;113(Suppl):S72–S77.
6. Wood JD, Enser M, Fisher AV, et al. Fat deposition, fatty acid composition and meat quality: a review. *Meat Sci.* 2008;78(4):343–358.
7. Neeland IJ, Turer AT, Ayers CR, et al. Dysfunctional adiposity and the risk of prediabetes and type 2 diabetes in obese adults. *JAMA.* 2012;308(11):1150–1159.
8. Pawar AS, Zhu XY, Eirin A, et al. Adipose tissue remodeling in a novel domestic porcine model of diet-induced obesity. *Obesity (Silver Spring).* 2015;23(2):399–407.

9. Koopmans SJ, Schuurman T. Considerations on pig models for appetite, metabolic syndrome and obese type 2 diabetes: from food intake to metabolic disease. *Eur J Pharmacol.* 2015;759:231–239.
10. Osto M, Lutz TA. Translational value of animal models of obesity-focus on dogs and cats. *Eur J Pharmacol.* 2015;759:240–252.
11. Lund EM, Armstrong PJ, Kirk CA, Klausner JS. Prevalence and risk factors for obesity in adult dogs from private US veterinary practices. *Intern J Appl Res Vet Med.* 2006;4(2):177–186.
12. Mao J, Xia Z, Chen J, Yu J. Prevalence and risk factors for canine obesity surveyed in veterinary practices in Beijing. *China Prev Vet Med.* 2013;112(3–4):438–442.
13. Wakshlag JJ, Struble AM, Levine CB, Bushey JJ, Laflamme DP, Long GM. The effects of weight loss on adipokines and markers of inflammation in dogs. *Br J Nutr.* 2011;106(Suppl 1):S11–S14.
14. Park HJ, Lee SE, Oh JH, Seo KW, Song KH. Leptin, adiponectin and serotonin levels in lean and obese dogs. *BMC Vet Res.* 2014;10:113.
15. Frank L, Mann S, Levine CB, Cummings BP, Wakshlag JJ. Increasing body condition score is positively associated interleukin-6 and monocyte chemoattractant protein-1 in Labrador retrievers. *Vet Immunol Immunopathol.* 2015;167(3–4):104–109.
16. Switonski M, Stachowiak M, Cieslak J, Bartz M, Grzes M. Genetics of fat tissue accumulation in pigs: a comparative approach. *J Appl Genet.* 2010;51(2):153–168.
17. Switonski M, Mankowska M. Dog obesity—the need for identifying predisposing genetic markers. *Res Vet Sci.* 2013;95(3):831–836.
18. de Godoy MR, Swanson KS. Companion Animals Symposium: nutrigenomics: using gene expression and molecular biology data to understand pet obesity. *J Anim Sci.* 2013;91(6):2949–2964.
19. Egerszegi I, Ratky J, Solti L, Brüssow KP. Mangalica—an indigenous swine breed from Hungary. *Arch Tierz.* 2003;46(3):245–256.
20. Lopez-Bote CJ. Sustained utilization of the Iberian pig breed. *Meat Sci.* 1998;49S1:S17–S27.
21. Ai H, Ren J, Zhang Z, et al. Detection of quantitative trait loci for growth- and fatness-related traits in a large-scale White Duroc × Erhualian intercross pig population. *Anim Genet.* 2012;43(4):383–391.
22. Clark BA, Alloosh M, Wenzel JW, Sturek M, Kostrominova TY. Effect of diet-induced obesity and metabolic syndrome on skeletal muscles of Ossabaw miniature swine. *Am J Physiol Endocrinol Metab.* 2011;300(5):E848–E857.
23. Toedebusch RG, Roberts MD, Wells KD, et al. Unique transcriptomic signature of omental adipose tissue in Ossabaw swine: a model of childhood obesity. *Physiol Genomics.* 2014;46(10):362–375.
24. Zhang X, Li ZL, Eirin A, et al. Cardiac metabolic alterations in hypertensive obese pigs. *Hypertension.* 2015;66(2):430–436.
25. Newell-Fugate AE, Taibl JN, Alloosh M, et al. Effects of obesity and metabolic syndrome on steroidogenesis and folliculogenesis in the female Ossabaw Mini-Pig. *PLoS One.* 2015;10(6):e0128749.
26. Szabo A, Viski A, Egyhazi Z, Hazas Z, Horn P, Romvari R. Comparison of Mangalica and Hungarian Large White pigs at identical bodyweight: 1 Backfat histology. *Arch Tierz.* 2010;53:141–146.
27. Lebret B, Dourmad JY, Mourot J, Pollet PY, Gondret F. Production performance, carcass composition, and adipose tissue traits of heavy pigs: influence of breed and production system. *J Anim Sci.* 2014;92(8):3543–3556.
28. Nakajima I, Oe M, Ojima K, Muroya S, Shibata M, Chikuni K. Cellularity of developing subcutaneous adipose tissue in Landrace and Meishan pigs: adipocyte size differences between two breeds. *Anim Sci J.* 2011;82(1):144–149.

29. Jeusette I, Greco D, Aquino F, et al. Effect of breed on body composition and comparison between various methods to estimate body composition in dogs. *Res Vet Sci.* 2010;88(2):227–232.
30. Mawby DI, Bartges JW, d'Avignon A, Laflamme DP, Moyers TD, Cottrell T. Comparison of various methods for estimating body fat in dogs. *J Am Anim Hosp Assoc.* 2004;40(2):109–114.
31. Jo J, Gavrilova O, Pack S, et al. Hypertrophy and/or hyperplasia: dynamics of adipose tissue growth. *PLoS Comput Biol.* 2009;5(3):e1000324.
32. Szczerbal I, Chmurzynska A. Chromosomal localization of nine porcine genes encoding transcription factors involved in adipogenesis. *Cytogenet Genome Res.* 2008;121(1):50–54.
33. Szczerbal I, Chmurzynska A, Switonski M. Cytogenetic mapping of eight genes encoding fatty acid binding proteins (FABPs) in the pig genome. *Cytogenet Genome Res.* 2007;118(1):63–66.
34. Boone C, Gregoire F, Remacle C. Regulation of porcine adipogenesis in vitro, as compared with other species. *Domest Anim Endocrinol.* 1999;17(2–3):257–267.
35. Pang W, Wang Y, Wei N, et al. Sirt1 inhibits akt2-mediated porcine adipogenesis potentially by direct protein-protein interaction. *PLoS One.* 2013;8(8):e71576.
36. Pang WJ, Wei N, Wang Y, et al. Obese and lean porcine difference of FoxO1 and its regulation through C/EBPbeta and PI3 K/GSK3beta signaling pathway. *J Anim Sci.* 2014;92(5):1968–1979.
37. Jiang S, Wei H, Song T, et al. KLF13 promotes porcine adipocyte differentiation through PPARgamma activation. *Cell Biosci.* 2015;5:28.
38. Kilpinen H, Dermitzakis ET. Genetic and epigenetic contribution to complex traits. *Hum Mol Genet.* 2012;21(R1):R24–R28.
39. van Dijk SJ, Molloy PL, Varinli H, Morrison JL, Muhlhausler BS. Members of Epi S. Epigenetics and human obesity. *Int J Obes.* 2015;39(1):85–97.
40. van Dijk SJ, Tellam RL, Morrison JL, Beverly S, Muhlhausler BS, Molloy PL. Recent developments on the role of epigenetics in obesity and metabolic disease. *Clin Epigenetics.* 2015;7:66.
41. Ronn T, Volkov P, Davegardh C, et al. A six months exercise intervention influences the genome-wide DNA methylation pattern in human adipose tissue. *PLoS Genet.* 2013;9 (6):e1003572.
42. Bouchard L, Rabasa-Lhoret R, Faraj M, et al. Differential epigenomic and transcriptomic responses in subcutaneous adipose tissue between low and high responders to caloric restriction. *Am J Clin Nutr.* 2010;91(2):309–320.
43. Benton MC, Johnstone A, Eccles D, et al. An analysis of DNA methylation in human adipose tissue reveals differential modification of obesity genes before and after gastric bypass and weight loss. *Genome Biol.* 2015;16:8.
44. Li M, Wu H, Luo Z, et al. An atlas of DNA methylomes in porcine adipose and muscle tissues. *Nat Commun.* 2012;3:850.
45. Li M, Wu H, Wang T, et al. Co-methylated genes in different adipose depots of pig are associated with metabolic, inflammatory and immune processes. *Int J Biol Sci.* 2012;8 (6):831–837.
46. Zhou C, Zhang J, Ma J, et al. Gene expression profiling reveals distinct features of various porcine adipose tissues. *Lipids Health Dis.* 2013;12:75.
47. Li M, Wang T, Wu H, et al. Genome-wide DNA methylation changes between the superficial and deep backfat tissues of the pig. *Int J Mol Sci.* 2012;13(6):7098–7108.
48. Wang T, Jiang A, Guo Y, et al. Deep sequencing of the transcriptome reveals inflammatory features of porcine visceral adipose tissue. *Int J Biol Sci.* 2013;9(6):550–556.
49. Ma J, Jiang Z, He S, et al. Intrinsic features in microRNA transcriptomes link porcine visceral rather than subcutaneous adipose tissues to metabolic risk. *PLoS One.* 2013;8 (11):e80041.

50. Peng Y, Yu S, Li H, Xiang H, Peng J, Jiang S. MicroRNAs: emerging roles in adipogenesis and obesity. *Cell Signal.* 2014;26(9):1888–1896.
51. Guo Y, Mo D, Zhang Y, et al. MicroRNAome comparison between intramuscular and subcutaneous vascular stem cell adipogenesis. *PLoS One.* 2012;7(9):e45410.
52. Gaffo E, Zambonelli P, Bisognin A, Bortoluzzi S, Davoli R. miRNome of Italian Large White pig subcutaneous fat tissue: new miRNAs, isomiRs and moRNAs. *Anim Genet.* 2014;45(5):685–698.
53. Sun L, Goff LA, Trapnell C, et al. Long noncoding RNAs regulate adipogenesis. *Proc Natl Acad Sci USA.* 2013;110(9):3387–3392.
54. Wei N, Wang Y, Xu RX, et al. PU.1 antisense lncRNA against its mRNA translation promotes adipogenesis in porcine preadipocytes. *Anim Genet.* 2015;46(2):133–140.
55. Szczerbal I, Foster HA, Bridger JM. The spatial repositioning of adipogenesis genes is correlated with their expression status in a porcine mesenchymal stem cell adipogenesis model system. *Chromosoma.* 2009;118(5):647–663.
56. Szczerbal I, Bridger JM. Association of adipogenic genes with SC-35 domains during porcine adipogenesis. *Chromosome Res.* 2010;18(8):887–895.
57. Kociucka B, Cieslak J, Szczerbal I. Three-dimensional arrangement of genes involved in lipid metabolism in nuclei of porcine adipocytes and fibroblasts in relation to their transcription level. *Cytogenet Genome Res.* 2012;136(4):295–302.
58. Feil R, Fraga MF. Epigenetics and the environment: emerging patterns and implications. *Nat Rev Genet.* 2011;13(2):97–109.
59. Vickers MH. Developmental programming and transgenerational transmission of obesity. *Ann Nutr Metab.* 2014;64(Suppl 1):26–34.
60. Feeney A, Nilsson E, Skinner MK. Epigenetics and transgenerational inheritance in domesticated farm animals. *J Anim Sci Biotechnol.* 2014;5(1):48.
61. Braunschweig M, Jagannathan V, Gutzwiller A, Bee G. Investigations on transgenerational epigenetic response down the male line in F2 pigs. *PLoS One.* 2012;7(2):e30583.
62. Liu X, Wang J, Li R, et al. Maternal dietary protein affects transcriptional regulation of myostatin gene distinctively at weaning and finishing stages in skeletal muscle of Meishan pigs. *Epigenetics.* 2011;6(7):899–907.
63. Cong R, Jia Y, Li R, et al. Maternal low-protein diet causes epigenetic deregulation of HMGCR and CYP7alpha1 in the liver of weaning piglets. *J Nutr Biochem.* 2012;23(12):1647–1654.
64. Meidtner K, Wermter AK, Hinney A, Remschmidt H, Hebebrand J, Fries R. Association of the melanocortin 4 receptor with feed intake and daily gain in F2 Mangalitsa × Pietrain pigs. *Anim Genet.* 2006;37(3):245–247.
65. Yang B, Zhang W, Zhang Z, et al. Genome-wide association analyses for fatty acid composition in porcine muscle and abdominal fat tissues. *PLoS One.* 2013;8(6):e65554.
66. Lindblad-Toh K, Wade CM, Mikkelsen TS, et al. Genome sequence, comparative analysis and haplotype structure of the domestic dog. *Nature.* 2005;438(7069):803–819.
67. Groenen MA, Archibald AL, Uenishi H, et al. Analyses of pig genomes provide insight into porcine demography and evolution. *Nature.* 2012;491(7424):393–398.
68. Andersson L, Haley CS, Ellegren H, et al. Genetic mapping of quantitative trait loci for growth and fatness in pigs. *Science.* 1994;263(5154):1771–1774.
69. Qiao R, Gao J, Zhang Z, et al. Genome-wide association analyses reveal significant loci and strong candidate genes for growth and fatness traits in two pig populations. *Genet Sel Evol.* 2015;47:17.
70. Rubin CJ, Megens HJ, Martinez Barrio A, et al. Strong signatures of selection in the domestic pig genome. *Proc Natl Acad Sci USA.* 2012;109(48):19529–19536.
71. Gondret F, Riquet J, Tacher S, et al. Towards candidate genes affecting body fatness at the SSC7 QTL by expression analyses. *J Anim Breed Genet.* 2012;129(4):316–324.

72. Ma J, Gilbert H, Iannuccelli N, et al. Fine mapping of fatness QTL on porcine chromosome X and analyses of three positional candidate genes. *BMC Genet.* 2013;14:46.
73. Molnar J, Nagy T, Steger V, Toth G, Marincs F, Barta E. Genome sequencing and analysis of Mangalica, a fatty local pig of Hungary. *BMC Genomics.* 2014;15:761.
74. Locke AE, Kahali B, Berndt SI, et al. Genetic studies of body mass index yield new insights for obesity biology. *Nature.* 2015;518(7538):197–206.
75. Kim J, Lee T, Kim TH, Lee KT, Kim H. An integrated approach of comparative genomics and heritability analysis of pig and human on obesity trait: evidence for candidate genes on human chromosome 2. *BMC Genomics.* 2012;13:711.
76. Machiela MJ, Chanock SJ. GWAS is going to the dogs. *Genome Biol.* 2014;15(3):105.
77. Vaysse A, Ratnakumar A, Derrien T, et al. Identification of genomic regions associated with phenotypic variation between dog breeds using selection mapping. *PLoS Genet.* 2011;7(10):e1002316.
78. Sutter NB, Bustamante CD, Chase K, et al. A single IGF1 allele is a major determinant of small size in dogs. *Science.* 2007;316(5821):112–115.
79. Fontanesi L, Scotti E, Buttazzoni L, et al. Confirmed association between a single nucleotide polymorphism in the FTO gene and obesity-related traits in heavy pigs. *Mol Biol Rep.* 2010;37(1):461–466.
80. Dvorakova V, Bartenschlager H, Stratil A, et al. Association between polymorphism in the FTO gene and growth and carcass traits in pig crosses. *Genet Sel Evol.* 2012;44:13.
81. Szydlowski M, Salamon S, Grzes M, Switonski M. SNP in the 5′ flanking region of the pig FTO gene is associated with fatness in Polish Landrace. *Livestock Sci.* 2012;150:397–400.
82. Fu Y, Li L, Ren S. Effect of FTO expression and polymorphism on fat deposition in Suzhong pigs. *Asian-Australas J Anim Sci.* 2013;26(10):1365–1373.
83. Munoz G, Alcazar E, Fernandez A, et al. Effects of porcine MC4R and LEPR polymorphisms, gender and Duroc sire line on economic traits in Duroc × Iberian crossbred pigs. *Meat Sci.* 2011;88(1):169–173.
84. Davoli R, Braglia S, Valastro V, et al. Analysis of MC4R polymorphism in Italian Large White and Italian Duroc pigs: association with carcass traits. *Meat Sci.* 2012;90(4):887–892.
85. Chao Z, Wang F, Deng CY, et al. Distribution and linkage disequilibrium analysis of polymorphisms of MC4R, LEP, H-FABP genes in the different populations of pigs, associated with economic traits in DIV2 line. *Mol Biol Rep.* 2012;39(5):6329–6335.
86. Rohrer GA, Nonneman DJ, Miller RK, Zerby H, Moeller SJ. Association of single nucleotide polymorphism (SNP) markers in candidate genes and QTL regions with pork quality traits in commercial pigs. *Meat Sci.* 2012;92(4):511–518.
87. Hirose K, Ito T, Fukawa K, et al. Evaluation of effects of multiple candidate genes (LEP, LEPR, MC4R, PIK3C3, and VRTN) on production traits in Duroc pigs. *Anim Sci J.* 2014;85(3):198–206.
88. Cieslak J, Flisikowska T, Schnieke A, et al. Polymorphisms in the promoter region of the adiponectin (ADIPOQ) gene are presumably associated with transcription level and carcass traits in pigs. *Anim Genet.* 2013;44(3):340–343.
89. Perez-Montarelo D, Fernandez A, Folch JM, et al. Joint effects of porcine leptin and leptin receptor polymorphisms on productivity and quality traits. *Anim Genet.* 2012;43 (6):805–809.
90. Mankowska M, Szydlowski M, Salamon S, Bartz M, Switonski M. Novel polymorphisms in porcine 3′UTR of the leptin gene, including a rare variant within target sequence for MIR-9 gene in Duroc breed, not associated with production traits. *Anim Biotechnol.* 2015;26(2):156–163.
91. Uemoto Y, Kikuchi T, Nakano H, et al. Effects of porcine leptin receptor gene polymorphisms on backfat thickness, fat area ratios by image analysis, and serum leptin concentrations in a Duroc purebred population. *Anim Sci J.* 2012;83(5):375–385.

92. Cepica S, Ovilo C, Masopust M, et al. Four genes located on a SSC2 meat quality QTL region are associated with different meat quality traits in Landrace × Chinese-European crossbred population. *Anim Genet.* 2012;43(3):333–336.
93. Stachowiak M, Nowacka-Woszuk J, Szydlowski M, Switonski M. The ACACA and SREBF1 genes are promising markers for pig carcass and performance traits, but not for fatty acid content in the longissimus dorsi muscle and adipose tissue. *Meat Sci.* 2013;95 (1):64–71.
94. Rusc A, Sieczkowska H, Krzecio E, et al. The association between acyl-CoA synthetase (ACSL4) polymorphism and intramuscular fat content in (Landrace × Yorkshire) × Duroc pigs. *Meat Sci.* 2011;89(4):440–443.
95. Shao GC, Luo LF, Jiang SW, Deng CY, Xiong YZ, Li FE. A C/T mutation in microRNA target sites in BMP5 gene is potentially associated with fatness in pigs. *Meat Sci.* 2011;87(3):299–303.
96. Zhu X, Mo D, Wang C, et al. Analysis of sequence variability in the pig CART gene and association of polymorphism with fatness traits in a F2 population. *MolBiolRep.* 2014;41 (8):5229–5236.
97. Oczkowicz M, Ropka-Molik K, Tyra M. Analysis of the associations between polymorphisms in GNAS complex locus and growth, carcass and meat quality traits in pigs. *Mol Biol Rep.* 2013;40(11):6419–6427.
98. Fan B, Du ZQ, Rothschild MF. The Hepatocyte nuclear factor-1 alpha (HNF1A) gene is associated with fatness and loin muscle area in the pig. *Animal.* 2010;4(10):1619–1627.
99. Bartz M, Kociucka B, Mankowska M, Switonski M, Szydlowski M. Transcript level of the porcine ME1 gene is affected by SNP in its 3′UTR, which is also associated with subcutaneous fat thickness. *J Anim Breed Genet.* 2014;131(4):271–278.
100. Tu PA, Lo LL, Chen YC, et al. Polymorphisms in the promoter region of myostatin gene are associated with carcass traits in pigs. *J Anim Breed Genet.* 2014;131(2):116–122.
101. Cepica S, Bartenschlager H, Ovilo C, et al. Porcine NAMPT gene: search for polymorphism, mapping and association studies. *Anim Genet.* 2010;41(6):646–651.
102. Lim KS, Kim JM, Lee EA, Choe JH, Hong KC. A candidate single nucleotide polymorphism in the 3′ untranslated region of stearoyl-CoA desaturase gene for fatness quality and the gene expression in Berkshire pigs. *Asian-Australas J Anim Sci.* 2015;28 (2):151–157.
103. Bartz M, Szydlowski M, Kociucka B, Salamon S, Jelen HH, Switonski M. Transcript abundance of the pig stearoyl-CoA desaturase gene has no effect on fatty acid composition in muscle and fat tissues, but its polymorphism within the putative microRNA target site is associated with daily body weight gain and feed conversion ratio. *J Anim Sci.* 2013;91(1):10–19.
104. Fontanesi L, Bertolini F, Scotti E, et al. Polymorphisms in an obesity-related gene (PCSK1) are associated with fat deposition and production traits in Italian heavy pigs. *Animal.* 2012;6(12):1913–1924.
105. Stachowiak M, Szydlowski M, Flisikowski K, et al. Polymorphism in 3′ untranslated region of the pig PPARA gene influences its transcript level and is associated with adipose tissue accumulation. *J Anim Sci.* 2014;92(6):2363–2371.
106. Wang H, Xiong K, Sun W, et al. Two completely linked polymorphisms in the PPARG transcriptional regulatory region significantly affect gene expression and intramuscular fat deposition in the longissimus dorsi muscle of Erhualian pigs. *Anim Genet.* 2013;44 (4):458–462.
107. Ryan MT, Hamill RM, O'Halloran AM, et al. SNP variation in the promoter of the PRKAG3 gene and association with meat quality traits in pig. *BMC Genet.* 2012;13:66.
108. Galve A, Burgos C, Varona L, Carrodeguas JA, Canovas A, Lopez-Buesa P. Allelic frequencies of PRKAG3 in several pig breeds and its technological consequences on a Duroc × Landrace-Large White cross. *J Anim Breed Genet.* 2013;130(5):382–393.

109. Sadkowski S, Molinska-Glura M, Molinski K, Szczepankiewicz D, Switonski M, Szydlowski M. A well-known mutation in RYR1 alters distribution of adipose tissue in gilts. *Anim Sci Pap Rep.* 2015;33(2):147–154.
110. Smemo S, Tena JJ, Kim KH, et al. Obesity-associated variants within FTO form long-range functional connections with IRX3. *Nature.* 2014;507(7492):371–375.
111. Fontanesi L, Schiavo G, Scotti E, et al. A retrospective analysis of allele frequency changes of major genes during 20 years of selection in the Italian Large White pig breed. *J Anim Breed Genet.* 2015;132(3):239–246.
112. Fontanesi L, Russo V. Nucleotide variability and haplotype heterogeneity at the porcine fat mass and obesity-associated (FTO) gene. *Anim Genet.* 2013;44(1):96–100.
113. Tao X, Men XM, Deng B, Xu ZW. Effects of breed, postnatal development, and nutrition on mRNA expression of the FTO gene in porcine muscle and its relationship with intramuscular fat deposition. *Czech J Anim Sci.* 2013;58:381–388.
114. Tao YX. Mutations in melanocortin-4 receptor and human obesity. *Prog Mol Biol Transl Sci.* 2009;88:173–204.
115. Switonski M, Mankowska M, Salamon S. Family of melanocortin receptor (MCR) genes in mammals-mutations, polymorphisms and phenotypic effects. *J Appl Genet.* 2013;54(4):461–472.
116. Fontanesi L, Galimberti G, Calo DG, et al. Identification and association analysis of several hundred single nucleotide polymorphisms within candidate genes for back fat thickness in Italian Large White pigs using a selective genotyping approach. *J Anim Sci.* 2012;90(8):2450–2464.
117. Van Laere AS, Nguyen M, Braunschweig M, et al. A regulatory mutation in IGF2 causes a major QTL effect on muscle growth in the pig. *Nature.* 2003;425(6960): 832–836.
118. Perez-Montarelo D, Rodriguez MC, Fernandez A, et al. Haplotypic diversity of porcine LEP and LEPR genes involved in growth and fatness regulation. *J Appl Genet.* 2015;56 (4):525–533.
119. Ovilo C, Fernandez A, Fernandez AI, et al. Hypothalamic expression of porcine leptin receptor (LEPR), neuropeptide Y (NPY), and cocaine- and amphetamine-regulated transcript (CART) genes is influenced by LEPR genotype. *Mamm Genome.* 2010;21 (11–12):583–591.
120. Bender N, Allemann N, Marek D, et al. Association between variants of the leptin receptor gene (LEPR) and overweight: a systematic review and an analysis of the CoLaus study. *PLoS One.* 2011;6(10):e26157.
121. Dubern B, Clement K. Leptin and leptin receptor-related monogenic obesity. *Biochimie.* 2012;94(10):2111–2115.
122. Fu Y, Sun W, Xu C, et al. Genetic variants in KDR transcriptional regulatory region affect promoter activity and intramuscular fat deposition in Erhualian pigs. *Anim Genet.* 2014;45(3):373–380.
123. Grzes M, Szczerbal I, Fijak-Nowak H, Szydlowski M, Switonski M. Two candidate genes (FTO and INSIG2) for fat accumulation in four canids: chromosome mapping, gene polymorphisms and association studies of body and skin weight of red foxes. *Cytogenet Genome Res.* 2011;135(1):25–32.
124. Skorczyk A, Stachowiak M, Szczerbal I, et al. Polymorphism and chromosomal location of the MC4R (melanocortin-4 receptor) gene in the dog and red fox. *Gene.* 2007;392 (1–2):247–252.
125. van den Berg L, van den Berg SM, Martens EE, et al. Analysis of variation in the melanocortin-4 receptor gene (mc4r) in Golden Retriever dogs. *Anim Genet.* 2010;41 (5):557.
126. Zeng R, Zhang Y, Du P. SNPs of melanocortin 4 receptor (MC4R) associated with body weight in Beagle dogs. *Exp Anim.* 2014;63(1):73–78.

127. Miyabe M, Gin A, Onozawa E, et al. Genetic variants of the unsaturated fatty acid receptor GPR120 relating to obesity in dogs. *J Vet Med Sci.* 2015;77(10): 1201–1206.
128. Skorczyk A, Flisikowski K, Szydlowski M, Cieslak J, Fries R, Switonski M. Association of MC3R gene polymorphisms with body weight in the red fox and comparative gene organization in four canids. *Anim Genet.* 2011;42(1):104–107.
129. Nishii N, Takasu M, Soe OK, et al. Cloning, expression and investigation for polymorphisms of canine peroxisome proliferator-activated receptors. *Comp Biochem Physiol B.* 2007;147(4):690–697.
130. Mankowska M, Stachowiak M, Graczyk A, et al. Sequence analysis of three canine adipokine genes revealed an association between TNF polymorphisms and obesity in Labrador dogs. *Anim Genet*, 2016. doi:10.1111/age.12390
131. Feuk L, Carson AR, Scherer SW. Structural variation in the human genome. *Nat Rev Genet.* 2006;7(2):85–97.
132. Zarrei M, MacDonald JR, Merico D, Scherer SW. A copy number variation map of the human genome. *Nat Rev Genet.* 2015;16(3):172–183.
133. Almal SH, Padh H. Implications of gene copy-number variation in health and diseases. *J Hum Genet.* 2012;57(1):6–13.
134. Bochukova EG, Huang N, Keogh J, et al. Large, rare chromosomal deletions associated with severe early-onset obesity. *Nature.* 2010;463(7281):666–670.
135. Walters RG, Jacquemont S, Valsesia A, et al. A new highly penetrant form of obesity due to deletions on chromosome 16p11.2. *Nature.* 2010;463(7281):671–675.
136. Sha BY, Yang TL, Zhao LJ, et al. Genome-wide association study suggested copy number variation may be associated with body mass index in the Chinese population. *J Hum Genet.* 2009;54(4):199–202.
137. Viljakainen H, Andersson-Assarsson JC, Armenio M, et al. Low copy number of the AMY1 locus is associated with early-onset female obesity in Finland. *PLoS One.* 2015;10(7):e0131883.
138. Wang J, Jiang J, Wang H, Kang H, Zhang Q, Liu JF. Improved detection and characterization of copy number variations among diverse pig breeds by Array CGH. *G3.* 2015;5(6):1253–1261.
139. Chen C, Qiao R, Wei R, et al. A comprehensive survey of copy number variation in 18 diverse pig populations and identification of candidate copy number variable genes associated with complex traits. *BMC Genomics.* 2012;13:733.
140. Fowler KE, Pong-Wong R, Bauer J, et al. Genome wide analysis reveals single nucleotide polymorphisms associated with fatness and putative novel copy number variants in three pig breeds. *BMC Genomics.* 2013;14:784.
141. Jiang J, Wang J, Wang H, et al. Global copy number analyses by next generation sequencing provide insight into pig genome variation. *BMC Genomics.* 2014;15:593.
142. Wang Y, Tang Z, Sun Y, et al. Analysis of genome-wide copy number variations in Chinese indigenous and western pig breeds by 60 K SNP genotyping arrays. *PLoS One.* 2014;9(9):e106780.
143. Schiavo G, Dolezal MA, Scotti E, et al. Copy number variants in Italian Large White pigs detected using high-density single nucleotide polymorphisms and their association with back fat thickness. *Anim Genet.* 2014;45(5):745–749.
144. Glessner JT, Bradfield JP, Wang K, et al. A genome-wide study reveals copy number variants exclusive to childhood obesity cases. *Am J Hum Genet.* 2010;87(5):661–666.
145. Paudel Y, Madsen O, Megens HJ, et al. Evolutionary dynamics of copy number variation in pig genomes in the context of adaptation and domestication. *BMC Genomics.* 2013;14:449.
146. Ramirez O, Olalde I, Berglund J, et al. Analysis of structural diversity in wolf-like canids reveals post-domestication variants. *BMC Genomics.* 2014;15:465.

147. Jin SL, Richard FJ, Kuo WP, D'Ercole AJ, Conti M. Impaired growth and fertility of cAMP-specific phosphodiesterase PDE4D-deficient mice. *Proc Natl Acad Sci USA.* 1999;96(21):11998–12003.
148. Song Y, Altarejos J, Goodarzi MO, et al. CRTC3 links catecholamine signalling to energy balance. *Nature.* 2010;468(7326):933–939.
149. Ou Z, Wang G, Li Q, Ma Z, Dai M, Zou F. CRTC3 polymorphisms were associated with the plasma level of total cholesterol and the risks of overweight and hypertriglyceridemia in a Chinese Han population. *Mol Biol Rep.* 2014;41(1):125–130.
150. Axelsson E, Ratnakumar A, Arendt ML, et al. The genomic signature of dog domestication reveals adaptation to a starch-rich diet. *Nature.* 2013;495(7441):360–364.
151. Arendt M, Fall T, Lindblad-Toh K, Axelsson E. Amylase activity is associated with AMY2B copy numbers in dog: implications for dog domestication, diet and diabetes. *Anim Genet.* 2014;45(5):716–722.
152. Nakajima K, Nemoto T, Muneyuki T, Kakei M, Fuchigami H, Munakata H. Low serum amylase in association with metabolic syndrome and diabetes: a community-based study. *Cardiovasc Diabetol.* 2011;10:34.

INDEX

A

F

G

H

I

J

L

M

O

Q

R

S

T

U

V

W

X

Z

(A)

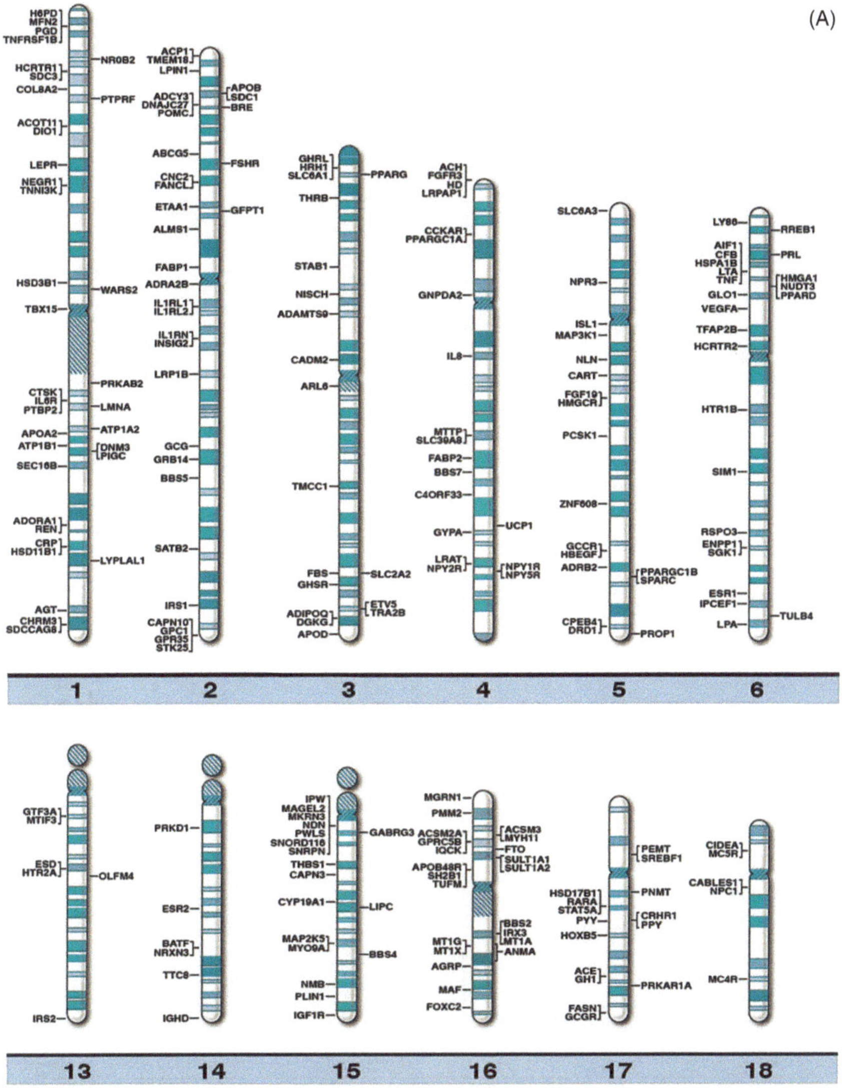

M.G. Butler, Figure 1 (A) Obesity gene ideogram part A. High resolution human chromosome ideograms (850 band level) with symbols representing recognized genes for obesity positioned at the chromosome band location. The upper "p" and lower "q" arms for each chromosome are separated by the centromere area highlighted in black.[16]

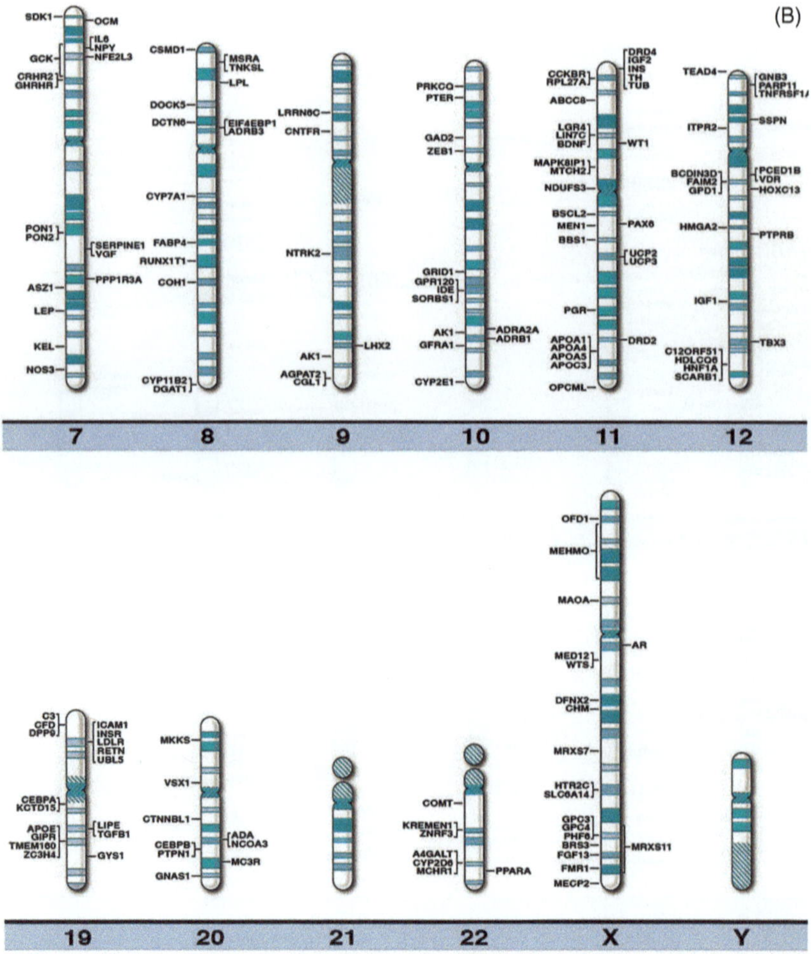

M.G. Butler, Figure 1 (B) Obesity gene ideogram part B. High resolution human chromosome ideograms (850 band level) with symbols representing recognized genes for obesity positioned at the chromosome band location. The upper "p" and lower "q" arms for each chromosome are separated by the centromere area highlighted in black.[16]

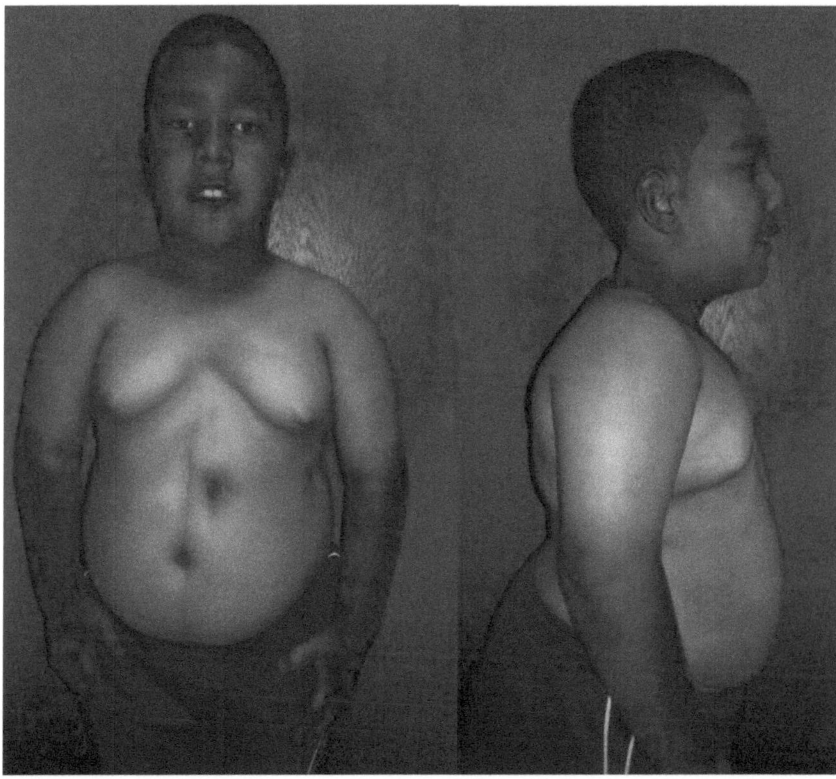

M.G. Butler, Figure 2 Frontal and profile view of an 8-year-old male with PWS showing the typical facial features, central obesity, and gastrostomy-tube site located on the abdomen.

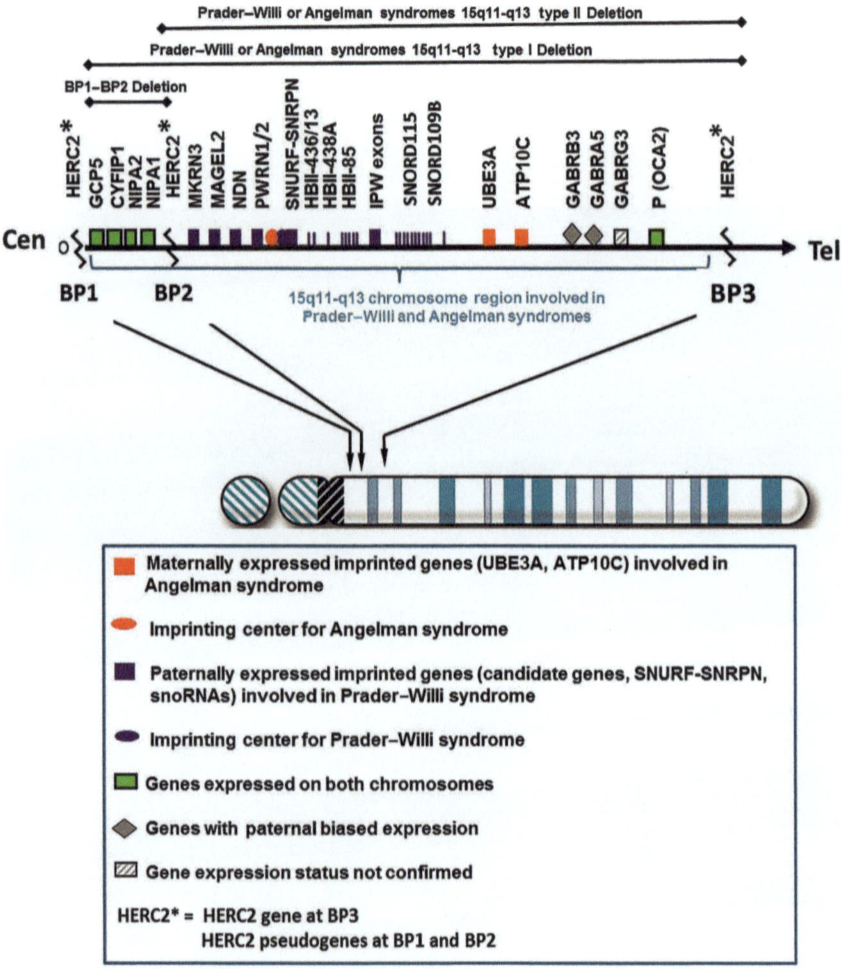

M.G. Butler, Figure 3 Chromosome 15 ideogram showing the location of genes in the 15q11-q13 region. BP1, BP2, and BP3 are the three common chromosome 15 breakpoints in the region at the site of breakage leading to the larger typical Type I deletion between BP1 and BP3 and the smaller Type II deletion between BP2 and BP3. The *dark gray* (blue in the web version) *colored rectangle*-shaped symbols represent paternally expressed genes (eg, MAGEL2) which when disturbed leads to PWS. The *gray* (red in the web version) *colored square*-shaped symbols represent maternally expressed genes and the UBA3E gene when disturbed causes Angelman syndrome. The *light gray* (green in the web version) *colored rectangle*-shaped symbols represent genes (eg, CYFIP1) that are expressed on both the maternal and paternal chromosome 15s.

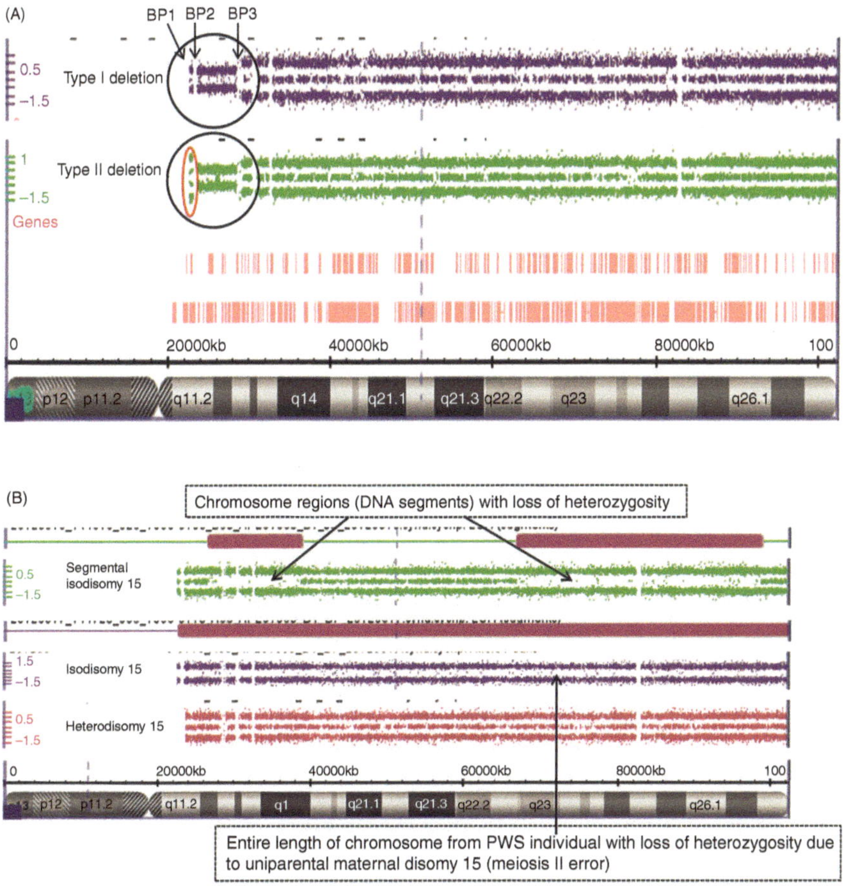

M.G. Butler, Figure 4 (A) Chromosome microarray results for PWS deletion subtypes. High resolution chromosome microarray using CNV and SNP probes to identify typical 15q11-q13 deletions in PWS classified as Type I involving breakpoints BP1 and BP3 and Type II involving breakpoints BP2 and BP3. (B) Chromosome Microarray Results for Uniparental Maternal Disomy 15. High resolution chromosome microarray using CNV and SNP probes to identify the maternal disomy 15 subtype (segmental isodisomy 15, isodisomy 15, and heterodisomy 15).

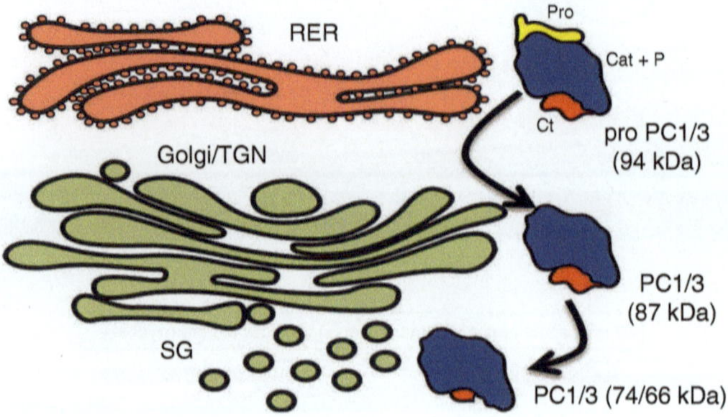

B. Ramos-Molina *et al.*, Figure 1 PC1/3 trafficking in the regulated secretory pathway. During translocation the signal peptide is removed, resulting in the inactive zymogen proPC1/3. In the ER, proPC1/3 is folded, N-glycosylated and undergoes rapid initial prodomain cleavage. In the Golgi compartments, the protein undergoes additional posttranslational modifications, including further addition glycosylation, sugar trimming; and sulfation. Once in the TGN, PC1/3 is sorted into dense-core granules, and the Ct domain is autocatalytically cleaved, resulting in complete maturation. *RER*, rough endoplasmic reticulum; *TGN*, trans-Golgi network; *SG*, secretory granules; *Pro*, prodomain; *Cat*, catalytic domain; *Ct*, carboxy-terminal domain.

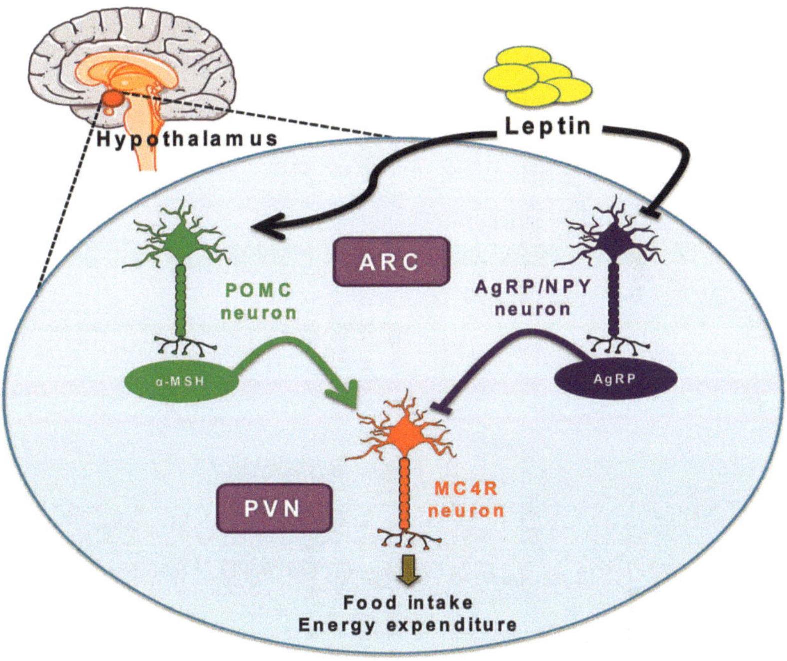

B. Ramos-Molina *et al.*, Figure 2 PC1/3 activity is essential to the leptin–melanocortin pathway. The hypothalamus is a key regulator of food intake and energy expenditure, responding to the action of peripheral hormones. Leptin is an anorexigenic hormone, produced and secreted by white adipose tissue, whose circulating levels are proportional to the mass of body fat. Leptin binds to leptin receptors on AgRP/NPY- and POMC-expressing neurons in the ARC of the hypothalamus. Leptin binding inhibits the synthesis of AgRP and stimulates the synthesis of POMC. Acting together, PC1/3 and PC2 accomplish the proteolytic processing of POMC to α-MSH within POMC-expressing neurons. PC1/3 is specifically required for the processing of proAgRP in NPY/AgRP-producing neurons. α-MSH and AgRP compete for the MC4R in the paraventricular nucleus *(PVN)* of the hypothalamus. Whereas increased receptor activity by α-MSH binding generates an anorexigenic signal, AgRP reduces receptor activity, generating an orexigenic signal.

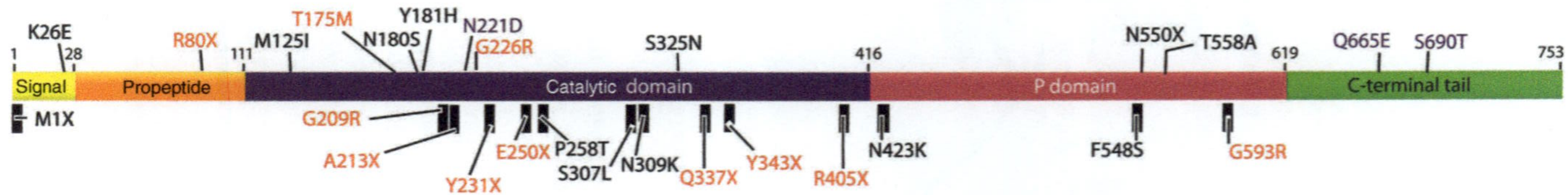

B. Ramos-Molina *et al.*, Figure 3 Human PC1/3 mutations and variations identified to date. To date, over two dozen different missense and nonsense mutations in *PCSK1* have been identified in infants with diarrhea and endocrine patients (shown below panel)[6,8,9,57] or in obese populations[81] (shown above panel). The three most frequent polymorphisms (N221D, Q665E/S690T) are shown in dark gray (blue in the web version); note that these may coexist with other mutations. Mutant proteins that are not secreted and likely are ER-retained are shown in light gray (red in the web version)

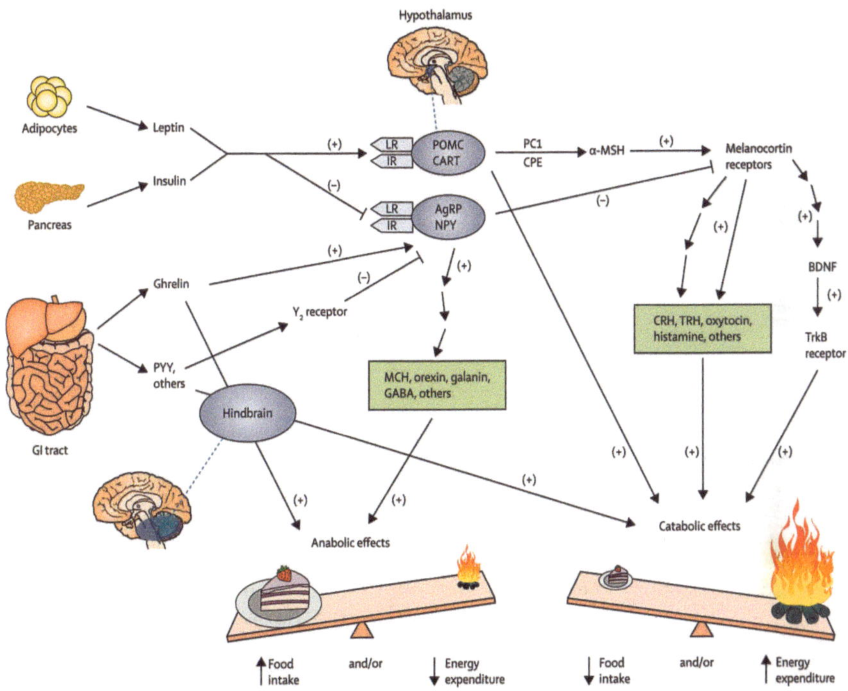

J.C. Han, Figure 1 Simplified overview of the leptin pathway. *Lines with arrowheads* show stimulatory action. *Lines with perpendicular endblocks* show inhibitory action. *AgRP*, agouti-related protein; *BDNF*, brain-derived neurotrophic factor; *CART*, cocaine–amphetamine related transcript; *CPE*, carboxypeptidase E; *CRH*, corticotropin-releasing hormone; *GABA*, gamma amino butyric acid; *GI*, gastrointestinal; *IR*, insulin receptor; *LR*, leptin receptor; *MCH*, melanin-concentrating hormone; *MSH*, melanocyte-stimulating hormone; *NPY*, neuropeptide Y; *PC1*, prohormone convertase 1; *POMC*, proopiomelanocortin; *PYY*, peptide YY; *TRH*, thyrotropin-releasing hormone; *TrkB*, tropomyosin-related kinase B.

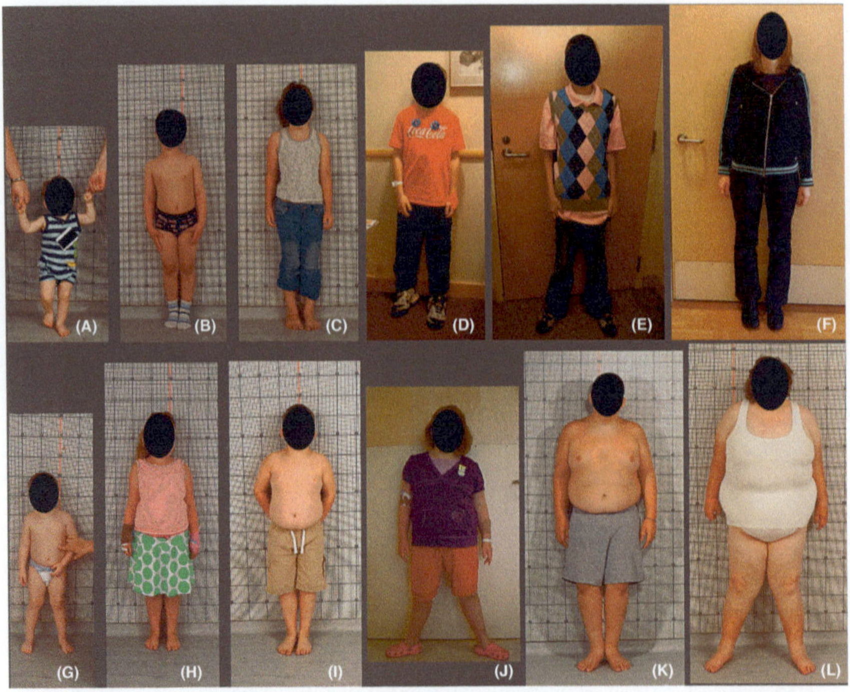

J.C. Han, Figure 2 Children and adults with WAGR syndrome. Patient characteristics are shown in Table 1. Patients (A–F), whose BMIs range from normal to overweight, have both alleles of *BDNF* intact. Patients (G–L) have heterozygous deletion of *BDNF*, and all have obesity.

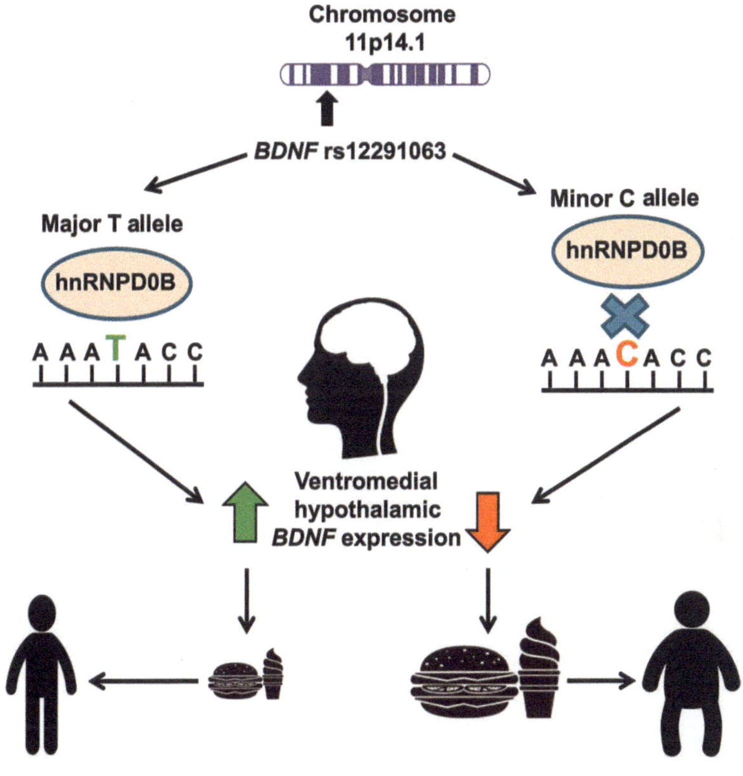

J.C. Han, Figure 3 Proposed mechanism for obesity caused by intronic *BDNF* SNP rs12291063.

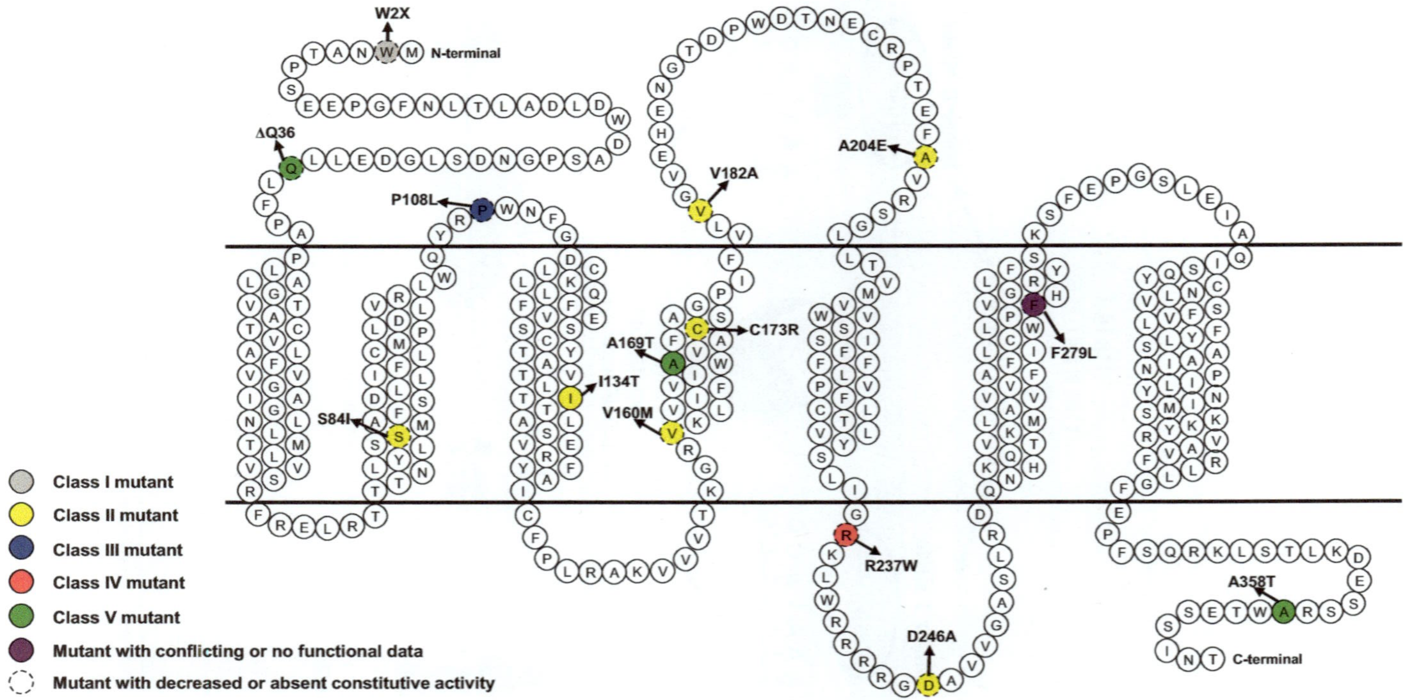

W. Yang and Y.-X. Tao, Figure 1 Schematic representation of naturally occurring *GHSR* mutations.

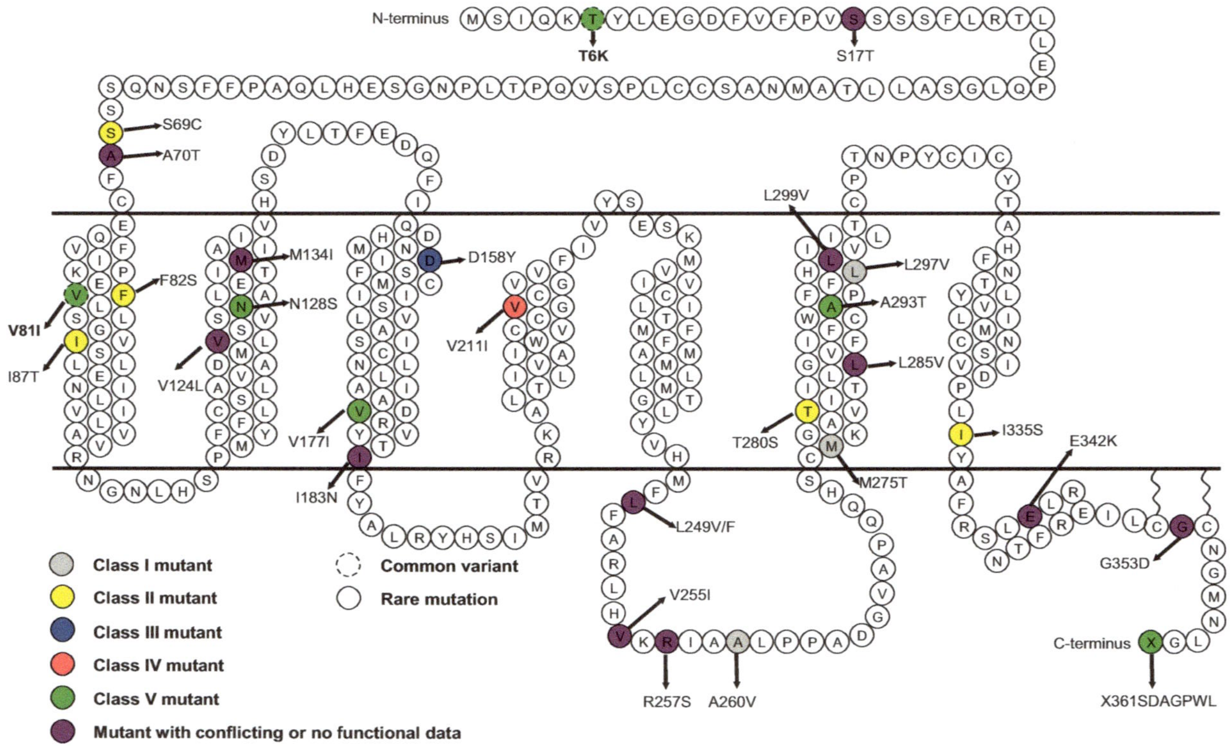

Z. Yang and Y.-X. Tao, Figure 1 Naturally occurring variants and mutations in the *MC3R*. Two common variants in the *MC3R*, Thr6Lys, and Val81Ile, are indicated. Mutants of different Classes are also indicated separately.

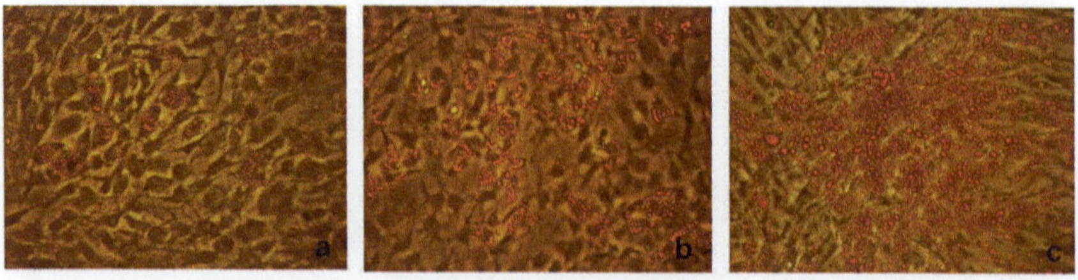

M. Stachowiak *et al.*, Figure 1 Accumulation of lipid droplets during porcine in vitro adipogenesis. Oil Red O staining was performed at days 3 (A), 7 (B), and 14 (C) of the differentiation. *Photo: I. Szczerbal.*

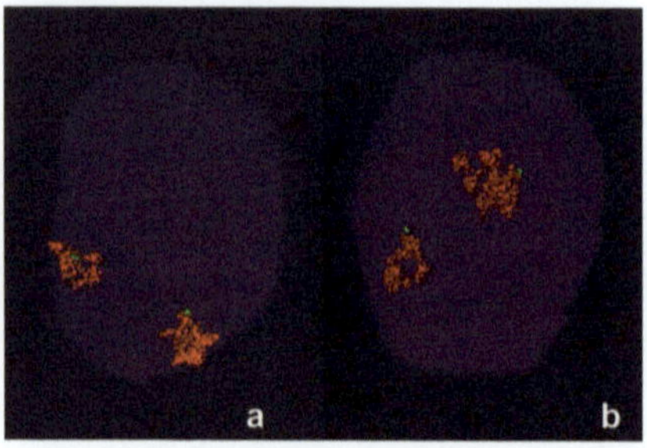

M. Stachowiak *et al.*, Figure 2 Changes of nuclear architecture during porcine in vitro adipogenesis. Distribution of the *FABP4* gene and SSC4 chromosome territories in a mesenchymal stem cell (A) and an adipocyte from day 14 of differentiation (B) are shown. *Photo: I. Szczerbal.*

Kolibri 360 GmbH
Ferdinand-Jühlke-Straße 7
99095 Erfurt, Deutschland
produktsicherheit@kolibri360.de